INTERNATIONAL CENTRE FOR MECHANICAL SCIENCES

COURSES AND LECTURES - No. 4

PROBABILISTIC METHODS IN GEOTECHNICAL ENGINEERING

EDITED BY

D. V. GRIFFITHS
COLORADO SCHOOL OF MINES, GOLDEN, USA

GORDON. A. FENTON
DALHOUSIE UNIVERSITY, HALIFAX, CANADA

SpringerWienNewYork

This volume contains 149 illustrations

SPIN 12048519

All contributions have been typeset by the authors.

ISBN 978-3-211-73365-3 SpringerWienNewYork

SpringerWienNewYork

CISM COURSES AND LECTURES

The series presents lecture notes, monographs, edited works and proceedings in the field of Mechanics, Engineering, Computer Science and Applied Mathematics.
Purpose of the series is to make known in the international scientific and technical community results obtained in some of the activities organized by CISM, the International Centre for Mechanical Sciences.

PREF CE

Soils and rocks are among the most variable of all engineering materials, and as such are highly amenable to a probabilistic treatment. The application of statistical and probabilistic concepts to geotechnical analysis is a rapidly growing area of interest for both academics and practitioners. The course is therefore aimed at students, researchers, and practitioners of geotechnical engineering who wish to keep abreast of developments in this evolving eld of study. The course content and delivery will assume no more that an introductory understanding of probability and statistics on the part of the course participants.

The main objective of the course is to present a state-of-the-art training on probabilistic techniques applied to geotechnical engineering in relation to both theory and practice. The course will include:

(a) discussion of potential bene ts of probabilistic approaches as opposed to the classical "Factor of Safety" methods, to review sources of uncertainty in geotechnical analysis and to introduce methods of LRFD and reliability concepts in Eurocode 7,

(b) review of relevant statistical theories needed to develop the methodologies and interpret the results of probabilistic analysis,

(c) examples of established probabilistic methods of analysis in geotechnical engineering, such as the First Order Second Moment (FOSM) method, the Point Estimate Method (PEM), the First and Second Order Reliability Methods (FORM/SORM) and Random Set (RS) theory.

(d) description of numerical methods of probabilistic analysis based on the -nite element method, such as the Stochastic Finite Element Method (SFEM) and recent developments on the Random Finite Element Method (RFEM),

(e) practical examples and case histories of probabilistic applications in geotechnical engineering.

Gordon A. Fenton
D. V. Griffiths

ONTENTS

Review of Probability Theory, Random Variables, and Random Fields

Gordon A. Fenton* and D.V. Griffiths†

* Department of Engineering Mathematics, Dalhousie University, Canada
† Division of Engineering, Colorado School of Mines, U.S.A.

Abstract Regulatory bodies are increasingly asking geotechnical engineers to provide rational risk assessments to accompany their designs. In order to provide these assessments, practicing geotechnical engineers need a good understanding of both basic probability theory and the more sophisticated, but realistic, random field soil models. This chapter lays the groundwork for this understanding. Starting with the basics of probability, the reader is lead through the theory of random variables and random fields and how they can be used to realistically model spatially variable soils.

1 Event Probabilities

The probability of an event A, denoted by $\mathrm{P}[A]$, is a number satisfying

$$0 \leq \mathrm{P}[A] \leq 1$$

Also, we assume that

$$\mathrm{P}[\emptyset] = 0, \quad \mathrm{P}[S] = 1.$$

Probabilities can sometimes be obtained using the counting rules discussed in the previous section. For example, if an experiment can result in any one of N different but equally likely outcomes, and if exactly m of these outcomes correspond to event A, then the probability of event A is $\mathrm{P}[A] = m/N$.

1.1 Additive Rules

Often we must compute the probability of some event which is expressed in terms of other events. For example, if A is the event that the company A requests your services and B is the event that company B requests your services, then the event that at least one of the two companies request your services is $A \cup B$. The probability of this is given by the following relationship;

If A and B are any two events, then

$$\mathrm{P}[A \cup B] = \mathrm{P}[A] + \mathrm{P}[B] - \mathrm{P}[A \cap B] \tag{1}$$

This relationship can be illustrated by the following Venn diagram. The desired quantity, $\mathrm{P}[A \cup B]$, is the area of $A \cup B$ which is shaded. If the shaded area is computed as the sum of the area of A, $\mathrm{P}[A]$, plus the area of B, $\mathrm{P}[B]$, then the intersection area, $\mathrm{P}[A \cap B]$, has been added twice. It must then be removed once to obtain the correct probability.

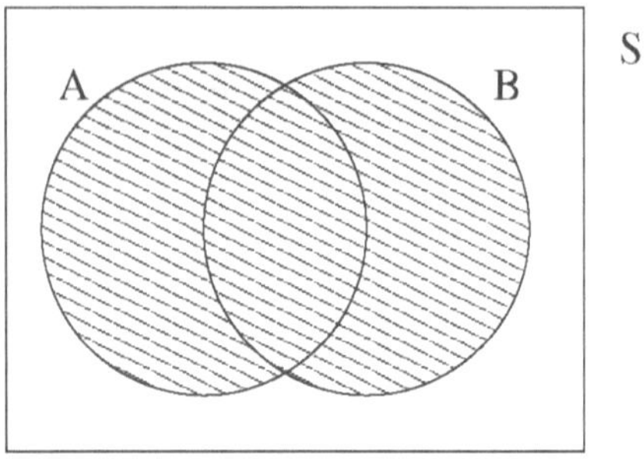

Figure 1. Venn diagram illustrating the union $A \cup B$.

Also,

If A and B are mutually exclusive, i.e. are disjoint and have no overlap, then

$$P[A \cup B] = P[A] + P[B] \tag{2}$$

If $A_1, A_2, ..., A_n$ are mutually exclusive, then

$$P[A_1 \cup ... \cup A_n] = P[A_1] + ... + P[A_n]. \tag{3}$$

Definition: We say that $A_1, A_2, ..., A_n$ is a **partition** of the sample space S if $A_1, A_2, ..., A_n$ are mutually exclusive and collectively exhaustive. Collectively exhaustive means that $A_1 \cup A_2 \cup ... \cup A_n = S$.
If $A_1, A_2, ..., A_n$ is a partition of the sample space S, then

$$P[A_1 \cup ... \cup A_n] = P[A_1] + ... + P[A_n] = P[S] = 1 \tag{4}$$

The above ideas can be extended to the union of more than two events. For example,
For any three events A, B, and C, we have

$$\begin{aligned} P[A \cup B \cup C] =& P[A] + P[B] + P[C] \\ & - P[A \cap B] - P[A \cap C] - P[B \cap C] \\ & + P[A \cap B \cap C] \end{aligned} \tag{5}$$

This can be seen by drawing a Venn diagram and keeping track of the areas which must be added and removed in order to get $P[A \cup B \cup C]$.

For the complementary events A and A^c, $P[A] + P[A^c] = 1$. *This is often used to compute* $P[A^c] = 1 - P[A]$.

2 Conditional Probability

The probability of an event is often affected by the occurrence of other events and/or the knowledge of information relevant to the event. Given two events, A and B, of an experiment, $P[B \mid A]$ is called the conditional probability of B given that A has already occurred. It is defined by

$$P[B \mid A] = \frac{P[A \cap B]}{P[A]} \tag{6}$$

That is, if we are given that event A has occurred, then A becomes our sample space. The probability that B has also occurred within this new sample space will be the ratio of the "area" of B within A to the "area" of A.

Sometimes we know $\mathrm{P}[B \mid A]$ and wish to compute $\mathrm{P}[A \cap B]$. If the events A and B can both occur, then

$$\mathrm{P}[A \cap B] = \mathrm{P}[B \mid A]\,\mathrm{P}[A] \tag{7}$$

2.1 Total Probability

Sometimes we know the probability of an event in terms of the occurrence of other events and want to compute the *unconditional* probability of the event. For example, when we want to compute the *total* probability of failure of a bridge, we can start by computing a series of simpler problems such as

1) the probability of bridge failure given a maximum static load,
2) the probability of bridge failure given a maximum dynamic traffic load,
3) the probability of bridge failure given an earthquake,
4) the probability of bridge failure given a flood,

etc. The *Total Probability Theorem* can be used to combine the above probabilities into the unconditional probability of network failure. We need to know the above conditional probabilities along with the probabilities that the 'conditions' occur (e.g. the probability that the maximum static load will occur during the design life, etc.).

The Total Probability Theorem is stated generally as follows;

Total Probability Theorem:

If the events $B_1, B_2, ..., B_k$ constitute a partition of the sample space S (i.e. are disjoint but collectively exhaustive) then for any event A in S,

$$\mathrm{P}[A] = \sum_{i=1}^{k} \mathrm{P}[B_i \cap A] = \sum_{i=1}^{k} \mathrm{P}[A \mid B_i]\,\mathrm{P}[B_i] \tag{8}$$

2.2 Bayes' Theorem

Sometimes we want to improve an estimate of a probability in light of additional information. Bayes' Theorem allows us to do this. It arises from the observation that $\mathrm{P}[A \cap B]$ can be written in two ways;

$$\begin{aligned} \mathrm{P}[A \cap B] &= \mathrm{P}[A \mid B] \cdot \mathrm{P}[B] \\ &= \mathrm{P}[B \mid A] \cdot \mathrm{P}[A] \end{aligned} \tag{9}$$

which implies that $\mathrm{P}[B \mid A] \cdot \mathrm{P}[A] = \mathrm{P}[A \mid B] \cdot \mathrm{P}[B]$, or

$$\mathrm{P}[B \mid A] = \frac{\mathrm{P}[A \mid B] \cdot \mathrm{P}[B]}{\mathrm{P}[A]} \tag{10}$$

Bayes' Theorem is stated formally as follows.

Bayes' Theorem:

If the events $B_1, B_2, ..., B_k$ constitute a partition of the sample space S (i.e. are disjoint and collectively exhaustive) then for any event A of S, such that $\mathrm{P}[A] \neq 0$,

$$\mathrm{P}[B_j \,|\, A] = \frac{\mathrm{P}[B_j \cap A]}{\sum_{i=1}^{k} \mathrm{P}[B_i \cap A]} = \frac{\mathrm{P}[A \,|\, B_j]\,\mathrm{P}[B_j]}{\sum_{i=1}^{k} \mathrm{P}[A \,|\, B_i]\,\mathrm{P}[B_i]} = \frac{\mathrm{P}[A \,|\, B_j]\,\mathrm{P}[B_j]}{\mathrm{P}[A]} \tag{11}$$

for any $j = 1, 2, ..., k$.

Bayes' Theorem is useful for revising or updating probabilities as more data and information becomes available. In the previous example on piezocones, there was an *initial* probability that a piezocone would have been manufactured at plant A: $\mathrm{P}[A] = 0.5$. This probability is referred to as the **prior** probability of A. That is, in the absence of any other information, a piezocone chosen at random has a probability of having been manufactured at plant A of 0.5. However, if a piezocone, chosen at random, is found to be defective (so that there is now more information on the piezocone), then its probability that it was manufactured at plant A reduces from 0.5 to 0.294. This latter probability is referred to as the **posterior** probability of A. **Bayesian** updating of probabilities is a very powerful tool in engineering reliability-based design.

For problems involving conditional probabilities, event trees are usually the easiest way to proceed. However, event trees are not always easy to draw, and the purely mathematical approach is sometimes necessary. As an example of a tree which is not quite straightforward, see if you can draw the event tree and answer the questions in the following *Exercise*. Remember that you must set up the tree in such a way that you can fill in most of the probabilities on the branches. If you are left with too many empty branches and no other given information, you are likely to have confused the order of the events - try reorganizing your tree.

2.3 Problem-Solving Methodology

Solving real-life problems (i.e. 'word problems') is not always easy. It is often not perfectly clear what is meant by a worded question. Two things improve one's chances of successfully solving problems which are expressed using words: (a) a systematic approach, and (b) *practice*. It is practice that allows you to identify those aspects of the question that need further clarification, if any. Below, a few basic recommendations are outlined.

1) Solving a word problem generally involves the computation of some quantity. Clearly identify this quantity at the beginning of the problem solution. Before starting any computations, it is good practice to write out your concluding sentence first. This forces you to concentrate on the essentials.
2) In any problem involving the probability of events, you should
   a) **clearly define your events.** Use the following guidelines:
      i) Keep events as simple as possible.
      ii) if your event definition includes the words "*and*", "*or*", "*given*", "*if*", "*when*", etc., then **it is NOT a good event definition.** Break your event into two (or more, if required) events and use " $\cap$ ", " $\cup$ ", or "|"

operators to express what you had originally intended. The complement is also a helpful operator, see (iii).

iii) You do not need to define separate events for, for example, "*an accident occurs*" and "*an accident does not occur*". In fact, this will often lead to confusion. Simply define A to be one of the events and use A^c when you want to refer to the other. This may also give you some hints as to how to proceed since you know that $\mathrm{P}[A^c] = 1 - \mathrm{P}[A]$.

b) Once your events are defined, you need to go through the worded problem to extract the given numerical information. Write this information down in the form of probabilities of the events that you defined above. For example, $\mathrm{P}[A] = 0.23$, $\mathrm{P}[B \mid A] = 0.6$, etc. Note that the conditional probabilities, are often difficult to unravel. Phrases such as

'*if ... occurs, the probability of ... doubles...*'
'*In the event that ... occurs, the probability of ... becomes 0.6*'
'*When ... occurs, the probability of ... becomes 0.43*'
'*Given that ... occurs, the probability of ... is 0.3*'

all translate into a probability statement of the form $\mathrm{P}[A \mid B]$. In this case, you will likely be using one of the conditional probability relationship ($\mathrm{P}[A \cap B] = \mathrm{P}[B \mid A]\,\mathrm{P}[A]$), the Total Probability Theorem, or Bayes' Theorem.

c) Now review the worded problem again and write down the probability that the question is asking for in terms of the events defined above. Although the question may be in worded form, you should be writing down something like $\mathrm{P}[A \cap B]$ or $\mathrm{P}[B \mid A]$, etc. Make sure that you can express the desired probability in terms of the events you defined above. If you can't, then you need to revise your original event definitions.

d) Finally, use the rules of combining probabilities (e.g. probabilities of unions, intersections, Bayes' Theorem, etc) to compute the desired probability.

3 Random Variables and Probability Distributions

Although probability theory is based on the idea of events and associated set theory, it becomes very unwieldy to treat random events like 'time to failure' using explicit event definitions. One would conceivably have to define a separate event for each possible time of failure and so would soon run out of symbols for the various events. For this reason, and also because they allow the use of a wealth of mathematical tools, *random variables* are used to represent a suite of possible events. In addition, since most engineering problems are expressed in terms of numerical quantities, random variables are particularly appropriate.

Definition: Consider a sample space S consisting of a set of outcomes $\{s_1, s_2, \ldots\}$. If X is a function that assigns a real number $X(s)$ to every outcome $s \in S$, then X is a *random variable*. Random variables will be denoted with upper case letters.

Now what does this mean in plain English? Essentially a random variable is a means of identifying events in numerical terms. For example, if the outcome s_1 means that an apple was selected and s_2 means that an orange was selected, then $X(s_1)$ could be set equal to 1 and $X(s_2)$ could be set equal to 0. $X > 0$ then means that an apple was selected. Now mathematics can be used on X, ie. if the fruit picking experiment is repeated n times and $x_1 = X_1(s)$ is the outcome of the first experiment, $x_2 = X_2(s)$ the outcome of the second, etc., then the total number of apples picked is $\sum_{i=1}^{n} x_i$. Note that mathematics could not be used on the actual outcomes themselves, e.g. picking an apple is a real event which knows nothing about mathematics nor can it be used in a mathematical expression without first mapping the event to a number.

For each outcome s, there is exactly one value of $x = X(s)$, but different values of s may lead to the same x.

The above discussion illustrates in a rather simple way one of the primary motivation for the use of random variables – simply so that mathematics can be used. One other thing might be noticed in the previous paragraph. After the 'experiment' has taken place and the outcome is known, it is referred to using the lower case, x_i. That is x_i has a known fixed value while X is unknown. In other words x is a realization of the random variable X. This is a rather subtle distinction, but it is important to remember that X is unknown. The most that we can say about X is to specify what its likelihoods of taking on certain values are – we cannot say exactly what the value of X is.

3.1 Discrete Random Variables

Discrete random variables are those that take on only discrete values $\{x_1, x_2, \ldots\}$, ie. have a countable number of outcomes. Note that countable just means that the outcomes can be numbered $1, 2, \ldots$, however there could still be an infinite number of them. For example, our experiment might be to count the number of soil tests performed before one yields a cohesion of 200 MPa. This is a discrete random variable since we outcome is one of $0, 1, \ldots$, but the number may be very large or even (in concept) infinite (implying that a soil sample with cohesion 200 MPa was never found).

Discrete Probability Distributions

As mentioned previously, we can never know for certain what the value of a random variable is (if we do measure it, it becomes a *realization* – presumably the next measurement is again uncertain until it is measured, and so on). The most that we can say about a random variable is what its probability is of assuming each of its possible values. The set of probabilities assigned to each possible value of X is called a **probability distribution.** The sum of these probabilities, over all possible values, must be 1.0.

Definition: The set of ordered pairs $(x, f_X(x))$ is the probability distribution of the discrete random variable X if, for each possible outcome x,

1) $0 \le f_X(x) \le 1$,
2) $\sum_{\text{all } x} f_X(x) = 1$,
3) $\mathrm{P}[X = x] = f_X(x)$

$f_X(x)$ is called the *probability mass function* of X. The subscript is used to indicate what random variable is being governed by the distribution. We shall see when we consider continuous random variables why we call this a probability 'mass' function.

Discrete Cumulative Distributions

An *equivalent* description of a random variable is the cumulative distribution function (CDF), which is defined as follows;

Definition: The cumulative distribution function, $F_X(x)$ of a discrete random variable X, with probability mass function $f_X(x)$, is defined by

$$F_X(x) = \mathrm{P}\left[X \leq x\right] = \sum_{t \leq x} f_X(t) \tag{12}$$

We say that this is equivalent to the probability mass function because one can be obtained from the other,

$$f_X(x_i) = F_X(x_i) - F_X(x_{i-1}) \tag{13}$$

3.2 Continuous Random Variables

Continuous random variables can take on an infinite number of possible outcomes – generally X takes values from the real line $\Re$. To illustrate the changes involved when we go from the discrete to the continuous case, consider the probability that a grain silo experiences a bearing capacity failure at *exactly* 4.3673458212... years from when it is installed. Clearly the probability that it fails at *exactly* that instant in time is essentially zero. In general the probability that it fails at any one instant in time is vanishingly small. In order to characterize probabilities for continuous random variables, we can't use probabilities directly (since they are all essentially zero) – we must use *relative likelihoods*. That is, we say that the probability that X lies in the small interval between x and $x + dx$ is $f_X(x)\,dx$, or

$$\mathrm{P}\left[x < X \leq x + dx\right] = f_X(x)\,dx \tag{14}$$

where $f_X(x)$ is now called the *probability density function* (pdf) of the random variable X. The word *density* is used because "density" must be multiplied by a length measure in order to get a "mass". Note that the above probability is vanishingly small because dx is vanishingly small. The function $f_X(x)$ is now the relative likelihood that X lies in a very small interval near x. Roughly speaking, we can think of this as $\mathrm{P}\left[X = x\right] = f_X(x)dx$.

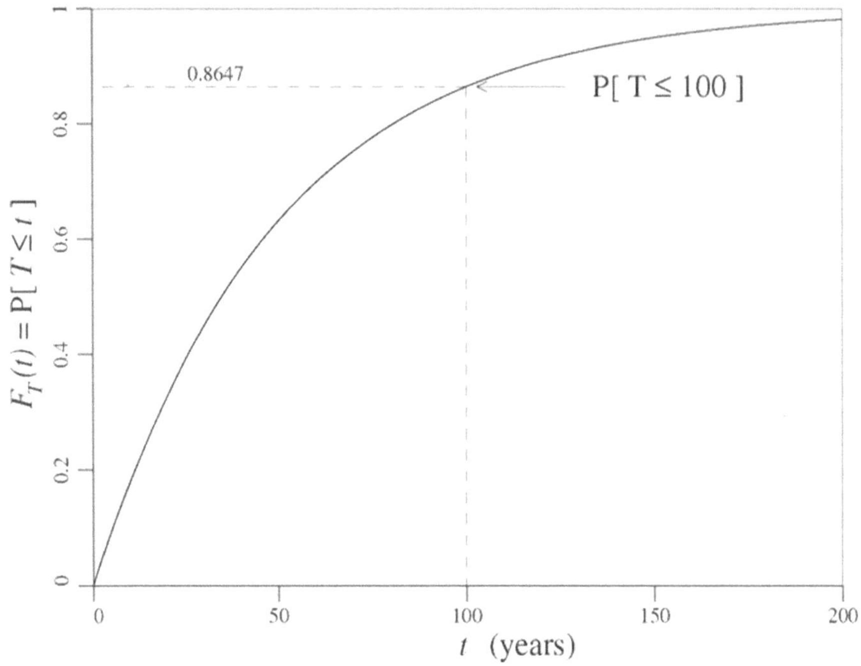

Figure 2. Cumulative distribution function for the exponential distribution.

Continuous Probability Distributions

Definition: The function $f_X(x)$ is a probability density function for the continuous random variable X, defined over the set of real numbers, if

1) $0 \leq f_X(x) < \infty$, for all $-\infty < x < +\infty$,
2) $\int_{-\infty}^{\infty} f_X(x)dx = 1$ (i.e. the area under the pdf is 1.0),
3) $\mathrm{P}[a < X < b] = \int_a^b f_X(x)dx$ (i.e. the area under $f_X(x)$ between a and b).

NOTE: it is important to recognize that, in the continuous case, $f_X(x)$ is no longer a probability. It has units of probability per unit length. In order to get probabilities, we have to find *areas* under the pdf, i.e. sum up values of $f_X(x)dx$.

Continuous Cumulative Distribution

The *cumulative distribution function* (cdf), for a continuous random variable is basically defined in the same way as it is for a discrete distribution.

Definition: The cumulative distribution function, $F_X(x)$, of a continuous random variable X having probability density function $f_X(x)$, is defined by the area under the density function to the left of x

$$F_X(x) = \mathrm{P}\left[X \leq x\right] = \int_{-\infty}^{x} f_X(t)dt \tag{15}$$

As in the discrete case, the cdf is equivalent to the pdf, in that one can be obtained from the other. It is simply another way of expressing the probabilities associcated with a random variable. Since the cdf is an integral of the pdf, the pdf can be obtained from the cdf as a derivative, ie.

$$f_X(x) = \frac{dF_X(x)}{dx} \tag{16}$$

4 Measures of Central Tendency, Variability, and Association

A random variable is completely described, as well as can be, if its probability distribution is specified. However, we will never know the precise distribution of any natural phenomenon. Nature cares not at all about our mathematical models and the 'truth' is usually far more complex than we are able to represent. So we very often have to describe a random variable using less complete, but more easily estimated, measures. The most important of these measures are *central tendency* and *variability*. Even if the complete probability distribution is known, these quantities remain useful because they convey information about the properties of the random variable that are of first importance in practical applications. Also, the parameters of the distribution are often derived as functions of these quantities, or they may be the parameters themselves.

The most common measures of central tendency and variability are the *mean* and the *variance*, respectively. In engineering, the variability of a random quantity is often expressed using the dimensionless *coefficient of variation* which is the ratio of the *standard deviation* over the *mean*. Also, when one has two random variables, X and Y, it is frequently of interest to measure how strongly they are related (or associated) to one another. A typical measure of the strength of the relationship between two random variables is their *covariance*. As we shall see, covariance depends on the units of the random variables involved and their individual variabilities, and so a more intuitive measure of the strength of the relationship between two random variables is the *correlation coefficient*, which is both dimensionless and bounded. All of these characteristics will be covered in this section.

4.1 Mean

The *mean* is the most important characteristic of a random variable, in that it tells us about its central tendency. It is defined mathematically as follows;

Definition: Let X be a random variable with probability density function $f(x)$. The *mean* or *expected value* of X, denoted μ_X, is defined by

$$\mu_X = \mathrm{E}\left[X\right] = \sum_x x f(x) \qquad \text{if } X \text{ is discrete} \tag{17a}$$

$$\mu_X = \mathrm{E}[X] = \int_{-\infty}^{\infty} x f(x)dx \qquad \text{if } X \text{ is continuous.} \qquad (17b)$$

where the subscript on μ, when present, denotes what μ is the mean of.

Expectation

The notation $\mathrm{E}[X]$ refers to a mathematical operation called *expectation*. The expectation of any random variable is a sum of all possible values of the random variable weighted by the probability of each value occurring. For example, if X is a random variable with probability (mass or density) function, $f_X(x)$, then the expected value of the random variable $g(X)$, where g is any function of X, is

$$\mu_{g(X)} = \mathrm{E}[g(X)] = \sum_x g(x) f_X(x) \qquad \text{if } X \text{ is discrete}$$

$$\mu_{g(X)} = \mathrm{E}[g(X)] = \int_{-\infty}^{\infty} g(x) f_X(x)dx \qquad \text{if } X \text{ is continuous.} \qquad (18)$$

If we have a sample of observations , $x_1, x_2, \ldots, x_n$, of some population X, then the population mean, μ_X, is estimated by the *sample mean*, $\bar{x}$, defined as

$$\boxed{\text{Sample Mean:}} \qquad \bar{x} = \frac{1}{n}\sum_{i=1}^{n} x_i$$

4.2 Median

The median is another measure of central tendency. We shall denote the median as $\tilde{\mu}$. It is the point which divides the distribution into two equal halves. Most commonly, $\tilde{\mu}$ is found by solving

$$F_X(\tilde{\mu}) = \mathrm{P}[X \le \tilde{\mu}] = 0.5$$

for $\tilde{\mu}$. For example, if $f_X(x) = \lambda e^{-\lambda x}$, then $F_X(x) = 1 - e^{-\lambda x}$, and we get

$$1 - e^{-\lambda\tilde{\mu}} = 0.5 \qquad \Longrightarrow \qquad \tilde{\mu}_X = -\frac{\ln(0.5)}{\lambda} = \frac{0.693}{\lambda}$$

While the mean is strongly affected by extremes in the distribution, the median is largely unaffected.

In general the mean and the median are not the same. If the distribution is positively skewed (or skewed right, which means a longer tail to the right than to the left), as are most soil properties, then the mean will be to the right of the median. Conversely, if the distribution is skewed left, then the mean will be to the left of the median. If the distribution is symmetric, then the mean and the median will coincide.

NOTE: the median is the point which divides the distribution in half.

If we have a sample of observations , $x_1, x_2, \ldots, x_n$, of some population X, then the population median, $\tilde{\mu}_X$, is estimated by the *sample median*, $\tilde{x}$. To define $\tilde{x}$, we must first order the observations from smallest to largest, $x_{(1)} \le x_{(2)} \le \cdots \le x_{(n)}$. When we have done so, the sample median is defined as

$$\boxed{\text{Sample Median}} \qquad \tilde{x} = \begin{cases} x_{((n+1)/2)} & \text{if } n \text{ is odd} \\ \frac{1}{2}\left(x_{(n/2)} + x_{((n+1)/2)}\right) & \text{if } n \text{ is even} \end{cases}$$

4.3 Variance

The mean (expected value) or median of the random variable X tells where the probability distribution is "centered". The next most important characteristic of a random variable is whether the distribution is "wide", "narrow", or somewhere in between. This distribution "variability" is commonly measured by a quantity call the variance of X.

Definition: Let X be a random variable with probability (mass or density) function $f_X(x)$ and mean μ_X. The **variance**, σ_X^2, of X is defined by

$$\sigma_X^2 = \text{Var}[X] = \text{E}\left[(X-\mu_X)^2\right] = \sum_x (x-\mu_X)^2 f_X(x) \tag{19a}$$

$$\sigma_X^2 = \text{Var}[X] = \text{E}\left[(X-\mu_X)^2\right] = \int_{-\infty}^{\infty} (x-\mu_X)^2 f_X(x)dx \tag{19b}$$

for the discrete and continous cases, respectively.

The variance of the random variable X is sometimes more easily computed as

$$\sigma_X^2 = \text{E}\left[X^2\right] - \text{E}^2[X] = \text{E}\left[X^2\right] - \mu_X^2 \tag{20}$$

The variance, σ_X^2, has units of X^2. The square root of the variance, σ_X, is called the *standard deviation* of X. Since the standard deviation has the same units as X, it is often preferable to report the standard deviation as a measure of variability.

Even though the standard deviation has the same units as the mean, it is often still not particularly informative. For example, a standard deviation of 1.0 may indicate significant variability when the mean is 1.0, but indicates virtually deterministic behaviour when the mean is one million. For example, an error of 1 m on a 1 m survey would be considered unacceptable, whereas an error of 1 m on a one thousand km survey might be considered quite accurate. A measure of variability which is both non-dimensional and delivers a relative sense of the magnitude of variability is the *coefficient of variation*, defined as

$$v = \frac{\sigma}{\mu} \tag{21}$$

Note that the coefficient of variation becomes undefined if the mean of X is zero. It is, however, quite popular as a way of expressing variability in engineering, particularly for material property and load variability, which generally have non-zero means.

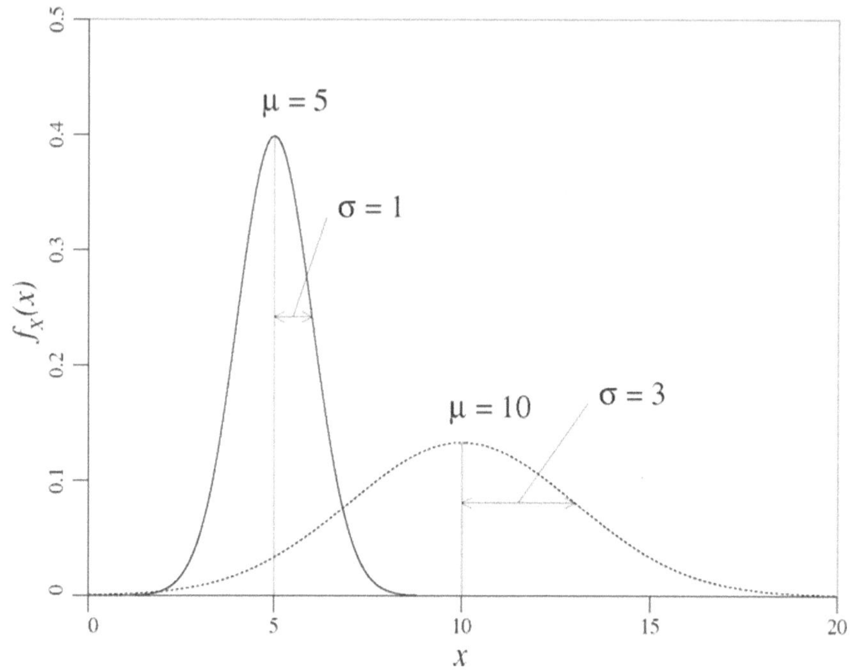

Figure 3. Two distributions illustrating how the position and shape changes with changes in mean and variance.

4.4 Covariance

Often one must consider more than one random variable at a time. For example, the two components of a drained soil's shear strength, $\tan(\phi')$ and c', will vary randomly from location to location in a soil. These two quantities can be modeled by two random variables, and since they may influence one another (or they may be jointly influenced by some other factor), they are characterized by a *bivariate distribution.*

Properties of the Bivariate Distribution:

Discrete:

$$f_{XY}(x,y) = \mathrm{P}\left[X = x \cap Y = y\right]$$

$$0 \leq f_{XY}(x,y) \leq 1$$

$$\sum_{\text{all } x} \sum_{\text{all } y} f_{XY}(x,y) = 1$$

Continuous:

$$f_{XY}(x,y)\,dx\,dy = \mathrm{P}\left[x < X \leq x + dx \cap y < Y \leq y + dy\right]$$

$$f_{XY}(x,y) \geq 0 \text{ for all } (x,y) \in \Re^2$$

$$\int_{-\infty}^{\infty} \int_{-\infty}^{\infty} f_{XY}(x,y)\,dx\,dy = 1$$

$$\mathrm{P}\left[x_1 < X \leq x_2 \cap y_1 < Y \leq y_2\right] = \int_{y_1}^{y_2} \int_{x_1}^{x_2} f_{XY}(x,y)\,dx\,dy$$

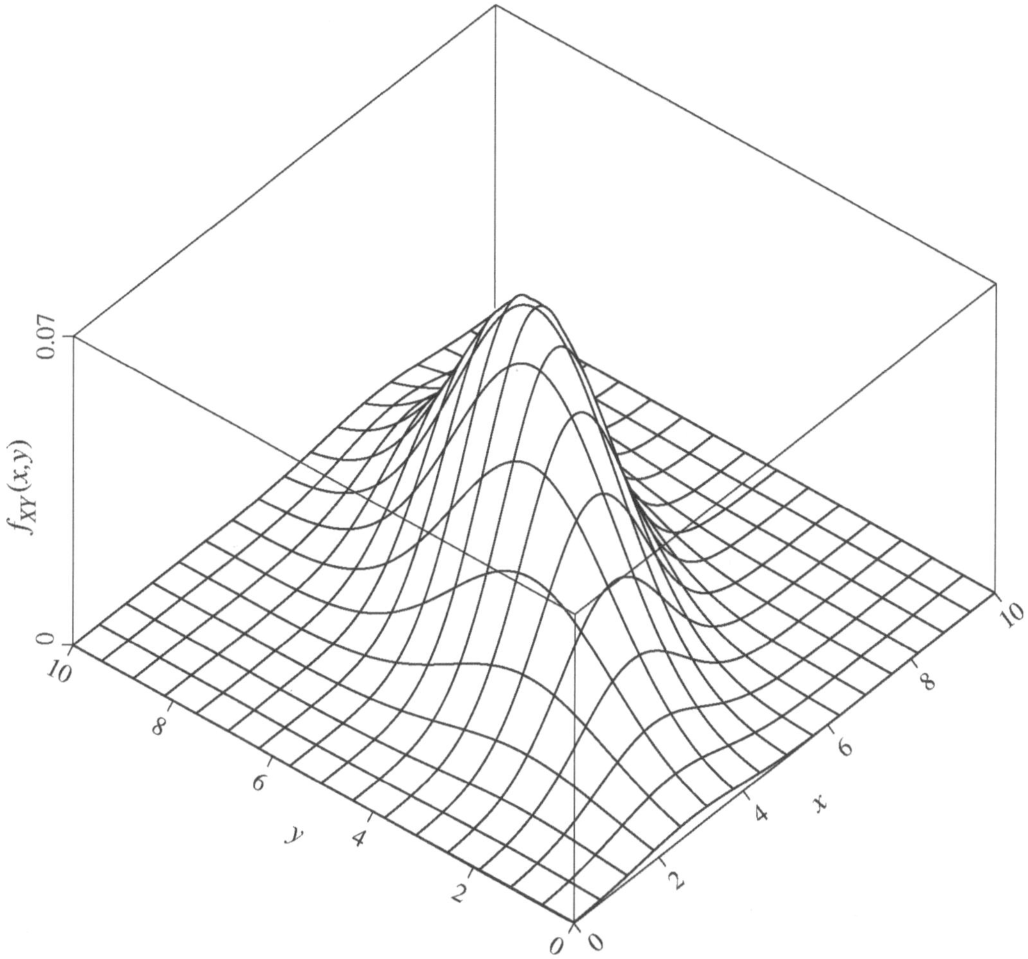

Figure 4. Example bivariate probability density function, $f_{XY}(x, y)$.

Definition: Let X and Y be random variables with joint probability distribution $f_{XY}(x, y)$. The *covariance* between X and Y is defined by

$$\begin{aligned}\operatorname{Cov}[X, Y] &= \mathrm{E}[(X - \mu_X)(Y - \mu_Y)] \\ &= \sum_x \sum_y (x - \mu_x)(y - \mu_Y) f_{XY}(x, y), \\ &= \int_{-\infty}^{\infty} \int_{-\infty}^{\infty} (x - \mu_X)(y - \mu_Y) f_{XY}(x, y)\, dx\, dy,\end{aligned} \tag{22}$$

for the discrete and continuous cases, respectively.

The covariance between two random variables X and Y, having means μ_X and μ_Y, respectively, may also be computed as

$$\mathrm{Cov}[X, Y] = \mathrm{E}[XY] - \mathrm{E}[X]\,\mathrm{E}[Y] = \mathrm{E}[XY] - \mu_X \mu_Y \tag{23}$$

Although the covariance between two random variables does give information regarding the nature of the relationship, the magnitude of $\mathrm{Cov}[X, Y]$ does not indicate anything regarding the strength of the relationship. This is because $\mathrm{Cov}[X, Y]$ depends on the units and variability of X and Y. A quantity which is both normalized and non-dimensional is the correlation coefficient, to be discussed next.

4.5 Correlation Coefficient

Definition: Let X and Y be random variables with joint probability distribution $f_{XY}(x, y)$. The **correlation coefficient** between X and Y is defined to be

$$\rho_{XY} = \frac{\mathrm{Cov}[X, Y]}{\sigma_X \sigma_Y}. \tag{24}$$

Figure 5 illustrates the effect that the correlation coefficient has on the shape of a bivariate probability density function, in this case for X and Y jointly normal. If $\rho_{XY} = 0$, then the contours form ovals with axes aligned with the cartesian axes (if the variances of X and Y are equal, then the ovals are circles). When $\rho_{XY} > 0$, the ovals become stretched and the major axis has a positive slope. What this means is that when Y is large, X will also tend to be large. For example, when $\rho_{XY} = 0.6$, as shown on the right plot of Figure 5, then when $Y = 8$, the most likely value X will take is around 7, since this is the peak of the distribution along the line $Y = 8$. Similarly, if $\rho_{XY} < 0$, then the ovals will be oriented so that the major axis has a negative slope. In this case, large values of Y will tend to give small values of X.

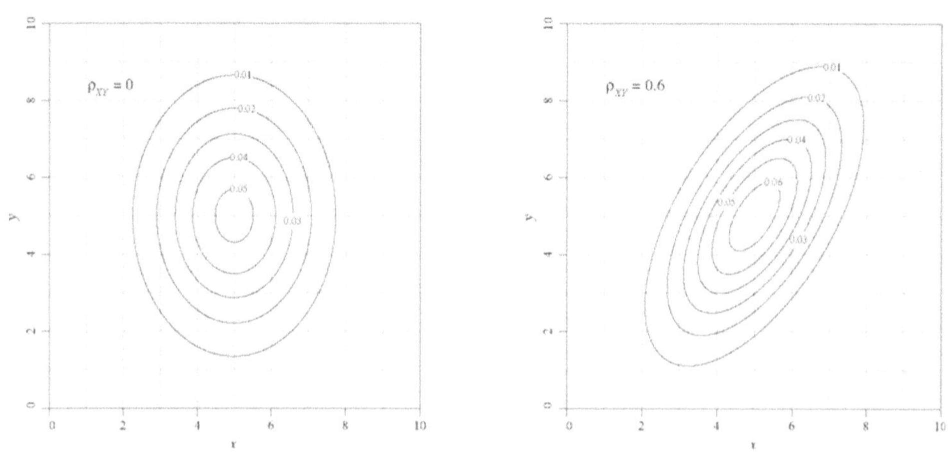

Figure 5. Effect of correlation coefficient, ρ_{XY}, on contours of a bivariate probability density function, $f_{XY}(x, y)$, having $\mu_X = \mu_Y = 5$ and $\sigma_X = 1.5$ and $\sigma_Y = 2.0$.

We can show that $-1 \leq \rho_{XY} \leq 1$ as follows: Consider two random variables X and Y having variances σ_X^2 and σ_Y^2, respectively, and correlation coefficient ρ_{XY}. Then

$$\begin{aligned}\text{Var}\left[\frac{X}{\sigma_X} + \frac{Y}{\sigma_Y}\right] &= \frac{\sigma_X^2}{\sigma_X^2} + \frac{\sigma_Y^2}{\sigma_Y^2} + 2\frac{\text{Cov}\left[X, Y\right]}{\sigma_X \sigma_Y} \\ &= 2\Big[1 + \rho_{XY}\Big] \\ &\geq 0\end{aligned}$$

which implies that $\rho_{XY} \geq -1$. Similarly,

$$\begin{aligned}\text{Var}\left[\frac{X}{\sigma_X} - \frac{Y}{\sigma_Y}\right] &= \frac{\sigma_X^2}{\sigma_X^2} + \frac{\sigma_Y^2}{\sigma_Y^2} - 2\frac{\text{Cov}\left[X, Y\right]}{\sigma_X \sigma_Y} \\ &= 2\Big[1 - \rho_{XY}\Big] \\ &\geq 0\end{aligned}$$

which implies that $\rho_{XY} \leq 1$. Taken together, these imply that $-1 \leq \rho_{XY} \leq 1$.

The correlation coefficient is a direct measure of the degree of *linear* dependence between X and Y. When the two variables are perfectly linearly related, ρ_{XY} will be either +1 or -1 (+1 if Y increases with X and -1 if Y decreases when X increases). When $|\rho_{XY}|$ is less that 1, the dependence between X and Y is not completely linear; however, there could still be a strong nonlinear dependence. If two random variables X and Y are independent, then their correlation coefficient will be 0. If the correlation coefficient between two random variables X and Y is 0, **it does not mean** that they are independent, only that they are uncorrelated. Independence is a much stronger statement than is $\rho_{XY} = 0$, since the latter only implies linear independence. For example, $Y = X^2$ may be linearly independent of X (this depends on the range of X), but clearly Y and X are completely (non-linearly) dependent.

5 Common Discrete Probability Distributions

Many engineered systems have the same statistical behaviour – we generally only need a handful of probability distributions to characterize most naturally occurring phenomena. In this section, the most common discrete distribution will be reviewed (the next section looks at the most comment continous distributions). These are the *Bernoulli family* of distributions, since they all derive from the first,

1) Bernoulli
2) Binomial
3) Geometric
4) Negative Binomial
5) Poisson
6) Exponential
7) Gamma

The Poisson, Exponential, and Gamma are the continuous-time analogs of the Binomial, Geometric, and Negative Binomial, respectively, arising when each instant in time is viewed as an independent Bernoulli trial. In this section we consider the *discrete* members of the Bernoulli family, which are the first five members listed above, looking briefly at the main characteristics of each of these distributions and describing how they are most commonly used in practice.

For a more complete description of these distributions, the interested reader should consult an introductory textbook on probability and statistics, such as Law and Kelton (2000) or Devore (2003).

5.1 Bernoulli Trials

All of the discrete distributions considered in this section (and the first two in the next section) are derived from the idea of *Bernoulli Trials*. A Bernoulli trial is an experiment which has only two possible outcomes, *success* or *failure* (or [1, 0], or [true, false], or [$<5, \geq 5$], etc). If a sequence of Bernoulli trials are mutually independent with constant (stationary) probability, p, of success, then the sequence is called a *Bernoulli Process*. There are many examples of Bernoulli processes: one might model the failures of earth dams using a Bernoulli process. The success or failure of each of a sequence of bids made by a company might be a Bernoulli process. The failure of piles to support the load applied on them might be a Bernoulli process if it can be assumed that the piles fail (or survive) independently and with constant probability. However, if the failure of one pile is dependent on the failure of adjacent piles, as might be the case if the soil structures are similar and load transfer takes place, the Bernoulli model may not be appropriate and a more complex, 'dependent', model may be required, e.g. random field modeling of the soil and finite element analysis of the structural response within a Monte Carlo simulation. Evidently, when we depart from satisfying the assumptions underlying the simple models, such as those required for the Bernoulli model, the required models rapidly become very much more complicated. In some cases, applying the simple model to the more complex problem will yield a ballpark estimate, or at least a bound on the probability, and so it may be appropriate to proceed with a Bernoulli model taking care to treat the results as approximate. The degree of approximation depends very much on the degree of dependence between 'trials' and the 'stationarity' of the probability of 'success', p.

If we let

$$X_j = \begin{cases} 1 & \text{if the } j^{th} \text{ trial results in a success,} \\ 0 & \text{if the } j^{th} \text{ trial results in a failure} \end{cases} \tag{25}$$

then the Bernoulli distribution, or probability mass function, is given by

$$\begin{aligned} \mathrm{P}\left[X_j = 1\right] &= p \\ \mathrm{P}\left[X_j = 0\right] &= 1 - p = q \end{aligned} \tag{26}$$

for all $j = 1, 2, \ldots$. Note that we commonly denote $1 - p$ as q for simplicity.

For a single Bernoulli trial the following results hold

$$\mathrm{E}\left[X_j\right] = \sum_{i=0}^{1} i \cdot \mathrm{P}\left[X_j = i\right] = 0(1-p) + 1(p) = p \tag{27a}$$

$$\mathrm{E}\left[X_j^2\right] = \sum_{i=0}^{1} i^2 \cdot \mathrm{P}\left[X_j = i\right] = 0^2(1-p) + 1^2(p) = p$$

$$\text{Var}[X_j] = \text{E}\left[X_j^2\right] - \text{E}^2[X_j] = p - p^2 = pq \tag{27b}$$

For a sequence of trials, the assumption of independence between the trials means that

$$\text{P}[X_1 = x_1 \cap X_2 = x_2 \cap \cdots X_n = x_n] = \text{P}[X_1 = x_1]\,\text{P}[X_2 = x_2]\cdots\text{P}[X_n = x_n] \tag{28}$$

The Maximum Likelihood Estimate of p is just the average of the set of observations, $x_1, x_2, \ldots, x_n$, of X,

$$\hat{p} = \frac{1}{n}\sum_{i=1}^{n} x_i \tag{29}$$

Notice that we use a hat to indicate that this is just an *estimate* of the true parameter p. Since the next set of observations will likely give a different value for $\hat{p}$, we see that $\hat{p}$ is actually a random variable itself, rather than the true population parameter, which is non-random. The mean and variance of the sequence of $\hat{p}$ can be found by considering the random $\hat{P}$,

$$\hat{P} = \frac{1}{n}\sum_{i=1}^{n} X_i \tag{30}$$

obtained *prior* to observing the results of our Bernoulli trials. We get

$$\begin{aligned}\text{E}\left[\hat{P}\right] &= \text{E}\left[\frac{1}{n}\sum_{i=1}^{n} X_i\right] \\ &= \frac{1}{n}\sum_{i=1}^{n}\text{E}[X_i] = \frac{1}{n}(np) \\ &= p\end{aligned} \tag{31}$$

which means that the estimator given by Eq. (29) is *unbiased* (that is, the estimator is 'aimed' at its desired target on average).

The estimator variance is

$$\begin{aligned}\text{Var}\left[\hat{P}\right] &= \text{Var}\left[\frac{1}{n}\sum_{i=1}^{n} X_i\right] \\ &= \frac{1}{n^2}\sum_{i=1}^{n}\text{Var}[X_i] = \frac{1}{n^2}(npq) \\ &= \frac{pq}{n}\end{aligned} \tag{32}$$

where we made use of the fact that the variance of a sum is the sum of the variances **if the random variables are uncorrelated.** We are assuming that, since this is a Bernoulli process, not only are the random variables uncorrelated, they are completely independent (the probability of one occurring is not affected by the probability of other occurrences).

Note that the estimator variance depends on the true value of p on the right-hand-side of Eq. (32). Since we are estimating p, we obviously don't know the true value. The solution is to use our estimate of p to estimate its variance, so that

$$\sigma^2_{\hat{p}} \simeq \frac{\hat{p}\hat{q}}{n} \tag{33}$$

Once we have determined the estimator variance, we can compute its *standard error*, which is commonly taken to be equal to the standard deviation and which gives an indication of how accurate our estimate is,

$$\sigma_{\hat{p}} \simeq \sqrt{\frac{\hat{p}\hat{q}}{n}} \tag{34}$$

For example, if $\hat{p} = 0.01$, then we would prefer $\sigma_{\hat{p}}$ to be quite a bit smaller than 0.01 and we can adjust the number of observations, n, to achieve this goal.

Later in this book, we will be estimating the probability of failure, p_f, of various classic geotechnical problems using a technique called *Monte Carlo simulation*. The standard error given by Eq. (34) will allow us to estimate the accuracy of our failure probability estimates, assuming that each 'simulation' results in an independent failure/success 'trial'.

Applications

The classic Bernoulli trial is the toss of a coin, but many other experiments can lead to Bernoulli trials under the above conditions. Consider the following examples;

1) Soil anchors at a particular site have a 1% probability of pulling out. When an anchor is examined, it is classified as a *success* if it has not pulled out, or a *failure* if it has. This is a Bernoulli trial with $p = 0.99$ if the anchors fail independently and if the probability of success remains constant from trial to trial.
2) Suppose that each sample of soil at a site has a 10% chance of containing significant amounts of chromium. A sample is analyzed and classified as a *success* if it does not contain significant amounts of chromium, and a *failure* if it does. This is a Bernoulli trial with $p = 0.90$ if the samples are independent and if the probability of success remains constant from trial to trial.
3) A highway through a certain mountain range passes below a series of steep rock slopes. It is estimated that each rock slope has a 2% probability of failure (resulting in some amount of rock blocking the highway) over the next 10 year. If we define each rock slope as a trial which is a success if it does not fail in the next 10 years, then this can be modeled as a Bernoulli trial with $p = 0.98$ (assuming rock slopes fail independently – which might not be a good assumption if they generally fail due to earthquakes...).

5.2 Binomial Distribution

Let N_n be the number of successes in n Bernoulli trials, each with probability of success p. Then N_n follows a binomial distribution where

$$\mathrm{P}\left[N_n = k\right] = \binom{n}{k} p^k q^{n-k}, \quad k = 0, 1, 2, \ldots, n \tag{35}$$

The quantity $p^k q^{n-k}$ is the probability of obtaining k successes and $n-k$ failures in n trials and $\binom{n}{k}$ is the number of possible ways of arranging the k successes over the n trials.

For example, consider 8 trials, which can be represented as a series of 8 dashes:

___ ___ ___ ___ ___ ___ ___ ___ .

One possible realization of 3 successes in 8 trials might be:

F S F F S S F F

where successes are shown as S and failures as F. Another possible realization might be

S F F S F F F S

and so on. Clearly these involve 3 successes, which have probability p^3, and 5 failures, which have probability q^5. Combining these two probabilities with the fact that 3 successes in 8 trials can be arranged in $\binom{8}{3}$ different ways leads to

$$\mathrm{P}\left[N_8 = 3\right] = \binom{8}{3} p^3 q^{8-3}$$

which generalizes to the binomial distribution, for n trials and k successes, given above.

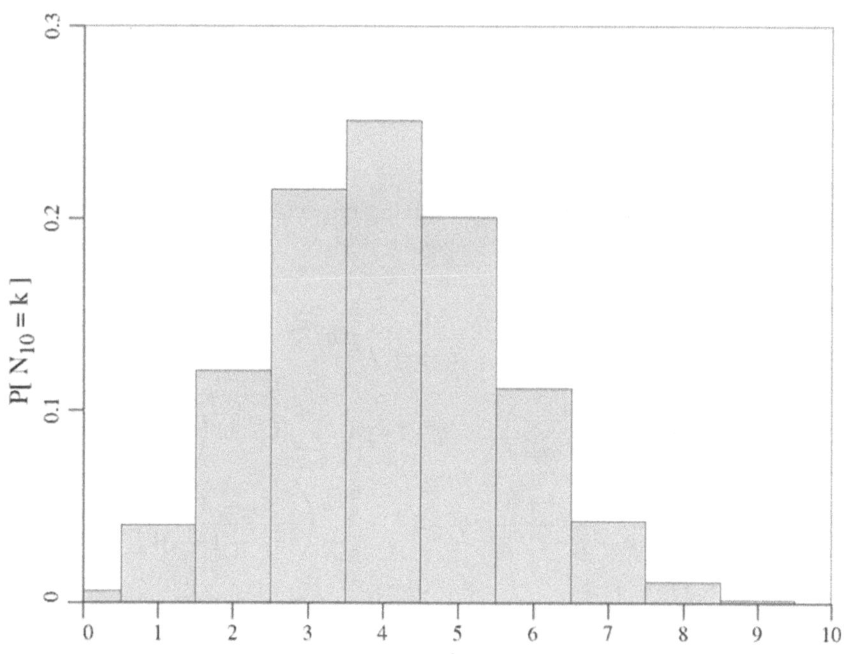

Figure 6. Binomial distribution for $n = 10$ and $p = 0.4$.

Properties:
In the following proofs, we make use of the binomial theorem, which states that

$$(\alpha+\beta)^n = \sum_{i=0}^{n} \binom{n}{i} \alpha^i \beta^{n-i} = \sum_{i=0}^{n} \frac{n!}{i!(n-i)!} \alpha^i \beta^{n-i} \tag{36}$$

The expected number of successes in n trials can be found directly from the definition of the discrete case expectation,

$$\begin{aligned}
\mathrm{E}[N_n] &= \sum_{i=0}^{n} i \binom{n}{i} p^i q^{n-i} = \sum_{i=0}^{n} i \left(\frac{n!}{i!(n-i)!} \right) p^i q^{n-i} \\
&= np \sum_{i=1}^{n} \frac{(n-1)!}{(i-1)!(n-i)!} p^{i-1} q^{n-i} = np \sum_{i=0}^{(n-1)} \frac{(n-1)!}{i!((n-1)-i)!} p^i q^{(n-1)-i} \\
&= np(p+q)^{n-1} \\
&= np
\end{aligned} \tag{37}$$

since $p+q=1$.
Alternatively, we could write

$$\begin{aligned}
\mathrm{E}[N_n] &= \mathrm{E}[X_1 + X_2 + \cdots + X_n] \\
&= \mathrm{E}[X_1] + \mathrm{E}[X_2] + \cdots + \mathrm{E}[X_n] \\
&= np
\end{aligned}$$

where X_i is a Bernoulli random variable, having expectation p.

To find the variance of N_n, we first need to find

$$\begin{aligned}
\mathrm{E}\left[N_n^2\right] &= \sum_{i=0}^{n} i^2 \binom{n}{i} p^i q^{n-i} = \sum_{i=1}^{n} i^2 \left(\frac{n!}{i!(n-i)!} \right) p^i q^{n-i} \\
&= np \sum_{i=1}^{n} i \left(\frac{(n-1)!}{(i-1)!(n-i)!} \right) p^{i-1} q^{n-i} = np \sum_{i=0}^{n-1} (i+1) \left(\frac{(n-1)!}{i!(n-1-i)!} \right) p^i q^{n-1-i} \\
&= np \left\{ \sum_{i=0}^{n-1} i \left(\frac{(n-1)!}{i!(n-1-i)!} \right) p^i q^{n-1-i} + \sum_{i=0}^{n-1} \left(\frac{(n-1)!}{i!(n-1-i)!} \right) p^i q^{n-1-i} \right\} \\
&= np\left\{(n-1)p+1\right\} \\
&= npq + n^2p^2
\end{aligned}$$

where for the first sum, we made use of the result given by Eq. 37. The variance is thus,

$$\mathrm{Var}[N_n] = \mathrm{E}\left[N_n^2\right] - \mathrm{E}^2[N_n] = npq + n^2p^2 - n^2p^2 = npq \tag{38}$$

The same result could have been obtained much more easily by considering the variance of a sum of independent random variables, since in this case, the variance of a sum is the sum of the variances;

$$\mathrm{Var}\left[N_n\right] = \mathrm{Var}\left[\sum_{i=1}^{n} X_i\right] = \sum_{i=1}^{n} Var X_i = npq$$

5.3 Geometric Distribution

Consider a Bernoulli process in which T_1 is the number of trials required to achieve the first success. Thus, if $T_1 = 3$, then we must have had 2 failures followed by a success (the value of T_1 fully prescribes the sequence of trials). This has probability

$$\mathrm{P}\left[T_1 = 3\right] = \mathrm{P}\left[\{\text{failure, failure, success}\}\right] = q^2 p$$

In general

$$\mathrm{P}\left[T_1 = k\right] = q^{k-1} p, \quad k = 1, 2, \ldots \tag{39}$$

Note that this is a valid probability mass function since

$$\sum_{k=1}^{\infty} q^{k-1} p = p \sum_{k=0}^{\infty} q^k = \frac{p}{1-q} = 1$$

where we used the fact that for any $|alpha| < 1$ (see, e.g., Gradshteyn and Ryzhik, 1980),

$$\sum_{k=0}^{\infty} \alpha^k = \frac{1}{1-\alpha} \tag{40}$$

As an example, in terms of the actual sequence of trials, the event that the first success occurs on the 8th trial appears as

$\underline{\text{ F }}\ \ \underline{\text{ F }}\ \ \underline{\text{ F }}\ \ \underline{\text{ F }}\ \ \underline{\text{ F }}\ \ \underline{\text{ F }}\ \ \underline{\text{ F }}\ \ \underline{\text{ S }}$

That is, the single success always occurs on the last trial. If $T_1 = 8$, then we have had 7 failures, having probability q^7, and 1 success, having probability p. Thus

$$\mathrm{P}\left[T_1 = 8\right] = q^7 p$$

Generalizing this for $T_1 = k$ leads to the geometric distribution shown above.

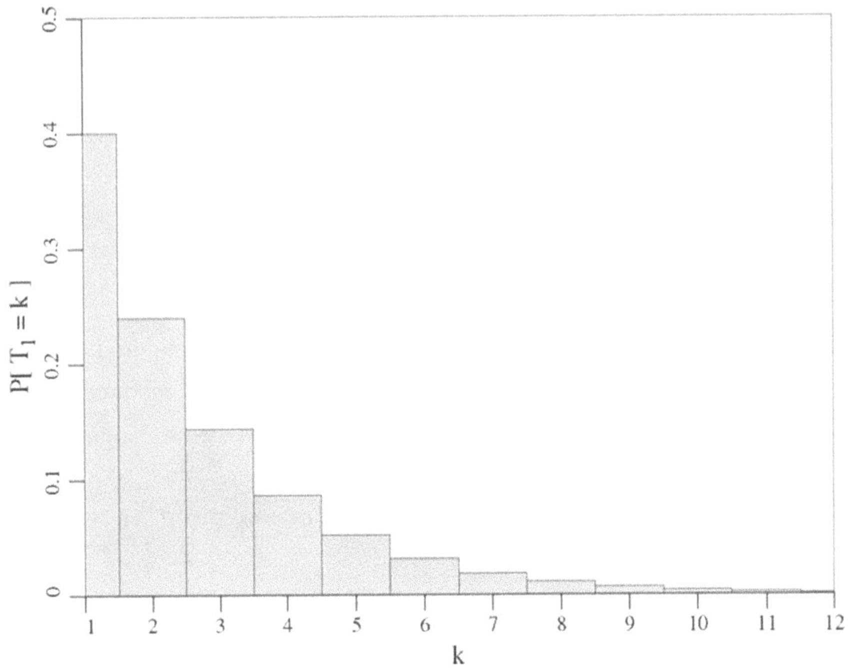

Figure 7. Geometric distribution for $p = 0.4$.

Because trials are assumed independent, the geometric distribution also models the number of trials *between* successes in a Bernoulli process. That is, suppose we observe the result of the Bernoulli process at trial number 1032. We will observe either a success or failure, but whichever is observed, it is now *known*. We can then ask a question such as: What is the probability that the next success occurs on trial 1040? To determine this, we start with trial 1032. Because we have observed it there is no uncertainty associated with trial 1032, so it does not enter into the probability problem. However, trials 1033, 1034, ..., 1040 are unknown. We are asking for the probability that trial 1040 is the first success after 1032. In order for this *event* to occur, trials 1033 to 1039 must be failures. Thus, the 8 trials, 1033 to 1040, must involve seven failures (q^7) followed by one success (p). The required probability is just the product

$$\mathrm{P}\left[T_1 = 8\right] = q^7 p$$

What this means is that the geometric distribution, by virtue of the independence between trials, is *memoryless*. It doesn't matter when you start looking at a Bernoulli process, the number of trials to the next 'success' is given by the geometric distribution (and is independent of the trial number).

Properties:

The mean of T_1, which is also sometimes refered to as the *return period* or the *mean recurrence time*, is determined as follows;

$$\begin{aligned} \mathrm{E}[T_1] &= \sum_{k=1}^{\infty} kpq^{k-1} = p\sum_{k=1}^{\infty} kq^{k-1} = p\frac{d}{dq}\sum_{k=1}^{\infty} q^k = p\frac{d}{dq}\left(\frac{q}{1-q}\right) \\ &= p\left(\frac{1}{(1-q)^2}\right) \\ &= \frac{1}{p} \end{aligned} \tag{41}$$

where we used Eq. 40 to evaluate the final sum above. We will use the second to last sum in the following proof.

The variance of T_1 is obtained from $\mathrm{Var}[T_1] = \mathrm{E}\left[T_1^2\right] - \mathrm{E}^2[T_1]$ as follows;

$$\begin{aligned} \mathrm{E}\left[T_1^2\right] &= \sum_{k=1}^{\infty} k^2pq^{k-1} = p\sum_{k=1}^{\infty} k^2q^{k-1} = p\frac{d}{dq}\sum_{k=1}^{\infty} kq^k \\ &= p\frac{d}{dq}\left(\frac{q}{(1-q)^2}\right) \\ &= \frac{1}{p} + \frac{2q}{p^2} \end{aligned}$$

thus

$$\begin{aligned} \mathrm{Var}[T_1] &= \mathrm{E}\left[T_1^2\right] - \mathrm{E}^2[T_1] \\ &= \frac{1}{p} + \frac{2q}{p^2} - \frac{1}{p^2} \\ &= \frac{q}{p^2} \end{aligned} \tag{42}$$

Aside: In engineering problems, we often reverse the meaning of 'success' and 'failure', and use the geometric distribution to model 'time' to failure, where 'time' is measured in discrete steps (trials).

5.4 Negative Binomial Distribution

Suppose we wish to know the number of trials ('time') in a Bernoulli process until the mth success. Letting T_m be the number of trials until the mth success, then

$$\mathrm{P}[T_m = k] = \binom{k-1}{m-1} p^m q^{k-m} \qquad \text{for } k = m, m+1, \ldots \tag{43}$$

which is the *negative binomial distribution.* Whereas a binomial distributed random variable is the number of successes in a fixed number of trials, a negative binomial distributed random variable is the number of trials for a fixed number of successes.

We note that the negative binomial is also often used to model the number of failures before the mth success, which results in a somewhat different distribution. We prefer the interpretation that the negative binomial distribution governs the number of trials until the mth success because it is a natural generalization of the geometric distribution and because it is then a discrete analog of the Gamma distribution considered in Section 1.6.2.

The name of the negative binomial distribution arises from the *negative binomial series*

$$(1-q)^{-m} = \sum_{k=m}^{\infty} \binom{k-1}{m-1} q^{k-m} \tag{44}$$

which converges for $|q| < 1$. This series can be used to show that the negative binomial distribution is a valid distribution, since

$$\sum_{k=m}^{\infty} \mathrm{P}\left[T_m = k\right] = \sum_{k=m}^{\infty} \binom{k-1}{m-1} p^m q^{k-m} = p^m \sum_{k=m}^{\infty} \binom{k-1}{m-1} q^{k-m} = p^m (1-q)^{-m} = 1 \tag{45}$$

as expected.

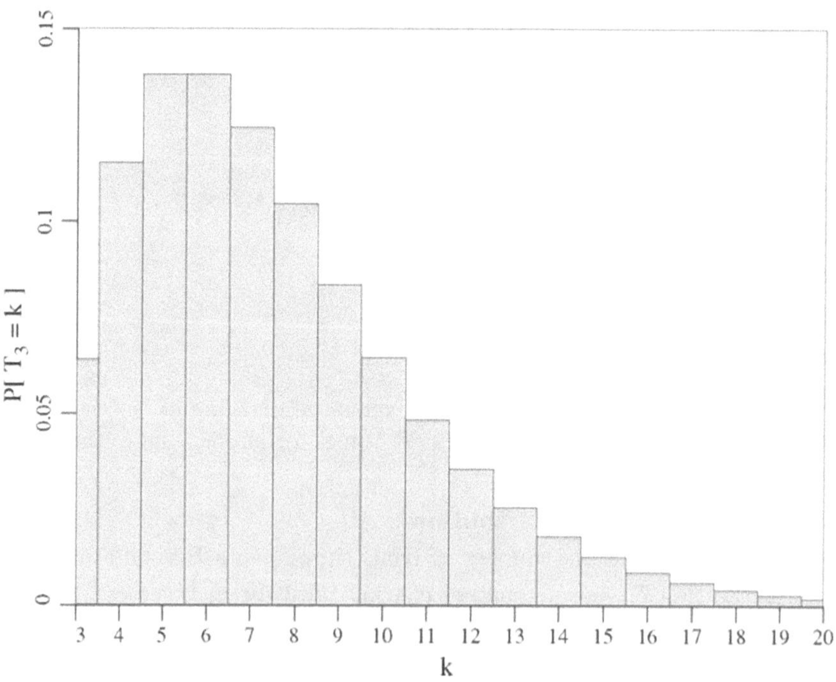

Figure 8. Negative binomial distribution for T_3 (ie. $m = 3$) and $p = 0.4$.

We see that the Geometric distribution is a special case of the Negative Binomial distribution with $m = 1$. The negative binomial distribution is often used to model 'time to the mth failure', where 'time' is measured in discrete steps, or trials. Consider one possible realization which has the 3rd success on the 8th trial;

F S S F F F F S

Another possible realization might be

F F F S F S F S

In both cases, the number of successes is 3, having probability p^3, and the number of failures is 5, having probability q^5. In terms of ordering, if $T_3 = 8$, then the 3rd success must occur on the 8'th trial (as shown above). Thus, the only other uncertainty is the ordering of the other two successes. This can occur in $\binom{7}{2}$ ways. The probability that the 3rd success occurs on the 8th trial is therefore given by

$$\mathrm{P}\left[T_3 = 8\right] = \binom{7}{2} p^3 q^5$$

Generalizing this for m successes and k trials leads to the negative binomial distribution shown above.

Properties:

Mean:

$$\begin{aligned}
\mathrm{E}\left[T_m\right] &= \sum_{j=m}^{\infty} j\mathrm{P}\left[T_m = j\right] = \sum_{j=m}^{\infty} j\binom{j-1}{m-1} p^m q^{j-m} = \sum_{j=m}^{\infty} j\left(\frac{(j-1)!}{(m-1)!(j-m)!}\right) p^m q^{j-m} \\
&= mp^m \sum_{j=m}^{\infty} \left(\frac{j!}{m!(j-m)!}\right) q^{j-m} \\
&= mp^m \left[1 + (m+1)q + \frac{(m+2)(m+1)}{2!}q^2 + \frac{(m+3)(m+2)(m+1)}{3!}q^3 + \cdots\right] \\
&= \frac{mp^m}{(1-q)^{m+1}} \\
&= \frac{m}{p}
\end{aligned} \tag{46}$$

which is just m times the mean of a single geometrically distributed random variable, T_1, as expected since the number of trials between successes follows a geometric distribution. In fact, this observation leads to the following alternative representation of T_m,

$$T_m = T_{1,1} + T_{1,2} + \cdots + T_{1,m} \tag{47}$$

where $T_{1,1}$ is the number of trials until the first success, $T_{1,2}$ is the number of trials after the first success until the second success, and so on. That is, the $T_{1,i}$ terms are just the 'times' between successes. Since all trials are independent, each of the $T_{1,i}$ terms will be independent geometrically distributed random variables, all having common probability of success, p. This leads to the following much simpler computation,

$$\mathrm{E}\left[T_m\right] = \mathrm{E}\left[T_{1,1}\right] + \mathrm{E}\left[T_{1,2}\right] + \cdots + \mathrm{E}\left[T_{1,m}\right] = \frac{m}{p} \tag{48}$$

since $\mathrm{E}[T_{1,i}] = 1/p$ for all $i = 1, 2, \ldots, m$. The mean in Figure 8 is $3/0.4 = 7.5$.

Variance:

To get the variance, $\mathrm{Var}[T_m]$, we'll again use Eq. 47. Due to independence of the $T_{1,i}$ terms, the variance of the sum is the sum of the variances,

$$\begin{aligned}\mathrm{Var}[T_m] &= \mathrm{Var}[T_{1,1}] + \mathrm{Var}[T_{1,2}] + \cdots + \mathrm{Var}[T_{1,m}] \\ &= m\mathrm{Var}[T_1] \\ &= \frac{mq}{p^2} \end{aligned} \tag{49}$$

which is just m times the variance of a single geometrically distributed random variable, T_1, as expected.

5.5 Poisson Distribution

The Poisson distribution governs many 'rate' dependent processes – for example, arrivals of vehicles at an intersection or number of points where a soil's cohesion exceeds some high threshold in a region. The Poisson is yet another distribution arising from the Bernoulli family and can be derived directly from the binomial distribution by letting each instant in time (or space) become an independent Bernoulli trial. For simplicity, we will talk about Poisson processes in time, but recognize that they can be equivalently applied over space, simply by replacing t with a distance (or area, volume, etc) measure.

For any non-zero time interval we have an infinite number of Bernoulli trials, since any time interval is made up of an infinite number of instants. Thus, the probability of success, p, in any one instant must go to zero – otherwise we would have an infinite number of successes in each time interval ($np \to \infty$ as $n \to \infty$). This means that we must abandon the probability of success, p, in favour of a *mean rate of success*, λ, which quantifies the mean number of successes per unit time.

The basic assumption on which the Poisson distribution rests is that each instant in time is a Bernoulli trial, with mean success (arrival) rate given by the parameter λ. This basic assumption leads to the following statements (which also define λ);

1) successes occur at random and at any point in time (or space),
2) the occurrence of a success in a given time (or space) interval is independent of successes occurring in all other disjoint intervals,
3) the probability of a success occurring in a small interval, Δt,is proportional to the size of Δt, ie., is $\lambda \Delta t$, where λ is the mean *rate* of occurrence.
4) for $\Delta t \to 0$, the probability of two or more successes in Δt is negligible (e.g. a Bernoulli trial can only have one success).

Now define N_t to be the number of successes ('arrivals' or 'occurrences') occurring in time t. If the above assumptions hold, then N_t is governed by the following distribution,

$$\mathrm{P}[N_t = k] = \frac{(\lambda t)^k}{k!} e^{-\lambda t}, \quad k = 0, 1, 2, \ldots \tag{50}$$

where λ is the mean rate of occurrence (λ has units of $1/\text{time}$).

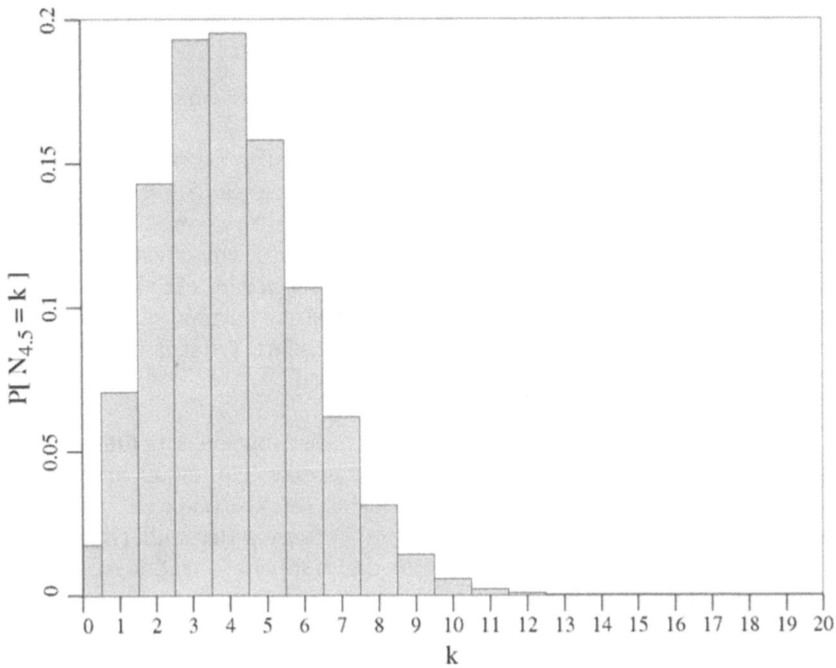

Figure 9. Poisson distribution for $t = 4.5$ and $\lambda = 0.9$.

Properties

Mean:

$$
\begin{aligned}
\mathrm{E}[N_t] &= \sum_{j=0}^{\infty} j\,\frac{(\lambda t)^j}{j!}\,e^{-\lambda t} = \lambda t e^{-\lambda t} \sum_{j=1}^{\infty} \frac{(\lambda t)^{j-1}}{(j-1)!} \\
&= \lambda t e^{-\lambda t} \sum_{j=0}^{\infty} \frac{(\lambda t)^j}{j!} \\
&= \lambda t
\end{aligned} \tag{51}
$$

The mean of the distribution shown in Figure 9 is $\mathrm{E}[N_{4.5}] = 0.9(4.5) = 4.05$.

Variance:

$$
\begin{aligned}
\mathrm{E}\left[N_t^2\right] &= \sum_{j=0}^{\infty} j^2\,\frac{(\lambda t)^j}{j!}\,e^{-\lambda t} = \lambda t e^{-\lambda t} \sum_{j=0}^{\infty} (j+1)\frac{(\lambda t)^j}{j!} \\
&= \lambda t e^{-\lambda t} \left[\sum_{j=0}^{\infty} j\,\frac{(\lambda t)^j}{j!} + \sum_{j=0}^{\infty} \frac{(\lambda t)^j}{j!} \right] \\
&= (\lambda t)^2 + (\lambda t)
\end{aligned}
$$

thus

$$\mathrm{Var}\left[N_t\right] = \mathrm{E}\left[N_t^2\right] - \mathrm{E}^2[N_t] = \lambda t \tag{52}$$

That is, the mean and variance of a Poisson process are the same.

Derivation from the Binomial Distribution

The Poisson distribution is often used to model 'arrival' processes. We shall see later that it is also useful to model so called 'excursion' processes (e.g. the number of times that a soil's property falls below some low threshold in a region). For the time being, however, we will consider a simple example of arrivals of customers at a bank, which should appeal to anyone who has had to line up for service in a bank.

Suppose that a financial engineer is monitoring customer arrivals at a bank and verifies that an average of 60 customers enter the bank per hour. What is the probability that 10 customers will arrive in the next 10-minute interval?

Solution:

In an attempt to model this using the binomial distribution, lets first divide time up into a series of intervals within each of which a customer can either arrive (success) or not (failure). As a starting point, let us assume that at most one customer can arrive in any 30 second interval. We make this assumption because a Bernoulli trial can only have two outcomes. If we wish to be able to count the number of customers, we must make these two possible outcomes that either 1 customer arrives (success) or 0 customers arrive (failure). If our 'trials' are 30 seconds in duration, then we have 120 trials per hour and the probability of a 'success' in each trial is the rate per hour divided by the number of trials per hour: $p = 60/120 = 0.5$. In our ten minute interval, we have $n = 600/30 = 20$ trials, so the probability we are looking for is

$$\mathrm{P}\left[10 \text{ customers in 10 minutes}\right] \simeq \binom{20}{10}(0.5)^{10}(0.5)^{20-10} = 0.176$$

Of course, we know that two or more customers could easily arrive within 30 seconds (the bank would approve of this). An improved solution is obtained by using a shorter 'trial' interval. If 10-second intervals were to be used then the probability of arrival of a customer in any interval becomes $p = 60/360 = \frac{1}{6}$ and the number of trials in 10 minutes (600 seconds) becomes $n = 600/10 = 60$

$$\mathrm{P}\left[10 \text{ customers in 10 minutes}\right] \simeq \binom{60}{10}(\tfrac{1}{6})^{10}(\tfrac{5}{6})^{50} = 0.137$$

In general, if time interval t is divided into n intervals then $p = \frac{\lambda t}{n}$ and

$$\mathrm{P}\left[N_t = k\right] = \binom{n}{k}\left(\frac{\lambda t}{n}\right)^k\left(1 - \frac{\lambda t}{n}\right)^{n-k}$$

where λt is the mean number of arrivals occurring in the time t. Now if arrivals pass through the bank's door 'instantaneously' and can arrive at any instant in time then

$$\mathrm{P}\left[N_t = k\right] = \lim_{n\to\infty}\binom{n}{k}\left(\frac{\lambda t}{n}\right)^k\left(1 - \frac{\lambda t}{n}\right)^{n-k}$$

$$= \lim_{n\to\infty} \left[\left\{ \frac{n}{n} \cdot \frac{n-1}{n} \cdots \frac{n-k+1}{n} \right\} \frac{(\lambda t)^k}{k!} \left(1 - \frac{\lambda t}{n}\right)^n \left(1 - \frac{\lambda t}{n}\right)^{-k} \right]$$

but since

$$\lim_{n\to\infty} \left\{ \frac{n}{n} \cdot \frac{n-1}{n} \cdots \frac{n-k+1}{n} \right\} = 1$$

$$\lim_{n\to\infty} \left(1 - \frac{\lambda t}{n}\right)^{-k} = 1$$

$$\lim_{n\to\infty} \left(1 - \frac{\lambda t}{n}\right)^{n} = e^{-\lambda t}$$

then our distribution simplifies to

$$\mathrm{P}\left[N_t = k\right] = \frac{(\lambda t)^k}{k!}\, e^{-\lambda t}$$

which is the Poisson distribution. In other words, the Poisson distribution is a limiting case of the Binomial distribution, obtained when the number of 'trials' goes to infinity, one for each instant in time, and p is replaced by the mean rate λ.

For our problem $\lambda = 1$ arrival/minute and $t = 10$ minutes so that, using the Poisson distribution

$$\mathrm{P}\left[N_{10} = 10\right] = \frac{(10)^{10}}{10!}\, e^{-10} = 0.125.$$

and we see that the binomial model using 10-second trial intervals gives a reasonably close result (with a relative error of less than 10%). We should also note that the Poisson model, in this case, is perhaps a bit too 'precise' since it is unlikely that any of us can pass through a door instantaneously, try as we might. Thus, the true probability will likely lie somewhere between 0.125 and 0.137.

6 Common Continuous Probability Distributions

Many naturally occurring, and continuous, random phenomena can be well modeled by a relatively small number of distributions. The following six continuous distributions are particularly common in engineering applications;

1) Exponential
2) Gamma
3) Uniform
4) Weibull
5) Rayleigh
6) Normal
7) Lognormal

As mentioned in the previous section, the Exponential and Gamma distributions are members of the *Bernoulli family*, deriving from the idea that each instant in time constitutes an independent Bernoulli trial. These are the continuous time analogs of the Geometric and Negative Binomial distributions.

Aside from the above list, there are certainly other continuous distributions which may be considered. Distributions which involve more than two parameters are generally difficult to justify because we rarely have enough data to estimate even two parameters with much accuracy. From a practical point of view what this means is that even if a geotechnical researcher has large volumes of data at a particular site, and can accurately estimate, for example, a modified six parameter Beta distribution, it is unlikely that anyone else will be able to do so at other sites. Thus, complex distributions, such as a six parameter Beta distribution, are of questionable at any site other than the site at which it was estimated.

As with the common discrete distributions, this section looks briefly at the main characteristics of each of these continous distributions and describes how they are most commonly used in practice. For a more complete description of these distributions, the interested reader should consult an introductory textbook on probability and statistics, such as Law and Kelton (2000) or Devore (2003).

6.1 Exponential Distribution

The exponential distribution is yet another distribution derived from the Bernoulli family – it is the continuous analog of the geometric distribution. Recall that the geometric distribution governs the number of trials until the first 'success' (or to the next success). If we imagine that each instant in time is now an independent trial, then the time until the first success is given by the exponential distribution (the mathematics associated with this transition from the geometric distribution involving 'discrete' trials to a 'continuous' sequence of trials is similar to that shown above for the transition from the binomial to the Poisson distribution and will not be repeated here).

As with the geometric distribution, the exponential distribution is often used to describe 'time-to-failure' type problems. It also governs the time between 'arrivals' of a Poisson process. If T_1 is the time to the occurrence (or failure) in question and T_1 is exponentially distributed, then its probability density function is

$$f_{T_1}(t) = \lambda e^{-\lambda t} \qquad t \geq 0 \tag{53}$$

where λ is the *mean rate* of occurrence (or failure). Its cumulative distribution function is

$$F_{T_1}(t) = \mathrm{P}\left[T_1 \leq t\right] = 1 - e^{-\lambda t} \qquad t \geq 0 \tag{54}$$

Properties:

$$\mathrm{E}\left[T_1\right] = \frac{1}{\lambda} \tag{55a}$$

$$\mathrm{Var}\left[T_1\right] = \frac{1}{\lambda^2} \tag{55b}$$

That is, the mean and standard deviation of an exponentially distributed random variable are equal.

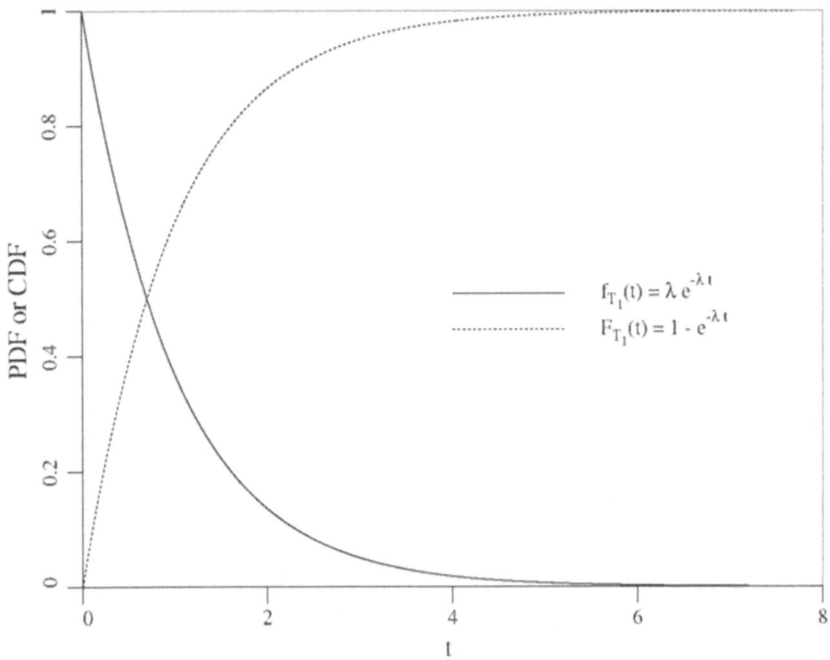

Figure 10. Exponential distribution for $\lambda = 1$.

Memoryless Property:

We will illustrate this property with an example: Let T_1 denote the time between arrivals of a customer at a bank. Assume that T_1 has an exponential distribution with a mean of 4 minutes. Thus, T_1 has mean 'arrival' rate of $\lambda = 1/4 = 0.25$ per minute. The probability that a customer arrives within the next 30 seconds is

$$\mathrm{P}\left[T_1 < 30 \text{ sec}\right] = \mathrm{P}\left[T_1 < 0.5 \text{ min}\right] = 1 - e^{-0.5 \times 0.25} = 0.1175$$

Now, suppose that the bank is empty at 11:00 am and 8 minutes pass without a customer arriving. What is the probability that a customer will arrive in the next 30 seconds? Because 8 minutes have gone by without an arrival, you might feel that an arrival is overdue and therefore more likely. That is, that the probability of an arrival in the next 30 seconds should be greater than 0.1175. However, for the exponential distribution, this is not the case, which is one of the features of the exponential distribution – the past is ignored. Each instant in time constitutes a trial which is independent of all other trials. In fact,

$$\mathrm{P}\left[T_1 < 8.5 \,|\, T_1 > 8\right] = \frac{\mathrm{P}\left[8 < T_1 < 8.5\right]}{\mathrm{P}\left[T_1 > 8\right]} = \frac{(1 - e^{-8.5 \times 0.25}) - (1 - e^{-8 \times 0.25})}{e^{-8 \times 0.25}} = 0.1175$$

Thus, after 8 minutes without an arrival, the probability of an arrival in the next 30 seconds is the same as the probability of an arrival in any 30 second interval. We found this same property existed in the Poisson process – indeed the times between arrivals in the Poisson process are exponentially distributed.

More generally, if T_1 is exponentially distributed with mean rate λ then

$$\begin{aligned} \mathrm{P}\left[T_1 > t+s \,|\, T_1 > t\right] &= \frac{\mathrm{P}\left[T_1 > t+s \cap T_1 > t\right]}{\mathrm{P}\left[T_1 > t\right]} = \frac{\mathrm{P}\left[T_1 > t+s\right]}{\mathrm{P}\left[T_1 > t\right]} = \frac{e^{-\lambda(t+s)}}{e^{-\lambda t}} \\ &= e^{-\lambda s} \\ &= \mathrm{P}\left[T_1 > s\right] \end{aligned} \tag{56}$$

Link to Poisson:
It was mentioned above that the exponential distribution governs the time between the occurrences of a Poisson process. This can be clearly seen through the following argument.

Let N_t be a Poisson distributed random variable with mean arrival rate λ. We wish to know the distribution of the time until the first 'arrival'. Let T_1 be the time to the first arrival. Then,

$$\mathrm{P}\left[T_1 > t\right] = \mathrm{P}\left[N_t = 0\right] = \frac{(\lambda t)^0}{0!} e^{\lambda t} = e^{-\lambda t}.$$

and so,

$$\mathrm{P}\left[T_1 \le t\right] = F_{T_1}(t) = 1 - e^{-\lambda t}.$$

But $1-e^{-\lambda t}$ is the cumulative distribution for the exponential probability density function $\lambda e^{-\lambda t}$. Consequently, T_1 must follow an exponential distribution with mean rate λ, ie. the time to the first occurrence in a Poisson process follows an exponential distribution with parameter λ which is equal to the Poisson rate λ. The same holds for the time *between* any occurrences of a Poisson process.

The link between all of Bernoulli Family distributions (binomial, geometric, negative binomial, Poisson, exponential, and Gamma) goes back to the idea of Bernoulli trials, where each trial is assumed to be independent. In the case of the Poisson, exponential, and Gamma (next subsection) distributions we are dealing with an infinite number of independent trials, one at each instant in time. The independence between trials means that it doesn't matter when we start looking – even if no one has arrived at our bank for 3 hours, the probability governing the time of the next arrival remains indifferent to the past. This makes sense if we assume that people go about their business entirely independently – that is, people don't (normally) plan their arrival at the bank around the banking plans of everyone else in the city.

6.2 Gamma Distribution

We consider here a particular form of the Gamma distribution, which is the continuous time analog of the negative binomial distribution. Specifically, if T_k is defined as the sum of k independent exponentially distributed random variables, E_i, each with parameter λ, that is $T_k = E_1 + E_2 + \cdots E_k$, then T_k has the following Gamma probability density function

$$f_{T_k}(t) = \frac{\lambda\,(\lambda t)^{k-1}}{(k-1)!} e^{-\lambda t} \qquad t \ge 0 \tag{57}$$

This form of the Gamma distribution is also refered to as the *k-Erlang distribution.* Note that $k = 1$ gives the exponential distribution, as expected. (The above distribution can be generalized to non-integer k if $(k-1)!$ is replaced by $\Gamma(k)$, which is the Gamma function – see Law and Kelton, 2000, for more information on the general Gamma distribution.)

To derive the cumulative distribution function, we integrate the above probability density function (by parts) to obtain, for integer k,

$$F_{T_k}(t) = \mathrm{P}\left[T \leq t\right] = 1 - e^{-\lambda t} \sum_{j=0}^{k-1} \frac{(\lambda t)^j}{j!} \tag{58}$$

The Gamma distribution, as defined here for integer k, is a member of the Bernoulli Family. That is, it derives from an infinite sequence of Bernoulli trials, one at each instant in time, with mean rate of success λ, and governs the time between every k'th occurrence of a 'success' in a Poisson process.

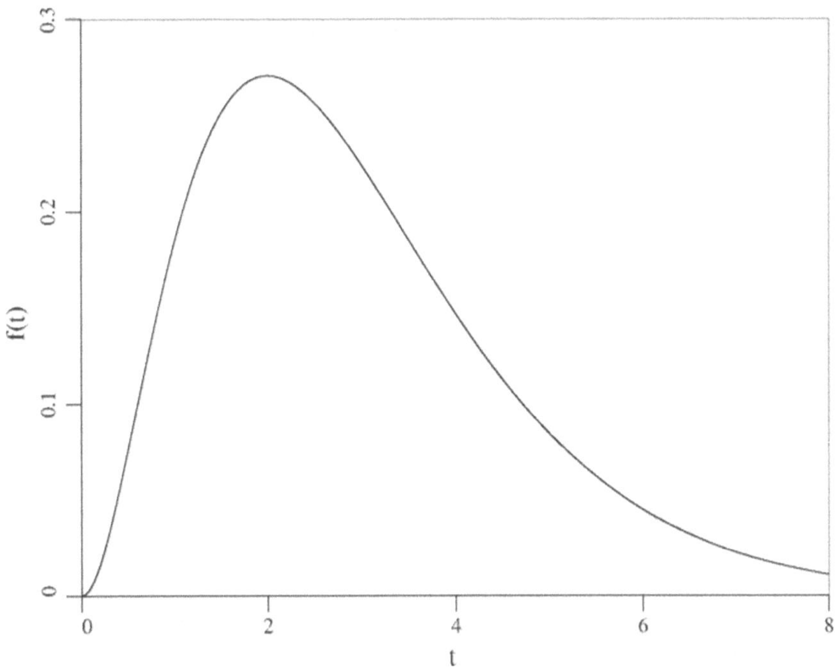

Figure 11. Gamma probability density function for $\lambda = 1$ and $k = 3$.

Properties:

$$\mathrm{E}\left[T_k\right] = \frac{k}{\lambda} \qquad \left(= k\mathrm{E}\left[E_i\right]\right) \tag{59a}$$

$$\mathrm{Var}\left[Y\right] = \frac{k}{\lambda^2} \qquad \left(= k\mathrm{Var}\left[E_i\right]\right) \tag{59b}$$

The Gamma distribution presented above is specialized to the sum of k independent and identically exponentially distributed random variables. It can be extended to other types of problems, so long as k is (at least approximately) a positive integer.

Although the Gamma distribution is not limited to integer values of k, the interpretation of the Gamma PDF as the distribution of a sum of independent, identically, and exponentially distributed random variables is lost if k is not an integer. The more general Gamma distribution has the form

$$f_X(x) = \frac{\lambda\,(\lambda x)^{k-1}}{\Gamma(k)} e^{-\lambda x} \qquad x \geq 0 \tag{60}$$

which is valid for any $k > 0$ and $\lambda > 0$. The *Gamma Function*, $\Gamma(k)$, for $k > 0$, is defined by the integral

$$\Gamma(k) = \int_0^\infty x^{k-1} e^{-x}\, dx \tag{61}$$

Tabulations of the Gamma Function can be found in Abramowitz and Stegun (1970), for example. However, when k is an integer, $\Gamma(k) = (k-1)!$.

6.3 Uniform Distribution

The continuous uniform distribution is the simplest of all continuous distribution since its density function is constant (over a range). Its general definition is

$$f(x) = \frac{1}{\beta - \alpha} \qquad \alpha \leq x \leq \beta$$

and its cumulative distribution is

$$F(x) = \mathrm{P}\left[X \leq x\right] = \frac{x - \alpha}{\beta - \alpha} \qquad \alpha \leq x \leq \beta \tag{62}$$

The uniform distribution is useful in representing random variables which have known *upper* and *lower* bounds and which have equal likelihood of occurring anywhere between these bounds. Another way of looking at the uniform distribution is that it is *non-informative* or *non-presumptive*. That is, if you know nothing else about the relative likelihood of a random variable, aside from its upper and lower bounds, then the uniform distribution is appropriate – it makes no assumptions regarding preferential likelihood of the random variable since all possible values are equilikely.

Properties:

$$\mathrm{E}\left[X\right] = \int_\alpha^\beta \frac{x\, dx}{\beta - \alpha} = \frac{\alpha + \beta}{2} \qquad \text{(this is the midpoint)} \tag{63a}$$

$$\mathrm{Var}\left[X\right] = \int_\alpha^\beta \frac{x^2}{\beta - \alpha}\, dx - \mathrm{E}^2[X] = \frac{(\beta - \alpha)^2}{12} \tag{63b}$$

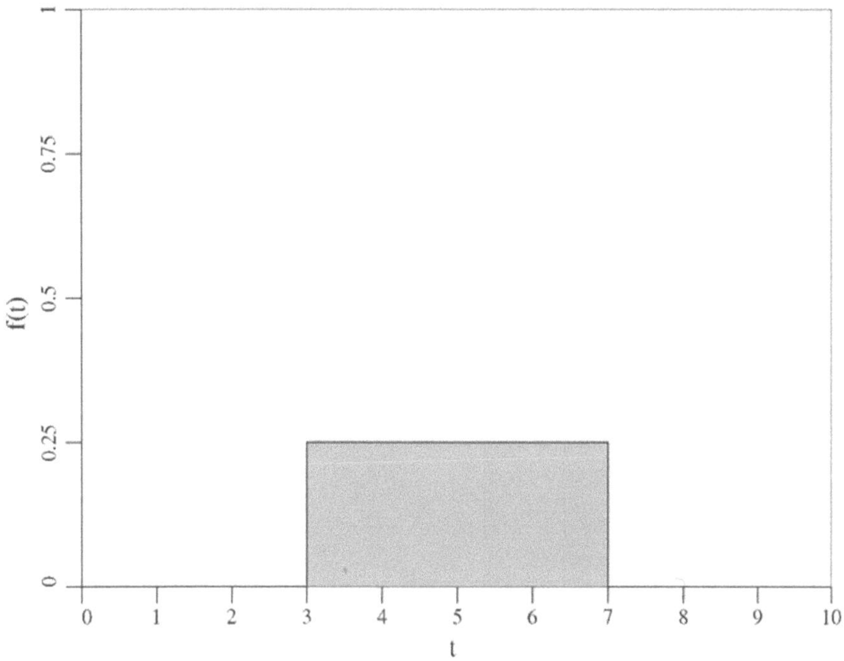

Figure 12. Uniform distribution for $\alpha = 3$ and $\beta = 7$.

6.4 Weibull Distribution

Often, engineers are concerned with the strength properties of materials and the lifetimes of manufactured devices. The Weibull distribution has become very popular in describing these types of problems. One of the attractive features of the Weibull distribution is that its cumulative distribution function is quite simple.

If a continuous random variable X has a Weibull distribution, then it has probability density function

$$f(x) = \frac{\beta}{x}(\lambda x)^{\beta} e^{-(\lambda x)^{\beta}}, \qquad \text{for } x > 0 \tag{64}$$

having parameters $\lambda > 0$ and $\beta > 0$. The Weibull has a particularly simple cumulative distribution function

$$F(x) = 1 - e^{-(\lambda x)^{\beta}} \qquad \text{if } x \geq 0 \tag{65}$$

Note that the exponential distribution is a special case of the Weibull distribution (simply set $\beta = 1$). While the exponential distribution has constant, memoryless, failure rate, the Weibull allows for a failure rate that decreases with time ($\beta < 1$) or a failure rate that increases with time ($\beta > 1$). This gives increased flexibility for modeling lifetimes of systems that improve with time (e.g. a good red wine might have $\beta < 1$) or degrade with time (e.g. reinforced concrete bridge decks subjected to salt might have $\beta > 1$).

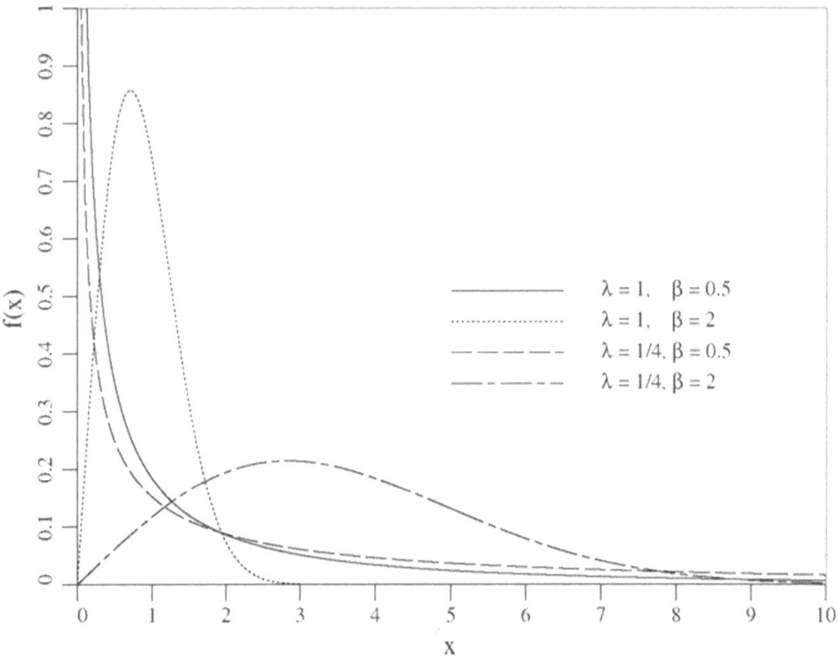

Figure 13. The Weibull distribution.

The mean and variance of a Weibull distributed random variable are

$$\mu = \frac{1}{\lambda\beta}\Gamma\left(\frac{1}{\beta}\right) \tag{66a}$$

$$\sigma^2 = \frac{1}{\lambda^2\beta}\left\{2\Gamma\left(\frac{2}{\beta}\right) - \frac{1}{\beta}\left[\Gamma\left(\frac{1}{\beta}\right)\right]^2\right\} \tag{66b}$$

where Γ is the Gamma function, which is commonly tabulated in math tables.

6.5 Rayleigh Distribution

The Rayleigh distribution is a non-negative distribution which finds application in the simulation of normally distributed random processes. In particular, consider the two orthogonal components, τ_1 and τ_2, of the vector $\underset{\sim}{\tau}$ in two-dimensional space. If the two components are independent and identically normally distributed random variables with zero means and common variance s^2, then the vector length, $|\tau| = \sqrt{\tau_1^2 + \tau_2^2}$, will be Rayleigh distributed with probability density function

$$f(x) = \frac{x}{s^2}\exp\left\{-\frac{x^2}{2s^2}\right\}, \qquad x \geq 0 \tag{67}$$

and cumulative distribution function

$$F(x) = 1 - e^{-\frac{1}{2}(x/s)^2} \qquad \text{if } x \geq 0 \tag{68}$$

which is actually a special case of the Weibull distribution ($\beta = 2$ and $\lambda = \frac{1}{s\sqrt{2}}$).

The mean and variance of a Rayleigh distributed random variable are

$$\mu = s\sqrt{\frac{\pi}{2}}$$
$$\sigma^2 = (2 - \pi/2)s^2$$

6.6 Normal Distribution

The *normal distribution* is probably the single most important distribution in use today. This is largely because sums of random variables tend to a normal distribution, as was proven by the *central limit theorem* – a theorem to be discussed shortly. Many natural 'additive' type phenomena, or phenomena involving many accumulating factors, therefore tend to have a normal distribution. For example, the cohesive strength of a soil is due to the sum of a very large number of electro-chemical interactions taking place at the molecular level – thus, the normal distribution has been widely used to represent the distribution of cohesion (its main competitor as a representative distribution is the lognormal distribution, discussed next).

A random variable X follows a normal (or *Gaussian*) distribution if its probability density function has the form

$$f(x) = \frac{1}{\sigma\sqrt{2\pi}} e^{-\frac{1}{2}\left(\frac{x-\mu}{\sigma}\right)^2} \qquad \text{on } -\infty < x < \infty. \tag{69}$$

The notation $X \sim N(\mu, \sigma^2)$ will be used to mean that X follows a normal distribution with mean μ and variance σ^2.

Properties:

1) the distribution is symmetric about the mean μ (which means that μ is also equal to the median),
2) the maximum point, or *mode*, of the distribution occurs at μ,
3) the inflection points of $f(x)$ occur at $x = \mu \pm \sigma$.

Characteristics:

$$\mathrm{E}[X] = \mu$$
$$\mathrm{Var}[X] = \sigma^2$$

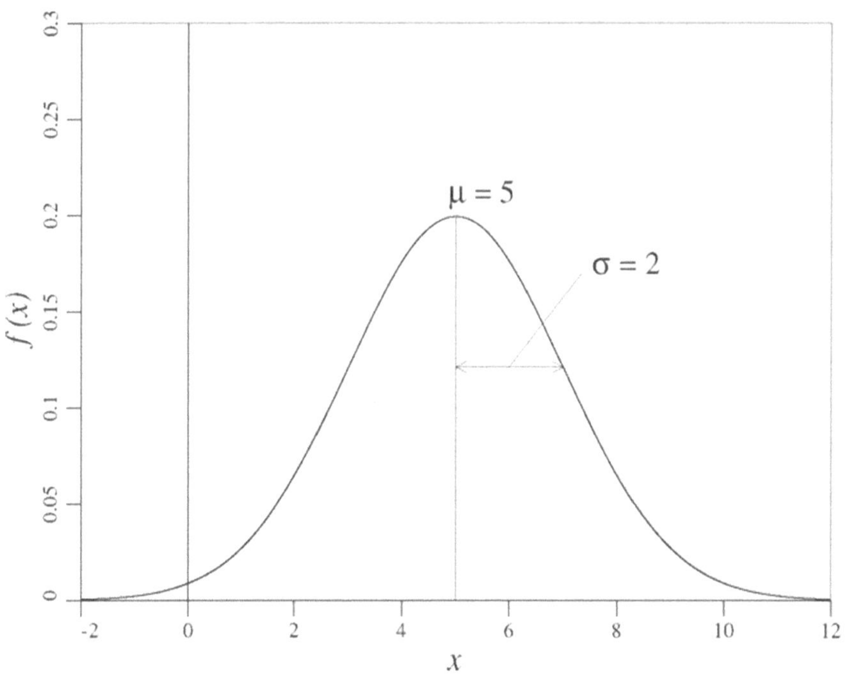

Figure 14. Normal distribution with $\mu = 5$ and $\sigma = 2$.

The Standard Normal

Unfortunately, no closed form solution exists for the integral of the normal probability density function. Probabilities associated with the normal distribution must be obtained by numerical integration. Traditionally, this has meant that normal probabilities have had to be obtained by consulting tables presented in manuals and textbooks. Of course, no book is big enough to contain the complete set of tables necessary for all possible values of μ and σ, so some way of encapsulating the tables is necessary. As it turns out, if the random variable X is transformed by subtracting its mean and dividing by its standard deviation,

$$Z = \frac{X - \mu}{\sigma} \tag{70}$$

then the resulting random variable, Z, has mean zero and unit variance. If a probability table is developed for Z, which is called the *standard normal* variate, then probabilities for all other normally distributed random variables can be obtained by performing the above normalizing transformation. That is, probabilities for any normally distributed random variable can be obtained by performing the above transformation and then consulting the single standard normal probability table.

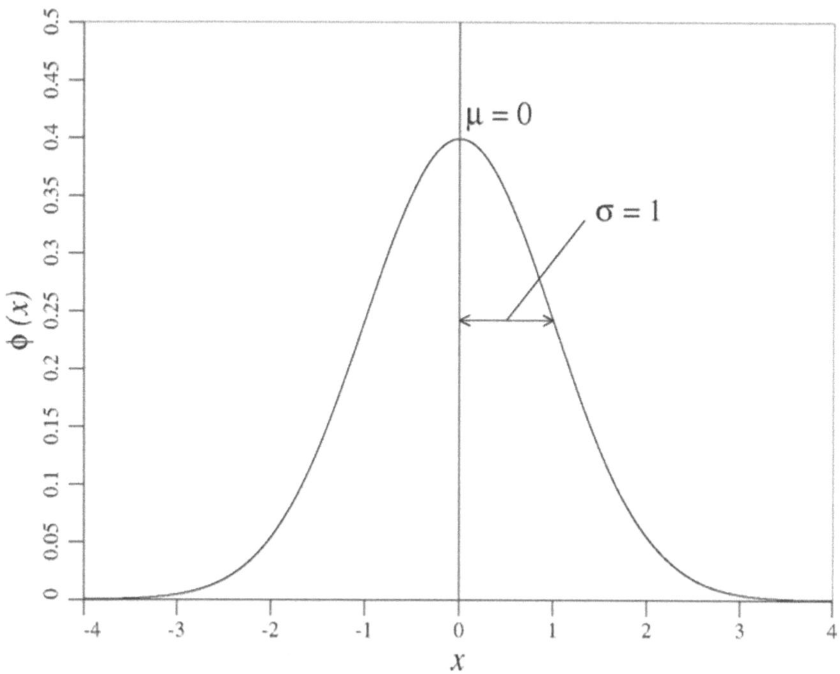

Figure 15. The standard normal distribution.

The distribution of the standard normal, Z, is given the special symbol $\phi(z)$, rather than $f(z)$, because of its importance in probability modeling and is defined by

$$\phi(z) = \frac{1}{\sqrt{2\pi}} e^{-\frac{1}{2}z^2} \qquad \text{on } -\infty < z < \infty. \tag{71}$$

The cumulative distribution function of the standard normal also has a special symbol, $\Phi(z)$, rather than $F(z)$, again because of its importance. Tables of $\Phi(z)$ are commonly included in textbooks. Computing probabilities for any normally distributed random variables proceeds by *standardization*, that is by subtracting the mean and dividing by the standard deviation on both sides of the inequality in the following;

$$\begin{aligned} \mathrm{P}\left[X < x\right] &= \mathrm{P}\left[\frac{X-\mu}{\sigma} < \frac{x-\mu}{\sigma}\right] \\ &= \mathrm{P}\left[Z < \frac{x-\mu}{\sigma}\right] \\ &= \Phi\left(\frac{x-\mu}{\sigma}\right) \\ &= \Phi(z) \end{aligned} \tag{72}$$

At which point, a standard normal table of probabilities can be consulted, with $z = (x-\mu)/\sigma$, to obtain the desired probability.

6.6.1 The Central Limit Theorem par

If $X_1, X_2, \ldots, X_n$ are independent random variables having arbitrary distributions, then the random variable

$$Y = X_1 + X_2 + \cdots + X_n \tag{73}$$

has a normal distribution as $n \to \infty$ if all the X's have about the same mean and variance (ie. none is dominant). See Papoulis (1991) for a proof of this theorem. In addition, if the X's are all normally distributed then Y is normally distributed for any n.

Specifically we will find the following result useful. If

$$\bar{X}_n = \frac{1}{n}\sum_{i=1}^{n} X_i$$

where $X_1, X_2, \ldots, X_n$ are independent samples taken from population X having mean μ and variance σ^2 (any distribution), then

$$\lim_{n\to\infty} \mathrm{P}\left[\frac{(\bar{X}_n - \mu)}{\sigma/\sqrt{n}} \le x\right] = \Phi(x) \tag{74}$$

Implications:

1) the sum of normal variates is normal (for any n) as mentioned above,
2) if the distributions of the X's are well-behaved (almost normal), then $n \ge 4$ gives a good approximation to the normal distribution,
3) if the distributions of the X's are uniform (or almost so), then $n \ge 6$ yields a reasonably good approximation to the normal distribution (out to at least about 3 standard deviations from the mean),
4) for poorly-behaved distributions, you may need $n > 100$ before the distribution begins to look reasonably normal. This happens, for example, with distributions whose tails fall off very slowly.

Thus for n sufficiently large and $X_1, X_2, \ldots, X_n$ independent and identically distributed (iid), then

$$Y = X_1 + X_2 + \cdots + X_n$$

is approximately normally distributed with

$$\mu_Y = \mathrm{E}[Y] = n\mathrm{E}[X_i] \tag{75a}$$

$$\sigma_Y^2 = \mathrm{Var}[Y] = n\mathrm{Var}[X_i] \tag{75b}$$

If the X's are *not* identically distributed, but are still independent, then

$$\mu_Y = \sum_{i=1}^{n} \mathrm{E}[X_i] \tag{76a}$$

$$\sigma_Y^2 = \sum_{i=1}^{n} \mathrm{Var}[X_i] \tag{76b}$$

6.6.2 Multivariate Normal Distribution par

The normal distribution is also popular as a distribution governing multiple random variables because it is simply defined knowing only the mean and variance of each random variable and the covariances acting between them. Consider two random variables, X and Y; these follow a bivariate normal distribution if their joint distribution has the form

$$f_{XY}(x,y) = \frac{1}{2\pi\sigma_X\sigma_Y\sqrt{1-\rho^2}} \exp\left\{\frac{-1}{2(1-\rho^2)}\left[\left(\frac{x-\mu_X}{\sigma_X}\right)^2 - 2\rho\left(\frac{x-\mu_X}{\sigma_X}\right)\left(\frac{y-\mu_Y}{\sigma_Y}\right) + \left(\frac{y-\mu_Y}{\sigma_Y}\right)^2\right]\right\} \tag{77}$$

for $-\infty < x, y < \infty$, where ρ is the correlation coefficient between X and Y and μ_X, μ_Y, and σ_X, σ_Y are the means and standard deviations of X and Y, respectively. Figures 4 and 5 illustrate the bivariate normal distribution.

If X and Y follow a bivariate normal distribution, then their marginal probability density functions, defined as

$$f_X(x) = \int_{-\infty}^{\infty} f_{XY}(x,y)\,dy \tag{78a}$$

$$f_Y(y) = \int_{-\infty}^{\infty} f_{XY}(x,y)\,dx \tag{78b}$$

are also normal distributions. For example, the marginal distribution of X is a normal distribution with mean μ_X and standard deviation σ_X, and similarly for the marginal distribution of Y. That is,

$$f_X(x) = \frac{1}{\sigma_X\sqrt{2\pi}} \exp\left\{-\frac{1}{2}\left(\frac{x-\mu_X}{\sigma_X}\right)^2\right\} \tag{79a}$$

$$f_Y(y) = \frac{1}{\sigma_Y\sqrt{2\pi}} \exp\left\{-\frac{1}{2}\left(\frac{y-\mu_Y}{\sigma_Y}\right)^2\right\} \tag{79b}$$

Recall that the conditional probability of A given B is

$$\mathrm{P}[A \mid B] = \frac{\mathrm{P}[A \cap B]}{\mathrm{P}[B]}$$

From this, we get the following result for *conditional distributions*

$$f_{X|Y}(x \mid y) = \frac{f_{XY}(x,y)}{f_Y(y)} \tag{80}$$

In particular, if X and Y follow a bivariate normal distribution, then it can be shown that

$$f_{X|Y}(x \mid y) = \frac{1}{\sigma_X\sqrt{1-\rho^2}\sqrt{2\pi}} \exp\left\{-\frac{1}{2}\left[\frac{x-\mu_X-\rho(y-\mu_Y)\sigma_X/\sigma_Y}{\sigma_X\sqrt{1-\rho^2}}\right]^2\right\} \tag{81}$$

It can be seen from this that the *conditional* distribution of X for a given $Y = y$ also follows a normal distribution with mean and standard deviation

$$\mu_{X|Y} = \mu_X + \rho(y - \mu_Y)\sigma_X/\sigma_Y \tag{82a}$$

$$\sigma_{X|Y} = \sigma_X \sqrt{(1-\rho^2)} \tag{82b}$$

To extend the multivariate normal distribution to more than two random variables, it is useful to use vector-matrix notation. Define

$$\underset{\sim}{\mu} = \begin{Bmatrix} \mu_1 \\ \mu_2 \\ \cdot \\ \cdot \\ \cdot \\ \mu_n \end{Bmatrix} \tag{83}$$

to be the vector of mean of the sequence of n random variables, $\underset{\sim}{X} = \{X_1, X_2, \dots, X_n\}$ and

$$\underset{\approx}{C} = \begin{bmatrix} C_{11} & C_{12} & \cdot & \cdot & \cdot & C_{1n} \\ C_{21} & C_{22} & \cdot & \cdot & \cdot & C_{2n} \\ \cdot & \cdot & \cdot & & & \cdot \\ \cdot & \cdot & & \cdot & & \cdot \\ \cdot & \cdot & & & \cdot & \cdot \\ C_{n1} & C_{n2} & \cdot & \cdot & \cdot & C_{nn} \end{bmatrix} \tag{84}$$

to be the matrix of covariances between X_i and X_j, $i = 1, 2, \dots, n$ and $j = 1, 2, \dots, n$. Each element of the covariance matrix is defined as

$$\begin{aligned} C_{ij} &= \mathrm{Cov}\,[X_i, X_j] = \rho_{ij}\sigma_i\sigma_j && \text{if } i \neq j \\ &= \mathrm{Var}\,[X_i] = \sigma_i^2 && \text{if } i = j \end{aligned}$$

Note that if the X_i's are uncorrelated, then the covariance matrix is diagonal

$$\underset{\approx}{C} = \begin{bmatrix} \sigma_1^2 & 0 & \cdot & \cdot & \cdot & 0 \\ 0 & \sigma_2^2 & \cdot & \cdot & \cdot & 0 \\ \cdot & \cdot & \cdot & & & \cdot \\ \cdot & \cdot & & \cdot & & \cdot \\ \cdot & \cdot & & & \cdot & \cdot \\ 0 & 0 & \cdot & \cdot & \cdot & \sigma_n^2 \end{bmatrix}$$

Using these definitions, the joint normal distribution of $\underset{\sim}{X} = \{X_1, X_2, \dots, X_n\}$ is

$$f_{\underset{\sim}{X}}(\underset{\sim}{x}) = \frac{1}{(2\pi)^{n/2}\sqrt{|\underset{\approx}{C}|}} \exp\left\{-\tfrac{1}{2}(\underset{\sim}{x} - \underset{\sim}{\mu})^T \underset{\approx}{C}^{-1}(\underset{\sim}{x} - \underset{\sim}{\mu})\right\} \tag{85}$$

where $|\underset{\approx}{C}|$ is the determinant of $\underset{\approx}{C}$ and superscript T means the transpose.

As in the bivariate case, all marginal distributions are also normally distributed;

$$f_{X_i}(x_i) = \frac{1}{\sigma_i\sqrt{2\pi}} \exp\left\{-\frac{1}{2}\left(\frac{x_i-\mu_i}{\sigma_i}\right)^2\right\} \tag{86}$$

The conditional distributions may be obtained by partitioning the vector $\underset{\sim}{X}$ into two parts (Vanmarcke, 1984): $\underset{\sim}{X}_a$ and $\underset{\sim}{X}_b$ of size n_a and n_b, where $n_a + n_b = n$, that is

$$\underset{\sim}{X} = \begin{Bmatrix} X_1 \\ \cdot \\ \cdot \\ \cdot \\ X_{n_a} \\ X_{n_a+1} \\ \cdot \\ \cdot \\ \cdot \\ X_n \end{Bmatrix} = \begin{Bmatrix} \underset{\sim}{X}_a \\ \underset{\sim}{X}_b \end{Bmatrix} \tag{87}$$

having mean vectors

$$\underset{\sim}{\mu}_a = \begin{Bmatrix} \mu_1 \\ \cdot \\ \cdot \\ \cdot \\ \mu_{n_a} \end{Bmatrix}, \qquad \underset{\sim}{\mu}_b = \begin{Bmatrix} \mu_{n_a+1} \\ \cdot \\ \cdot \\ \cdot \\ \mu_n \end{Bmatrix} \tag{88}$$

Using this partition, the covariance matrix can be split up into four submatrices;

$$\underset{\approx}{C} = \begin{pmatrix} \underset{\approx}{C}_{aa} & \underset{\approx}{C}_{ab} \\ \underset{\approx}{C}_{ba} & \underset{\approx}{C}_{bb} \end{pmatrix} \tag{89}$$

where $\underset{\approx}{C}_{ba} = \underset{\approx}{C}^T_{ab}$. Using these partitions, the conditional mean of the vector $\underset{\sim}{X}_a$ given the vector $\underset{\sim}{X}_b$ can be obtained from

$$\underset{\sim}{\mu}_{a\,|\,b} = \underset{\sim}{\mu}_a + \underset{\approx}{C}_{ab}\underset{\approx}{C}_{bb}^{-1}(\underset{\sim}{X}_b - \underset{\sim}{\mu}_b) \tag{90}$$

Similarly, the conditional covariance matrix is

$$\underset{\approx}{C}_{a\,|\,b} = \underset{\approx}{C}_{aa} - \underset{\approx}{C}_{ab}\underset{\approx}{C}_{bb}^{-1}\underset{\approx}{C}^T_{ab} \tag{91}$$

With these results, the conditional distribution of $\underset{\sim}{X}_a$ given $\underset{\sim}{X}_b$ is

$$f_{\underset{\sim}{X}_a\,|\,\underset{\sim}{X}_b}(\underset{\sim}{x}_a \,|\, \underset{\sim}{x}_b) = \frac{1}{(2\pi)^{n_a/2}\sqrt{|\underset{\approx}{C}_{a\,|\,b}|}} \exp\left\{-\tfrac{1}{2}(\underset{\sim}{x}_a - \underset{\sim}{\mu}_{a\,|\,b})^T \underset{\approx}{C}_{a\,|\,b}^{-1}(\underset{\sim}{x}_a - \underset{\sim}{\mu}_{a\,|\,b})\right\} \tag{92}$$

6.7 Lognormal Distribution

From the point of view of modeling material properties and loads in engineering, which are generally non-negative, the normal distribution suffers from the disadvantage of allowing negative values. For example, if a soil's elastic modulus were to be modeled using a normal distribution, then there would be a non-zero probability of obtaining a negative elastic modulus. Since a negative elastic modulus does not occur in practice, the normal cannot be its true distribution.

As an approximation, the normal is nevertheless often used to represent material properties. The error incurred may be slight when the coefficient of variation, v, is small. For example, if $v \leq 0.3$, then $\mathrm{P}[X < 0] \leq 0.0004$, which may be fine unless it is at these extremes that failure is initiated. A simple way to avoid such problems is to fit a non-negative distribution to the population in question, and one such candidate is the *lognormal distribution*. The lognormal distribution arises from the normal distribution through a simple, albeit non-linear, transformation. In particular, if G is a normally distributed random variable, having range $-\infty < g < +\infty$, then $X = \exp\{G\}$ will have range $0 \leq x < \infty$. We say that the resulting random variable, X, is *lognormally* distributed – note that its natural logarithm is normally distributed.

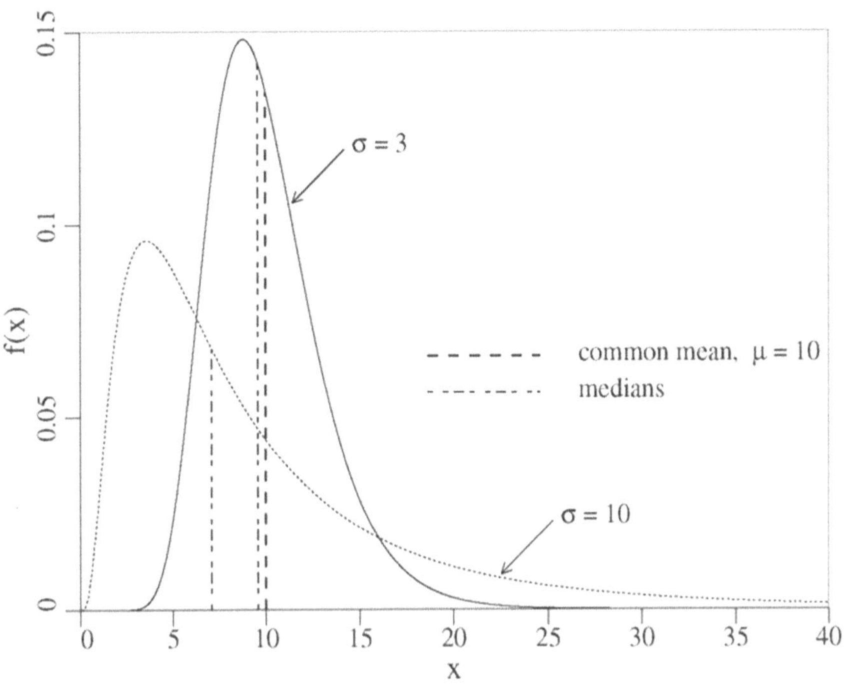

Figure 16. Two lognormal distributions illustrating the effect of changing variance.

The random variable X is lognormally distributed if $\ln(X)$ is normally distributed. If this is true, then X has probability density function

$$f(x) = \frac{1}{x\sigma_{\ln x}\sqrt{2\pi}} \exp\left\{-\frac{1}{2}\left(\frac{\ln x - \mu_{\ln x}}{\sigma_{\ln x}}\right)^2\right\} \qquad 0 \leq x < \infty \tag{93}$$

Note that this distribution is strictly non-negative and so is popular as a distribution of non-negative engineering properties, such as cohesion, elastic modulus, the tangent of the friction angle, and so on. (Aside, I suppose that it is possible to have a negative cohesion, but the authors have never heard of a soil possessing this property.) The two parameters of the distribution

$$\mu_{\ln x} = \mathrm{E}[\ln X]$$
$$\sigma^2_{\ln x} = \mathrm{Var}[\ln X]$$

are the mean and variance of the underlying normally distributed random variable, $\ln X$.

Computing Probabilities

In order to compute probabilities from the lognormal distribution, we must make use of the fact that $\ln(X)$ is normally distributed so that we can use the standard normal table. That is, in a probability expression, we take logarithms on both sides of the inequality, then standardize by subtracting the mean and dividing by the standard deviation of $\ln X$,

$$\begin{aligned} \mathrm{P}[X \leq a] &= \mathrm{P}[\ln(X) < \ln(a)] = \mathrm{P}\left[\frac{\ln(X) - \mu_{\ln x}}{\sigma_{\ln x}} < \frac{\ln(a) - \mu_{\ln x}}{\sigma_{\ln x}}\right] \\ &= \mathrm{P}\left[Z < \frac{\ln(a) - \mu_{\ln x}}{\sigma_{\ln x}}\right] = \Phi\left(\frac{\ln(a) - \mu_{\ln x}}{\sigma_{\ln x}}\right) \end{aligned} \tag{94}$$

where, as before, Z is the standard normal random variate.

Mean and Variance

The mean and variance of X are obtained by transforming the two parameters of the lognormal distribution,

$$\mu_X = \mathrm{E}[X] = e^{\mu_{\ln x} + \frac{1}{2}\sigma^2_{\ln x}} \tag{95a}$$

$$\sigma^2_X = \mathrm{Var}[X] = \mu^2_X\left(e^{\sigma^2_{\ln x}} - 1\right) \tag{95b}$$

Alternatively, if you are given μ_X and σ^2_X, you can obtain the parameters $\mu_{\ln x}$ and $\sigma^2_{\ln x}$ as follows;

$$\sigma^2_{\ln x} = \ln\left(1 + \frac{\sigma^2_X}{\mu^2_X}\right) \tag{96a}$$

$$\mu_{\ln x} = \ln(\mu_X) - \frac{1}{2}\sigma^2_{\ln x} \tag{96b}$$

Characteristics and Moments

$$\text{mode} = e^{\mu_{\ln X} - \sigma^2_{\ln X}} \tag{97a}$$

$$\text{median} = e^{\mu_{\ln X}} \tag{97b}$$

$$\text{mean} = e^{\mu_{\ln X} + \frac{1}{2}\sigma^2_{\ln X}} \tag{97c}$$

$$\mathrm{E}\left[X^k\right] = e^{k\mu_{\ln X} + \frac{1}{2}k^2\sigma^2_{\ln X}} \tag{97d}$$

Note that the mode < median < mean, and thus the lognormal distribution has *positive skew*. A distribution is skewed if one of its tails is longer than the other, and, by tradition, the sign of the skew indicates the direction of the longer tail.

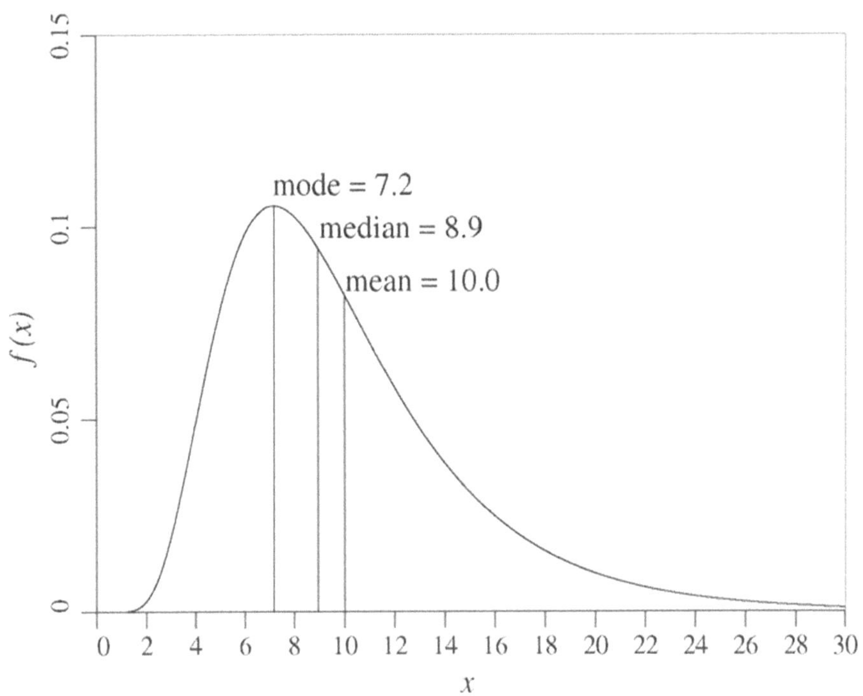

Figure 17. Location of mode, median, and mean in a lognormal distribution for $\mu_X = 10$ and $\sigma_X = 5$.

Figure 17 illustrates the relative locations of the mode, median, and mean for the non-symmetric lognormal distribution. Because of the positive skewed, or 'skewed right', shape of the distribution, with the long distribution tail to the right, realizations from the lognormal distribution will have very large values every now and then. This results in the mean being drawn to the right (for example, the arithmetic average is affected by very large values in the sum). Often, for the lognormal distribution, the median is actually viewed as the primary characteristic of the distribution, since it divides the distribution into equal halves.

It is worth digressing slightly at this point and consider the median of a lognormal distribution in a bit more detail, especially with respect to its estimation. Suppose that we have taken several observations, $x_1, x_2, \ldots, x_n$ of a lognormally distributed random variable, X. An estimate of the mean of $\ln(X)$ is just the average of $\ln(x_1), \ln(x_2), \ldots, \ln(x_n)$,

$$\hat{\mu}_{\ln X} = \frac{1}{n}\sum_{i=1}^{n} \ln(x_i) \tag{98}$$

where the 'hat' denotes that this is an estimate of $\mu_{\ln X}$. From this, an estimate of the median, $\tilde{x}$, is

$$\tilde{x} = \exp\{\hat{\mu}_{\ln X}\} = \exp\left\{\frac{1}{n}\sum_{i=1}^{n} \ln(x_i)\right\} \tag{99}$$

Alternatively, the geometric average, x_g, of a sequence of non-negative numbers is defined as the n'th root of the product of the n observations,

$$\begin{aligned} x_g &= (x_1 x_2 \cdots x_n)^{1/n} \\ &= \exp\left\{\ln\left((x_1 x_2 \cdots x_n)^{1/n}\right)\right\} \\ &= \exp\left\{\frac{1}{n}\sum_{i=1}^{n} \ln(x_i)\right\} \end{aligned} \tag{100}$$

which is identical to the equation for $\tilde{x}$, so we see that the geometric average is an estimate of the median of a lognormally distributed random variable. This also means that the median of a lognormal distribution is preserved under geometric averaging.

Multiplicative Property

If $X = Y_1 Y_2 \cdots Y_n$ and each Y_i are (positive) independent random variables of any distribution having about the same 'weight', then

$$\ln X = \ln Y_1 + \ln Y_2 + \cdots + \ln Y_n \tag{101}$$

and by the central limit theorem $\ln X$ tends to a normal distribution with

$$\mu_{\ln X} = \mu_{\ln Y_1} + \mu_{\ln Y_2} + \cdots + \mu_{\ln Y_n} \tag{102a}$$

$$\sigma^2_{\ln X} = \sigma^2_{\ln Y_1} + \sigma^2_{\ln Y_2} + \cdots + \sigma^2_{\ln Y_n} \tag{102b}$$

Thus X tends to a lognormal distribution with parameters $\mu_{\ln X}$ and $\sigma^2_{\ln X}$. This is a useful property since it can be used to approximate the distribution of many multiplicative functions.

In particular, if X is any multiplicative function, say

$$X = \frac{AB}{C} \qquad \Longrightarrow \qquad \ln X = \ln A + \ln B - \ln C \tag{103}$$

and A, B, and C are independent and lognormally distributed, then X is also lognormally distributed with

$$\mu_{\ln X} = \mu_{\ln A} + \mu_{\ln B} - \mu_{\ln C}$$
$$\sigma^2_{\ln X} = \sigma^2_{\ln A} + \sigma^2_{\ln B} + \sigma^2_{\ln C}$$

Recall that for variances, the coefficient of -1 appearing before the last term in Eq. 103 is squared, so that, in the case of independence, the variance of a sum is literally the sum of the variances. (If A, B, and C were correlated, then the covariance terms which would have to be added in to find $\sigma^2_{\ln X}$ would have sign dependendent on the signs appearing in the original sum.)

Consider again the geometric average, this time for random observations (i.e., before we have observed them),

$$X_g = (X_1 X_2 \cdots X_n)^{1/n} = X_1^{1/n} \times X_2^{1/n} \times \cdots \times X_n^{1/n}$$

which is a product of n random variables. By the central limit theorem, X_g will tend to a lognormal distribution so that

$$\ln(X_g) = \ln\left((X_1 X_2 \cdots X_n)^{1/n}\right) = \frac{1}{n}\sum_{i=1}^{n} \ln(X_i)$$

is normally distributed. As mentioned above, X_g is an estimate of the median of X, if X is lognormally distributed. However, even if X is not lognormally distributed, X_g, will tend to have a lognormal distribution, by the central limit theorem, if the X_i's are non-negative. This suggests that in a variety of cases, the lognormal distribution is a natural distribution for soil properties according to the central limit theorem.

6.7.1 Bivariate Lognormal Distribution par

Generally, the multivariate lognormal distribution is handled by directly considering the underlying multivariate normal distribution. That is, rather than considering the joint distribution between the lognormally distributed variated X_1, X_2, ..., we consider the joint distribution between $\ln X_1$, $\ln X_2$, ..., since these are all normally distributed and the results presented in the previous section can be used. However, we sometimes need to consider the lognormally distributed variates directly. Here we will present some results for two lognormally distributed random variables, X_1 and X_2.

If X_1 and X_2 are jointly lognormally distributed, then their bivariate distribution is

$$f_{X_1X_2}(x,y) = \frac{1}{2\pi\sigma_{\ln X_1}\sigma_{\ln X_2} r x y} \exp\left\{-\frac{1}{2r^2}\left[\Psi_1^2 - 2\rho_{\ln 12}\Psi_1\Psi_2 + \Psi_2^2\right]\right\}, \qquad x \geq 0,\ y \geq 0 \tag{104}$$

where $\Psi_1 = (\ln x - \mu_{\ln X_1}/\sigma_{\ln X_1}$, $\Psi_2 = (\ln y - \mu_{\ln X_2})/\sigma_{\ln X_2}$, $r^2 = 1 - \rho^2_{\ln 12}$, and $\rho_{\ln 12}$ is the correlation coefficient between $\ln X_1$ and $\ln X_2$.

In general, the parameters $\mu_{\ln X_1}$, $\sigma_{\ln X_1}$ can be obtained using the transformation equations given in the previous subsection from the parameters μ_{X_1}, σ_{X_1}, and so on. If we happen to have an estimate for the correlation coefficient, ρ_{12}, acting between X_1 and X_2, we can get $\rho_{\ln 12}$ from

$$\rho_{\ln 12} = \frac{\ln(1 + \rho_{12} v_{X_1} v_{X_2})}{\sqrt{\ln(1 + v_{X_1}^2)\ln(1 + v_{X_2}^2)}} \tag{105}$$

where $v_{X_i} = \sigma_{X_i}/\mu_{X_i}$ is the coefficient of variation of X_i. We can also invert this relationship to obtain an expression for ρ_{12},

$$\rho_{12} = \frac{\exp\{\rho_{\ln 12}\sigma_{\ln X_1}\sigma_{\ln X_2}\} - 1}{\sqrt{\left(\exp\{\sigma_{\ln X_1}^2\} - 1\right)\left(\exp\{\sigma_{\ln X_2}^2\} - 1\right)}} \tag{106}$$

7 Random Fields

We turn our attention in this section to random fields, $X(t)$, which consist of a sequence of continuous random variables at each point t. As an example of a one-dimensional random field, the following figure illustrates the tip resistance measured during a cone penetration test (CPT).

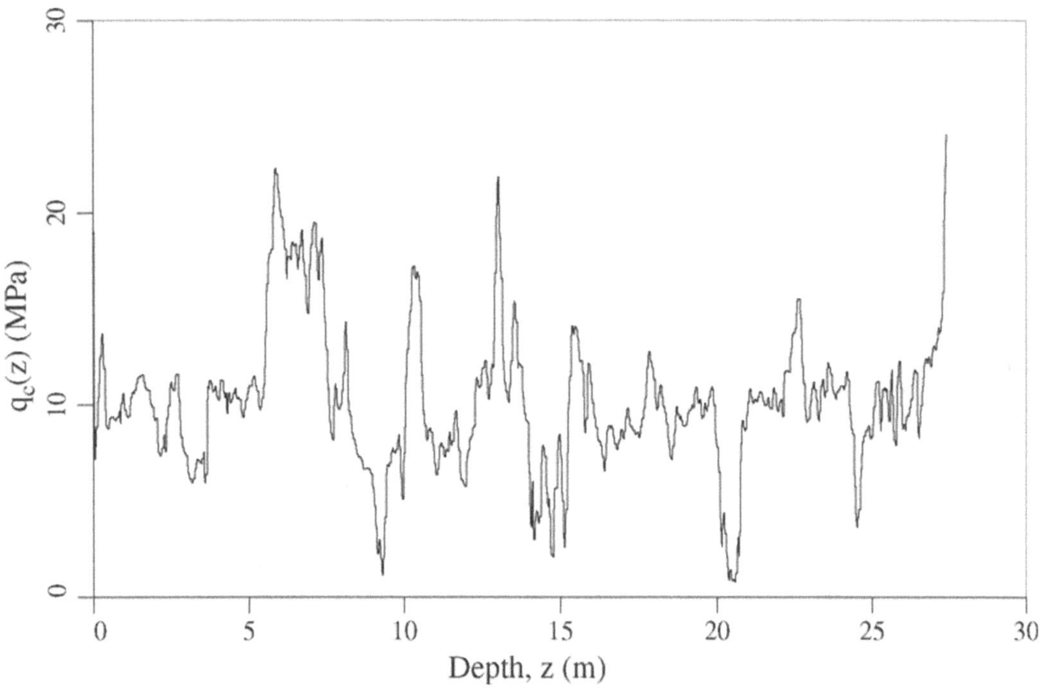

Figure 18. Tip resistance, $q_c(z)$, measured over depth z by a cone penetrometer

Aside from soil disturbance, measurement errors and problems with extracting engineering properties from CPT data, Figure 18 presumably gives a reasonably good idea about the soil properties at the location at which the CPT was taken. However, what can be said about the soil properties 1 or 5 metres away? The data presented in Figure 18 should be used to characterize the randomness (uncertainty) at locations which have not been sampled. But how do we characterize such a process? Some considerations involved in characterizing spatial variability are;

1) *variability at a point:* pick a specific position, t^*. At this point the process has a random value $X(t^*) = X^*$ which is governed by a probability density function, $f_{X^*}(x)$. If we picked another position, say t', then $X(t') = X'$ would have another, possibly different PDF, $f_{X'}(x)$. That is, the PDF's could evolve with position. In practice, evolving PDF's become quite difficult to estimate for anything beyond a simple trend in the mean or variance. An example where the point, or marginal, distribution evolves with time is earthquake ground motion where the motion variance increases drastically during the strong motion portion of the record.
2) *spatial dependence:* Consider again two positions, t^* and t' separated by distance $\tau = t' - t^*$. Presumably, the two random variables, $X(t')$ and $X(t^*)$ will exhibit some dependence on each other. For example, if X is cohesion, then we would expect $X(t')$ and $X(t^*)$ to be quite similar (highly dependent) when τ is small (e.g. a few centimetres) and possibly quite dissimilar when τ is large (e.g. tens or hundreds of metres). If $X(t^*)$ and $X(t')$ are independent for any two positions with separation $\tau = t' - t^* \neq 0$, then the process would be infinitely rough – points separated by vanishingly small lags could have quite different values. This is not physically realistic for most natural phenomena. Thus, $X(t^*)$ and $X(t')$ generally have some sort of dependence, that often decreases with separation distance. This interdependence results in a smoothing of the random process. That is, for small τ, nearby states of X are preferential – the random field is constrained by its neighbors. We characterize the interdependence with the joint distribution $f_{X^*X'}(x^*, x')$. If we extend this idea to the consideration of any three, or four, or five, ..., points then the complete probabilistic description of a random process is the infinite dimensional probability density function
$$f_{X_1X_2\ldots}(x_1, x_2, \ldots)$$
Again, such a PDF is difficult to use in practice, not only mathematically, but also in that it is difficult to estimate from real data.

To simplify the characterization problem, we introduce a number of assumptions which are commonly made;

1) *Gaussian process:* the joint PDF is a multivariate *normally distributed* random process. Such a process is also commonly refered to as a *Gaussian process* after Gauss, the father of the normal distribution. The great advantage to the multivariate normal distribution is that the complete distribution can be specified by just the mean vector and the covariance matrix. As we say earlier, the multivariate normal PDF has the form

$$f_{X_1X_2\ldots,X_k}(x_1, x_2, \ldots, x_k) = \frac{1}{(2\pi)^{k/2}} \frac{1}{|\underset{\approx}{C}|^{1/2}} \exp\left\{-\tfrac{1}{2}(\underset{\sim}{x} - \underset{\sim}{\mu})^T \underset{\approx}{C}^{-1} (\underset{\sim}{x} - \underset{\sim}{\mu})\right\}$$

where $\underset{\sim}{\mu}$ is the vector of mean values, one for each X_i, $\underset{\approx}{C}$ is the covariance matrix between the X's, and $|\underset{\approx}{C}|$ is its determinant. Specifically,

$$\underset{\sim}{\mu} = \mathrm{E}\left[\underset{\sim}{X}\right]$$
$$\underset{\approx}{C} = \mathrm{E}\left[(\underset{\sim}{X} - \underset{\sim}{\mu})(\underset{\sim}{X} - \underset{\sim}{\mu})^T\right]$$

where the superscript T means the transpose. The covariance matrix, $\underset{\approx}{C}$, is a $k \times k$ symmetric, positive definite, matrix. For a continuous random field, the dimensions of $\underset{\sim}{\mu}$ and $\underset{\approx}{C}$ are still infinite, since the random field is composed of an infinite number of X's, one for each point. However, we often quantify $\underset{\sim}{\mu}$ and $\underset{\approx}{C}$ using continuous functions of space. For example, in a one-dimensional random field (or random process), the mean may vary linearly;

$$\mu(t) = a + bt$$

and the covariance matrix can be expressed in terms of the standard deviations, which may vary with t, and the correlation function, ρ, as in

$$C(t_1, t_2) = \sigma(t_1)\sigma(t_2)\rho(t_1, t_2)$$

which specifies the covariance between $X(t_1)$ and $X(t_2)$. It the random field is stationary, then $\mu(t) = \mu$ is constant, Again, such a PDF is difficult to use in practice, not only mathematically, but also to estimate from real data.

2) *stationarity or statistical homogeneity:* the joint PDF is independent of spatial position, that is it depends just on *relative* positions of the points. This assumption implies that the mean, covariance, and higher order moments are constant in time (or space) and thus that the marginal, or point, PDF is also constant in time (or space). So called *weak stationarity* or *second order stationarity* just implies that the mean and variance are constant in space.
3) *isotropy:* in two and higher dimensional random fields, isotropy implies that the joint PDF is invariant under rotation. This condition implies stationarity. Isotropy just means that the correlation between two points only depends on the distance between the two points, not on their orientation relative to one another.

In this book, we will largely restrict ourselves to stationary Gaussian random fields, and to fields derived through simple transformations from Gaussian random fields (e.g. lognormally distributed random fields). Gaussian random fields are completely specified by their mean and covariance structure, ie. their first two moments. In practice, we are sometimes able to reasonably accurately estimate the mean, and sometimes a mean trend, of a soil property at a site. Estimating the variance and covariance requires considerably more data – we often need to resort to information provided by the literature in order to specify the variance and covariance structure. Because of this uncertainty in the basic parameters of even the covariance, there is often little point in adopting other joint distributions, which are more complicated and depend on higher moments, to govern the random fields representing soil properties, unless these distributions are suggested by mechanical or physical theory.

Under the simplifying assumptions that the random field is Gaussian and stationary, we need to know three things in order to characterize the field;

1) the field mean, μ_X,
2) the field variance, σ_X^2,
3) how rapidly the field varies in space

The last is characterized by the second moment of the field's joint distribution, which is captured equivalently by the covariance function, the spectral density function, or the variance function. These functions are discussed in the next few sections.

8 The Covariance Function

The second moment nature of a Gaussian random field can be expressed by the *covariance function*,

$$\begin{aligned} C(t', t^*) &= \mathrm{Cov}\left[X(t'), X(t^*)\right] = \mathrm{E}\left[\left(X(t') - \mu_X(t')\right)\left(X(t^*) - \mu_X(t^*)\right)\right] \\ &= \mathrm{E}\left[X(t')X(t^*)\right] - \mu_X(t')\mu_X(t^*) \end{aligned} \tag{107}$$

where $\mu_X(t)$ is the mean of X at the position t. Since the magnitude of the covariance depends on the size of the variance of $X(t')$ and $X(t^*)$, it tells us little about the degree of linear dependence between $X(t')$ and $X(t^*)$. A more meaningful measure, in this sense, is the *correlation function*,

$$\rho(t', t^*) = \frac{C(t', t^*)}{\sigma_X(t')\sigma_X(t^*)} \tag{108}$$

where $\sigma_X(t)$ is the standard deviation of X at the position t. As seen in Section 1.4.5, $-1 \leq \rho(t', t^*) \leq 1$, and when $\rho(t', t^*) = 0$ we say that $X(t')$ and $X(t^*)$ are uncorrelated. When X is Gaussian, being uncorrelated also implies independence. If $\rho(t', t^*) = \pm 1$, then $X(t')$ and $X(t^*)$ are perfectly linearly correlated, that is $X(t')$ can be expressed in terms of $X(t^*)$ as

$$X(t') = a \pm bX(t^*)$$

Furthermore, if $X(t')$ and $X(t^*)$ are perfectly correlated and the random field is stationary, then $X(t') = \pm X(t^*)$. The sign to use is the same as the sign of $\rho(t', t^*)$.

For stationary random fields, the mean and covariance are independent of position, so that

$$\begin{aligned} C(t', t^*) = C(t' - t^*) = C(\tau) &= \mathrm{Cov}\left[X(t), X(t+\tau)\right] = \mathrm{Cov}\left[X(0), X(\tau)\right] \\ &= \mathrm{E}\left[X(0)X(\tau)\right] - \mu_X^2 \end{aligned} \tag{109}$$

and the correlation function becomes

$$\rho(\tau) = \frac{C(\tau)}{C(0)} = \frac{C(\tau)}{\sigma_X^2}$$

Because $C(t', t^*) = C(t^*, t')$, we must have $C(\tau) = C(-\tau)$, when the field is stationary, and similarly $\rho(\tau) = \rho(-\tau)$.

Another property of the covariance function is that it is *positive definite*. To illustrate this property consider a linear combination of n of the random variables in the process $X(t)$, say $X_i = X(t_i)$ for any sequence of times, $t_1, t_2, \ldots, t_n$,

$$Y = a_1X_1 + a_2X_2 + \cdots + a_nX_n = \sum_{i=1}^{n} a_iX_i$$

where $a_1, a_2, \ldots, a_n$ are any set of coefficients. The variance of a linear combination is

$$\mathrm{Var}[Y] = \sum_{i=1}^{n}\sum_{j=1}^{n} a_i a_j \mathrm{Cov}[X_i, X_j]$$

Since Var $[Y]$ is also defined as E $\left[(Y-\mu_Y)^2\right]$ it cannot be negative. This means that the covariances between the X's must satisfy the following inequality for any a_i

$$\sum_{i=1}^{n}\sum_{j=1}^{n} a_i a_j \mathrm{Cov}[X_i, X_j] \geq 0 \tag{110}$$

which is the statement of positive definiteness. In the case of a stationary process where $\mathrm{Cov}[X_i, X_j] = \sigma_X^2 \rho(t_i - t_j) = \sigma_X^2 \rho_{ij}$, we see that the correlation function is also positive definite

$$\sum_{i=1}^{n}\sum_{j=1}^{n} a_i a_j \rho_{ij} \geq 0 \tag{111}$$

since $\sigma_X^2 \geq 0$.

One of the points of Eq. 110 and 111 is that not just any covariance and correlation function can be used to characterize the second moment of a random field. In particular, the following properties of the covariance function must be satisfied;

1) $|\mathrm{Cov}[X_i, X_j]| \leq \sigma_{X_i}\sigma_{X_j}$, which ensures that $-1 \leq \rho_{ij} \leq 1$,
2) $\mathrm{Cov}[X_i, X_j] = \mathrm{Cov}[X_j, X_i]$, and
3) $\sum_{i=1}^{n}\sum_{j=1}^{n} a_i a_j \mathrm{Cov}[X_i, X_j] \geq 0$.

If two covariance functions, $C_1(X_i, X_j)$ and $C_2(X_i, X_j)$, each satisfy the above conditions, then their sum $C(X_i, X_j) = C_1(X_i, X_j) + C_2(X_i, X_j)$ will also satisfy the above conditions and be a valid covariance function.

If the set of covariances $\mathrm{Cov}[X_i, X_j]$ is viewed as a matrix $\underset{\approx}{C} = [C_{ij}]$, with elements $C_{ij} = \mathrm{Cov}[X_i, X_j]$, then one of the results of positive definiteness is that the *square root* of $\underset{\approx}{C}$ will be real. The square root will be defined here as the lower triangular matrix, $\underset{\approx}{L}$, such that $\underset{\approx}{L}\underset{\approx}{L}^T = \underset{\approx}{C}$, where the superscript T denotes the matrix transpose. The lower triangular matrix $\underset{\approx}{L}$ has the form

$$\underset{\approx}{L} = \begin{bmatrix} \ell_{11} & 0 & 0 & \cdot & \cdot & \cdot & 0 \\ \ell_{21} & \ell_{22} & 0 & \cdot & \cdot & \cdot & 0 \\ \ell_{31} & \ell_{32} & \ell_{33} & \cdot & \cdot & \cdot & 0 \\ \cdot & \cdot & \cdot & \cdot & & & \cdot \\ \cdot & \cdot & \cdot & & \cdot & & \cdot \\ \cdot & \cdot & \cdot & & & \cdot & \cdot \\ \ell_{n1} & \ell_{n2} & \ell_{n3} & \cdot & \cdot & \cdot & \ell_{nn} \end{bmatrix} \tag{112}$$

We shall see how this matrix can be used to simulate a random field in Chapter 2.

A positive definite covariance matrix can also be decomposed into a matrix of eigenvectors $\underset{\approx}{Q}$ and positive eigenvalues $\underset{\approx}{\Psi}$ such that

$$\underset{\approx}{C} = \underset{\approx}{Q}^T \underset{\approx}{\Psi} \underset{\approx}{Q} \tag{113}$$

where $\underset{\approx}{\Psi}$ is a diagonal matrix whose elements are the eigenvalues, $\psi_1, \psi_2, \ldots, \psi_n$, of the covariance matrix $\underset{\approx}{C}$.

The eigenvectors composing each column of the matrix Q make up an *orthonormal basis*, which is a set of unit vectors which are mutually perpendicular. A property of orthonormal vectors is that $\underset{\approx}{Q}^T = \underset{\approx}{Q}^{-1}$. If we premultiply and postmultiply Eq. 113 by $\underset{\approx}{Q}$ and $\underset{\approx}{Q}^T$, respectively, we get

$$\underset{\approx}{Q}\underset{\approx}{C}\underset{\approx}{Q}^T = \underset{\approx}{\Psi} = \begin{bmatrix} \psi_1 & 0 & 0 & \cdot & \cdot & \cdot & 0 \\ 0 & \psi_2 & 0 & \cdot & \cdot & \cdot & 0 \\ 0 & 0 & \psi_3 & \cdot & \cdot & \cdot & 0 \\ \cdot & \cdot & \cdot & \cdot & & & \cdot \\ \cdot & \cdot & \cdot & & \cdot & & \cdot \\ \cdot & \cdot & \cdot & & & \cdot & \cdot \\ 0 & 0 & 0 & \cdot & \cdot & \cdot & \psi_n \end{bmatrix} \tag{114}$$

Now let us define the vector $\underset{\sim}{X} = \{X_1, X_2, \ldots, X_n\}^T$ which contains the sequence of $X(t)$ values discussed above, having covariance matrix $\underset{\approx}{C} = \mathrm{E}\left[(\underset{\sim}{X} - \mu_X)(\underset{\sim}{X} - \mu_X)^T\right]$. If we let

$$\underset{\sim}{Z} = \underset{\approx}{Q}\underset{\sim}{X} \tag{115}$$

be a sequence of random variables obtained by rotating the vector $\underset{\sim}{X}$ by the orthonormal basis, $\underset{\approx}{Q}$, then $\underset{\sim}{Z}$ is composed of uncorrelated random variables having variance $\psi_1, \psi_2, \ldots, \psi_n$. We can show this by computing the covariance matrix of $\underset{\sim}{Z}$. For this we will assume, without loss of generality and merely for simplicity, that $\mathrm{E}[X(t)] = 0$ so that $\mathrm{E}[\underset{\sim}{Z}] = \underset{\sim}{0}$. (The end result for a non-zero mean is exactly the same, just more complicated – try it.) The covariance matrix of $\underset{\sim}{Z}$, in this case, is given by

$$\begin{aligned} \underset{\approx}{C}_Z &= \mathrm{E}\left[\underset{\sim}{Z}\underset{\sim}{Z}^T\right] = \mathrm{E}\left[(\underset{\approx}{Q}\underset{\sim}{X})(\underset{\approx}{Q}\underset{\sim}{X})^T\right] = \mathrm{E}\left[\underset{\approx}{Q}\underset{\sim}{X}\underset{\sim}{X}^T\underset{\approx}{Q}^T\right] \\ &= \underset{\approx}{Q}\mathrm{E}\left[\underset{\sim}{X}\underset{\sim}{X}^T\right]\underset{\approx}{Q}^T \\ &= \underset{\approx}{Q}\underset{\approx}{C}\underset{\approx}{Q}^T \\ &= \underset{\approx}{\Psi} \end{aligned}$$

so that the matrix of eigenvectors, $\underset{\approx}{Q}$, can be viewed as a rotation matrix which transforms the set of correlated random variables $X_1, X_2, \ldots, X_n$ into a set of uncorrelated random variables $\underset{\sim}{Z} = Z_1, Z_2, \ldots, Z_n$ having variance $\psi_1, \psi_2, \ldots, \psi_n$, respectively.

8.1 Conditional Probabilities

We are often interested in conditional probabilities of the form; given that $X(t)$ has been observed to have some value x at time t, what is the probability distribution of $X(t+s)$? If $X(t)$ is a stationary Gaussian process, then the conditional distribution of $X(t+s)$ given $X(t) = x$ is also normally distributed with mean and variance

$$\mathrm{E}[X(t+s) \mid X(t) = x] = \mu_X + (x - \mu_X)\rho(s) \tag{116a}$$

$$\mathrm{Var}[X(t+s) \mid X(t) = x] = \sigma_X^2(1 - \rho^2(s)) \tag{116b}$$

where $\rho(s)$ is the correlation coefficient between $X(t+s)$ and $X(t)$.

9 The Spectral Density Function

We now turn our attention to an equivalent 2nd-moment description of a stationary random process, namely its *spectral representation*. We say 'equivalent' because the spectral representation, in the form of a *spectral density function*, contains the same information as the covariance function, just expressed in a different way. As we shall see, the spectral density function can be obtained from the covariance function and vice-versa. The two forms are merely transforms of one another.

Priestley (1981) shows that if $X(t)$ is a stationary random process, with $\rho(\tau)$ continuous at $\tau = 0$, then it can be expressed as a sum of sinusoids with mutually independent random amplitudes and phase angles,

$$X(t) = \mu_X + \sum_{k=-N}^{N} C_k \cos(\omega_k t + \Phi_k) \tag{117}$$
$$= \mu_X + \sum_{k=-N}^{N} \Big(A_k \cos(\omega_k t) + B_k \sin(\omega_k t) \Big)$$

where μ_X is the process mean, C_k is a random amplitude, and Φ_k is a random phase angle. The equivalent form involving A_k and B_k is obtained by setting $A_k = C_k \cos(\Phi_k)$ and $B_k = -C_k \sin(\Phi_k)$. If the random amplitudes A_k and B_k are normally distributed with zero means, then $X(t)$ will also be normally distributed with mean μ_X. For this to be true, C_k must be Raleigh distributed and Φ_k must be uniformly distributed on the interval $[0, 2\pi]$. Note that $X(t)$ will tend to a normal distribution anyhow, by virtue of the central limit theorem, for wide-band processes, so we will assume that $X(t)$ is normally distributed.

Consider the kth component of $X(t)$, and ignore μ_X for the time being,

$$X_k(t) = C_k \cos(\omega_k t + \Phi_k) \tag{118}$$

If C_k is independent of Φ_k, then $X_k(t)$ has mean

$$\mathrm{E}\left[X_k(t)\right] = \mathrm{E}\left[C_k \cos(\omega_k t + \Phi_k)\right] = \mathrm{E}\left[C_k\right] \mathrm{E}\left[\cos(\omega_k t + \Phi_k)\right] = 0$$

due to independence and the fact that for any t, $\mathrm{E}\left[\cos(\omega_k t + \Phi_k)\right] = 0$ since Φ_k is uniformly distributed on $[0, 2\pi]$. The variance of $X_k(t)$ is thus

$$\mathrm{Var}\left[X_k(t)\right] = \mathrm{E}\left[X_k^2(t)\right] = \mathrm{E}\left[C_k^2\right] \mathrm{E}\left[\cos^2(\omega_k t + \Phi_k)\right] = \tfrac{1}{2}\mathrm{E}\left[C_k^2\right] \tag{119}$$

Note that $\mathrm{E}\left[\cos^2(\omega_k t + \Phi_k)\right] = \frac{1}{2}$, which again uses the fact that Φ_k is uniformly distributed between 0 and 2π.

Priestley also shows that the component sinusoids are independent of one another, that is that $X_k(t)$ is independent of $X_j(t)$, for all $k \neq j$. Using this property, we can put the components back together to find the mean and variance of $X(t)$,

$$\mathrm{E}\left[X(t)\right] = \mu_X + \sum_{k=-N}^{N} \mathrm{E}\left[X_k(t)\right] = \mu_X \tag{120a}$$

$$\mathrm{Var}\left[X(t)\right] = \sum_{k=-N}^{N} \mathrm{Var}\left[X_k(t)\right] = \sum_{k=-N}^{N} \tfrac{1}{2}\mathrm{E}\left[C_k^2\right] \tag{120b}$$

In other words, the prescribed mean of $X(t)$ is preserved by the spectral representation and the variance of the sum is the sum of the variances of each component frequency, since the component sinusoids are independent. The amount that each component frequency contributes to the overall variance of $X(t)$ depends on the 'power' in the sinusoid amplitude, $\frac{1}{2}\mathrm{E}\left[C_k^2\right]$.

Now define the *two-sided spectral density function*, $S(\omega)$, such that

$$S(\omega_k)\Delta\omega = \mathrm{Var}\left[X_k(t)\right] = \mathrm{E}\left[X_k^2(t)\right] = \tfrac{1}{2}\mathrm{E}\left[C_k^2\right] \tag{121}$$

then the variance of $X(t)$ can be written as

$$\mathrm{Var}\left[X(t)\right] = \sum_{k=-N}^{N} S(\omega_k)\Delta\omega \tag{122}$$

In the limit as $\Delta\omega \to 0$ and $N \to \infty$, we get

$$\mathrm{Var}\left[X(t)\right] = \sigma_X^2 = \int_{-\infty}^{\infty} S(\omega)\, d\omega \tag{123}$$

which is to say, the variance of $X(t)$ is just the area under the two-sided spectral density function.

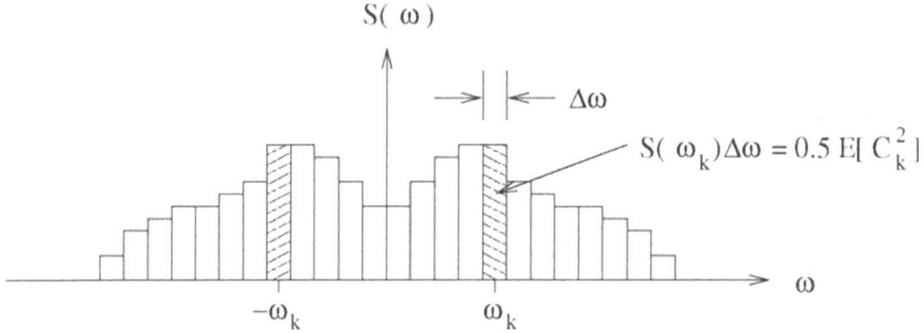

Figure 19. Two-sided spectral density function, $S(\omega)$.

9.1 Wiener-Khinchine Relations

We can use the spectral representation to express the covariance function, $C(\tau)$. Assuming that $\mu_X = 0$ for the time being to simplify the algebra (this is not a restriction, the end results are the same even if $\mu_X \neq 0$), we have

$$C(\tau) = \mathrm{Cov}\left[X(0), X(\tau)\right], \qquad \text{(due to stationarity)}$$

$$= \mathrm{E}\left[\sum_k X_k(0) \sum_j X_j(\tau)\right]$$
$$= \sum_k \sum_j \mathrm{E}\left[X_k(0) X_j(\tau)\right]$$
$$= \sum_k \mathrm{E}\left[X_k(0) X_k(\tau)\right], \qquad \text{(due to independence)}$$

Now, since $X_k(0) = C_k \cos(\Phi_k)$ and $X_k(\tau) = C_k \cos(\omega_k \tau + \Phi_k)$ we get

$$C(\tau) = \sum_k \mathrm{E}\left[C_k^2\right] \mathrm{E}\left[\cos(\Phi_k)\cos(\omega_k\tau + \Phi_k)\right]$$
$$= \sum_k \mathrm{E}\left[C_k^2\right] \mathrm{E}\left[\tfrac{1}{2}\{\cos(\omega_k\tau + 2\Phi_k) + \cos(\omega_k\tau)\}\right]$$
$$= \sum_k \tfrac{1}{2}\mathrm{E}\left[C_k^2\right] \cos(\omega_k\tau)$$
$$= \sum_k S(\omega_k)\cos(\omega_k\tau)\Delta\omega$$

which, in the limit as $\Delta\omega \to 0$ gives

$$C(\tau) = \int_{-\infty}^{\infty} S(\omega)\cos(\omega\tau)\, d\omega \tag{124}$$

Thus, the covariance function $C(\tau)$ is the Fourier transform of the spectral density function, $S(\omega)$. The inverse transform can be applied to find $S(\omega)$ in terms of $C(\tau)$,

$$S(\omega) = \frac{1}{2\pi}\int_{-\infty}^{\infty} C(\tau)\cos(\omega\tau)\, d\tau \tag{125}$$

so that knowing either $C(\tau)$ or $S(\omega)$ allows the other to be found (and hence these are 'equivalent' in terms of information). Also, since $C(\tau) = C(-\tau)$, ie. that the covariance between one point and another is the same regardless of which point you consider first, and since $\cos(x) = \cos(-x)$, we see that

$$S(\omega) = S(-\omega) \tag{126}$$

In other words, the two-sided spectral density function is an even function (see Figure 19). The fact that $S(\omega)$ is symmetric about $\omega = 0$ means that we need only know the positive half in order to know the entire function. This motivates the introduction of the *one-sided spectral density function*, $G(\omega)$ defined as

$$G(\omega) = 2S(\omega), \qquad \omega \geq 0 \tag{127}$$

The factor of two is included to preserve the total variance when only positive frequencies are considered. Now the Wiener-Khinchine relations become

$$C(\tau) = \int_{0}^{\infty} G(\omega)\cos(\omega\tau)\, d\omega \tag{128a}$$

$$G(\omega) = \frac{1}{\pi} \int_{-\infty}^{\infty} C(\tau) \cos(\omega\tau) \, d\tau \tag{128b}$$

$$= \frac{2}{\pi} \int_{0}^{\infty} C(\tau) \cos(\omega\tau) \, d\tau \tag{128c}$$

and the variance of $X(t)$ is the area under $G(\omega)$ (set $\tau = 0$ in Eq. 128a to see this),

$$\sigma_X^2 = C(0) = \int_0^{\infty} G(\omega) \, d\omega \tag{129}$$

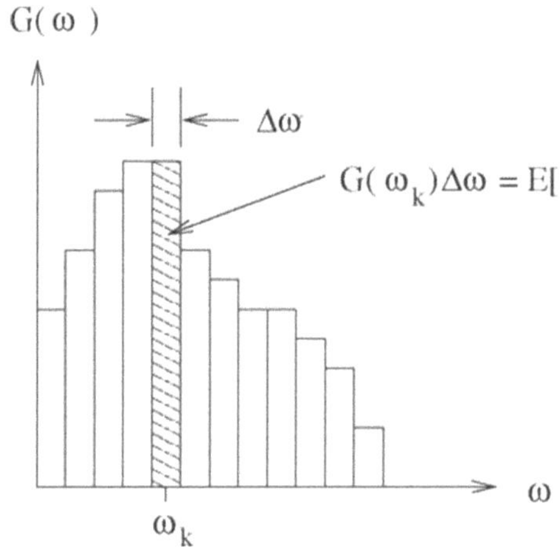

Figure 20. One-sided spectral density function, $G(\omega) = 2S(\omega)$ corresponding to Figure 19.

The spectral representation of a stationary Gaussian process is primarily used in situations where the frequency domain is an integral part of the problem being considered. For example, earthquake ground motions are often represented using the spectral density function because the motions are largely sinusoidal with frequency content dictated by resonance in the soil or rock through which the earthquake waves are traveling. In addition, the response of structures to earthquake motion is often performed using Fourier response 'modes', each having its own resonance frequency. Thus, if a structure has a 1 Hz primary response mode (single mass-and-spring oscillation), then it is of interest to see what 'power' the input ground motion has at 1 Hz. This is given by $G(\omega_k)\Delta\omega$ at $\omega_k = 1$ Hz.

In addition, the spectral representation provides a means to simulate a stationary Gaussian process, namely to simulate independent realizations of C_k and Φ_k, for $k = 0, 1, \ldots, N$, and then recombine using the spectral representation. We shall see more of this in Chapter 2.

10 The Variance Function

Virtually all engineering properties are actually properties of a local average of some sort. For example, the hydraulic conductivity of a soil is rarely measured at a point since, at the point level, we are either in a void having infinite conductivity, or in a solid, having negligible conductivity. Just as we rarely model soils at the microscopic, or particle, level for use in designs at the macroscopic level, the hydraulic conductivity is generally estimated using a laboratory sample of some volume, supplying a water pressure, and measuring the quantity of water which passes through the sample in some time interval. The paths that the water takes to migrate through the sample are not considered individually, rather it is the sum of these paths that are measured. This is a 'local average' over the laboratory sample. (As we shall see later there is more than one possible type of average to take, but for now we shall concentrate on the more common arithmetic average.)

Similarly, when the compressive strength of a material is determined, a load is applied to a finite sized sample until failure occurs. Failure takes place when the shear/tensile resistance of a large number of bonds are broken – the failure load is then a function of the average bond strength throughout the failure region.

Thus, it is of considerable engineering interest to investigate how averages of random fields behave. Consider the local average defined as

$$X_T(t) = \frac{1}{T}\int_{t-T/2}^{t+T/2} X(\xi)\, d\xi \tag{130}$$

which is a 'moving' local average. That is, $X_T(t)$ is the local average of $X(t)$ over a window of width T centered at t. As this window is moved along in time, the local average $X_T(t)$ changes more slowly.

For example, consider the 'boat in the water' example: if the motion of a piece of sawdust on the surface of the ocean is tracked, it is seen to have considerable variability in its elevation. In fact, it will have as much variability as the waves themselves. Now, replace the sawdust with an ocean liner. The liner does not bounce around with every wave, but rather it 'averages' out the wave motion over the area of the liner. It's vertical variability is drastically reduced.

In this example, it is also worth thinking about the spectral representation of the ocean waves. The piece of sawdust sees all of the waves, big and small, whereas the local averaging taking place over the ocean liner damps out the high frequency components leaving just the long wavelength components (wavelengths of the order of the size of the ship and longer). Thus, local averaging is a low-pass filter. If the ocean waves on the day that the sawdust and ocean liner are being observed are composed of just long wavelength swells, then the variability of the sawdust and liner will be the same. Conversely, if the ocean surface is just choppy without any swells, then the ocean liner may hardly move up and down at all. Both the sawdust and the ocean liner will have the same mean elevation in all cases.

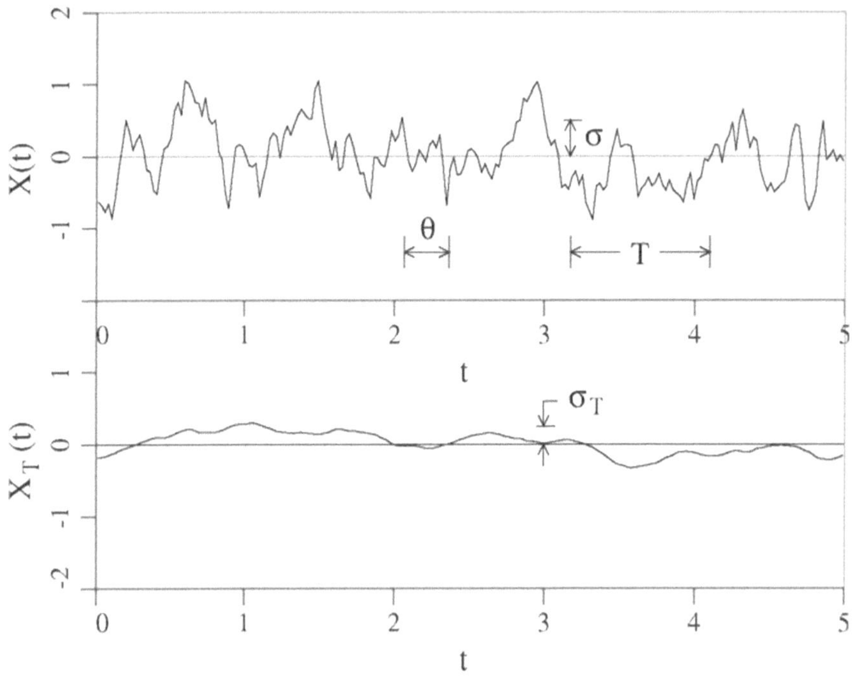

Figure 21. Illustration of the effect of local averaging on variance. T is the moving window length over which the top plot is averaged to get the lower plot.

The two main effects of local averaging are to reduce the variance and to damp the contribution from the high frequency components. The amount of variance reduction increases with increasing high-frequency content in the random field. An increased high-frequency content corresponds to increasing independence in the random field, so that another way of putting this is that variance reduction increases when the random field consists of more 'independence'. This is illustrated in Figure 21. A random process is shown in the upper plot which is then averaged within a moving window of width T to obtain the lower plot. Notice that averaging both smooths the process and reduces its variance.

Let us look in more detail at the moments of $X_T(t)$. Its mean is

$$\begin{aligned} \mathrm{E}\left[X_T(t)\right] &= \mathrm{E}\left[\frac{1}{T}\int_{t-T/2}^{t+T/2} X(\xi)\,d\xi\right] \\ &= \frac{1}{T}\int_{t-T/2}^{t+T/2} \mathrm{E}\left[X(\xi)\right]\,d\xi \\ &= \mathrm{E}\left[X\right] \end{aligned} \tag{131}$$

for stationary $X(t)$. That is, local averaging preserves the mean of the random field (the mean of an arithmetic average is just the mean of the process). Consider the variance,

$$\mathrm{Var}\left[X_T(t)\right] = \mathrm{E}\left[(X_T(t) - \mu_{X_T})^2\right] \tag{132}$$

where, since $\mu_{X_T} = \mu_X$,

$$X_T - \mu_{X_T} = \frac{1}{T}\int_{t-T/2}^{t+T/2} X(\xi)\,d\xi - \mu_X$$
$$= \frac{1}{T}\int_{t-T/2}^{t+T/2} [X(\xi) - \mu_X]\,d\xi$$

so that (due to stationarity, the bounds of the integral can be changed to any domain of length T without changing the expectation; we will use the domain $[0,T]$ for simplicity),

$$\begin{aligned}
\operatorname{Var}[X_T(t)] &= \mathrm{E}\left[\frac{1}{T}\int_0^T [X(\xi)-\mu_X]\,d\xi \cdot \frac{1}{T}\int_0^T [X(\eta)-\mu_X]\,d\eta\right] \\
&= \frac{1}{T^2}\int_0^T\int_0^T \mathrm{E}\left[(X(\xi)-\mu_X)(X(\eta)-\mu_X)\right]\,d\xi\,d\eta \\
&= \frac{1}{T^2}\int_0^T\int_0^T C_X(\xi-\eta)\,d\xi\,d\eta \\
&= \frac{\sigma_X^2}{T^2}\int_0^T\int_0^T \rho_X(\xi-\eta)\,d\xi\,d\eta \\
&= \sigma_X^2\,\gamma(T)
\end{aligned} \tag{133}$$

where $C_X(\tau)$ is the covariance function of $X(t)$, and $\rho_X(\tau)$ is the correlation function of $X(t)$, such that $C_X(\tau) = \sigma_X^2 \rho_X(\tau)$. In the final expression, $\gamma(T)$ is the so-called *variance function*, which gives the amount that the variance is reduced when $X(t)$ is averaged over the length T. The variance function has value 1.0 when $T = 0$, which is to say that $X_T(t) = X(t)$ when $T = 0$ and so the variance is not at all reduced. As T increases, the variance function decreases towards zero. It has mathematical definition

$$\gamma(T) = \frac{1}{T^2}\int_0^T\int_0^T \rho_X(\xi-\eta)\,d\xi\,d\eta \tag{134}$$

The variance function can be seen, in Eq. (134), to be an average of the correlation coefficient between every pair of points on the interval $[0,T]$. If the correlation function falls off rapidly, so that the correlation between pairs of points becomes rapidly smaller with separation distance, then $\gamma(T)$ will be small. On the other hand, if all points on the interval $[0,T]$ are perfectly correlated, having $\rho(\tau) = 1$ for all τ, then $\gamma(T)$ will be 1.0. Such a field displays no variance reduction under local averaging. (In fact, if the field is stationary, all points will have the same random value, $X(t) = X$.)

The integral, in Eq. 134, is over the square region $[0,T] \times [0,T]$ in (ξ,η) space. Evidently, $\rho_X(\xi - \eta)$ is constant along diagonal lines where $\xi - \eta =$ constant. The length of the main diagonal, where $\xi = \eta$, is $\sqrt{2}T$, and the other diagonal lines decrease linearly in length to zero in the corners. The double integral can be collapsed to a single integral by integrating in a direction perpendicular to the diagonals; each diagonal differential area has length $\sqrt{2}(T - |\tau|)$, width $d\tau/\sqrt{2}$, and 'height' equal to $\rho_X(\xi - \eta) = \rho_X(\tau)$. The integral can therefore be written as follows

$$\begin{aligned}\gamma(T) &= \frac{1}{T^2}\int_0^T\int_0^T \rho(\xi - \eta)\,d\xi\,d\eta \\ &= \frac{1}{T^2}\left[\int_{-T}^{0}\sqrt{2}(T - |\tau_1|)\rho(\tau_1)\frac{d\tau_1}{\sqrt{2}} + \int_0^T \sqrt{2}(T - |\tau_2|)\rho(\tau_2)\frac{d\tau_2}{\sqrt{2}}\right] \\ &= \frac{1}{T^2}\int_{-T}^{T}(T - |\tau|)\rho_X(\tau)\,d\tau \end{aligned} \tag{135}$$

Furthermore, since $\rho_X(\tau) = \rho_X(-\tau)$, the integrand is even which results in the further simplification

$$\gamma(T) = \frac{2}{T^2}\int_0^T (T - \tau)\rho_X(\tau)\,d\tau \tag{136}$$

The variance function is another 'equivalent' second-moment description of a random field, since it can be obtained through knowledge of the correlation function, which in turn can be obtained from the spectral density function. The inverse relationship between $\gamma(T)$ and $\rho(\tau)$ is obtained by differentiation;

$$\rho(\tau) = \frac{1}{2}\frac{d^2}{d\tau^2}[\tau^2\gamma(\tau)] \tag{137}$$

The variance function can also be obtained from the spectral density function (Vanmarcke, 1984)

$$\gamma(T) = \int_0^\infty \frac{G(\omega)}{\sigma_X^2}\left[\frac{\sin(\omega T/2)}{\omega T/2}\right]^2 d\omega \tag{138}$$

11 The Scale of Fluctuation

A convenient measure of the variability of a random field is the *scale of fluctuation*, θ, also sometimes refered to as the *correlation length*. Loosely speaking θ is the distance beyond which points are largely uncorrelated. Conversely, two points separated by a distance less than θ will be significantly correlated. Mathematically, θ is defined here as the area under the correlation function (Vanmarcke, 1984),

$$\theta = \int_{-\infty}^{\infty}\rho(\tau)\,d\tau = 2\int_0^\infty \rho(\tau)\,d\tau \tag{139}$$

The correlation length is sometimes defined without the factor of two showing on the right-hand-side of Eq. 139 (see, for example, Journel and Huijbregts, 1978)

Eq. 139 implies that if θ is to be finite then $\rho(\tau)$ must decrease sufficiently quickly to zero as τ increases. Not all correlation functions will satisfy this criteria, and for such random processes, $\theta = \infty$. An example of a process with infinite scale of fluctuation is a *fractal* process.

In addition, the scale of fluctuation is really only meaningful for strictly non-negative correlation functions. Since $-1 \leq \rho \leq 1$, one could conceivably have an oscillatory correlation function whose area is zero but which has significant correlations (positive or negative) over significant distances. An example of such a correlation function might be that governing wave heights in a body of water.

The scale of fluctuation can also be defined in terms of the spectral density function,

$$G(\omega) = \frac{2\sigma^2}{\pi} \int_0^\infty \rho(\tau) \cos(\omega\tau) \, d\tau \tag{140}$$

since, when $\omega = 0$,

$$G(0) = \frac{2\sigma^2}{\pi} \int_0^\infty \rho(\tau) \, d\tau = \frac{\sigma^2}{\pi}\theta \tag{141}$$

which means that

$$\theta = \frac{\pi G(0)}{\sigma^2} \tag{142}$$

What this means is that if the spectral density function is finite at the origin, then θ will also be finite. In practice $G(0)$ is quite difficult to estimate, since it requires data over an infinite distance ($\omega = 0$ corresponds to an infinite wavelength). Thus, Eq. 142 is of limited value in estimating the scale of fluctuation from real data.

The scale of fluctuation can also be defined in terms of the variance function as a limit (Vanmarcke, 1984);

$$\theta = \lim_{T \to \infty} T\gamma(T) \tag{143}$$

This implies that if the scale of fluctuation is finite, then the variance function has the following limiting form as the averaging region grows very large;

$$\lim_{T \to \infty} \gamma(T) = \frac{\theta}{T} \tag{144}$$

which, in turn, means that θ/T can be used as an approximation for $\gamma(T)$ when $T >> \theta$. A more extensive approximation for $\gamma(T)$, useful when the precise correlation structure of a random field is unknown, but for which θ is known (or estimated) is

$$\gamma(T) \simeq \frac{\theta}{\theta + |T|} \tag{145}$$

which has the correct limiting form for $T >> \theta$ and which has value 1.0 when $T = 0$, as expected. The correlation function corresponding to Eq. 145 is,

$$\rho(\tau) = \frac{\theta^3}{(\theta + \tau)^3} \tag{146}$$

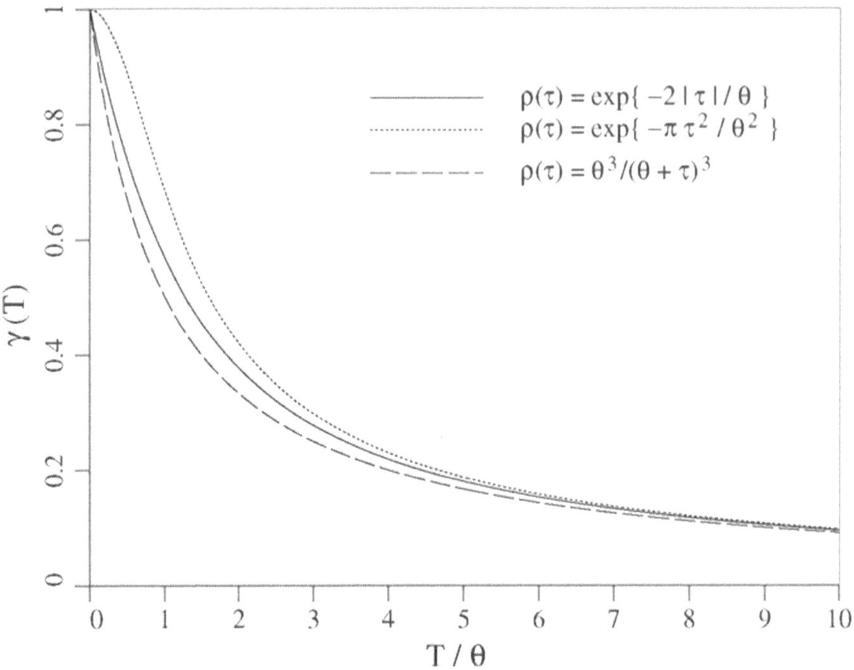

Figure 22. Variance function corresponding to three different correlation models.

The next figure shows two random field realizations. The field on the left has a small scale of fluctuation ($\theta = 0.04$) and can be seen to be quite rough. The field on the right has a large scale of fluctuation ($\theta = 2$) and can be seen to be more slowly varying.

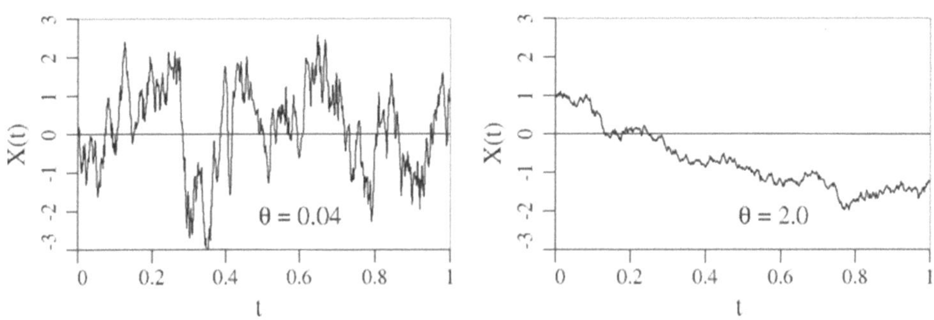

Figure 23. Sample realizations of $X(t)$ for two different scales of fluctuation.

12 Random Fields in Higher Dimensions

Figure 1.24 illustrates a two-dimensional random field, $X(t_1, t_2)$, where X varies randomly in two directions, rather than just along a line. The elevation of a soil's surface, or the thickness of a soil layer, at any point on the plan area of a site are examples of two-dimensional random fields. The cohesion of the soil at plan location (t_1, t_2) and depth t_3 is an example of a three-dimensional random field, $X(t_1, t_2, t_3)$. The coordinate labels t_1, t_2, and t_3 are often replaced by the more common Cartesian coordinates x, y, and z. We shall keep the current notation to remain consistent with that developed in the one-dimensional case.

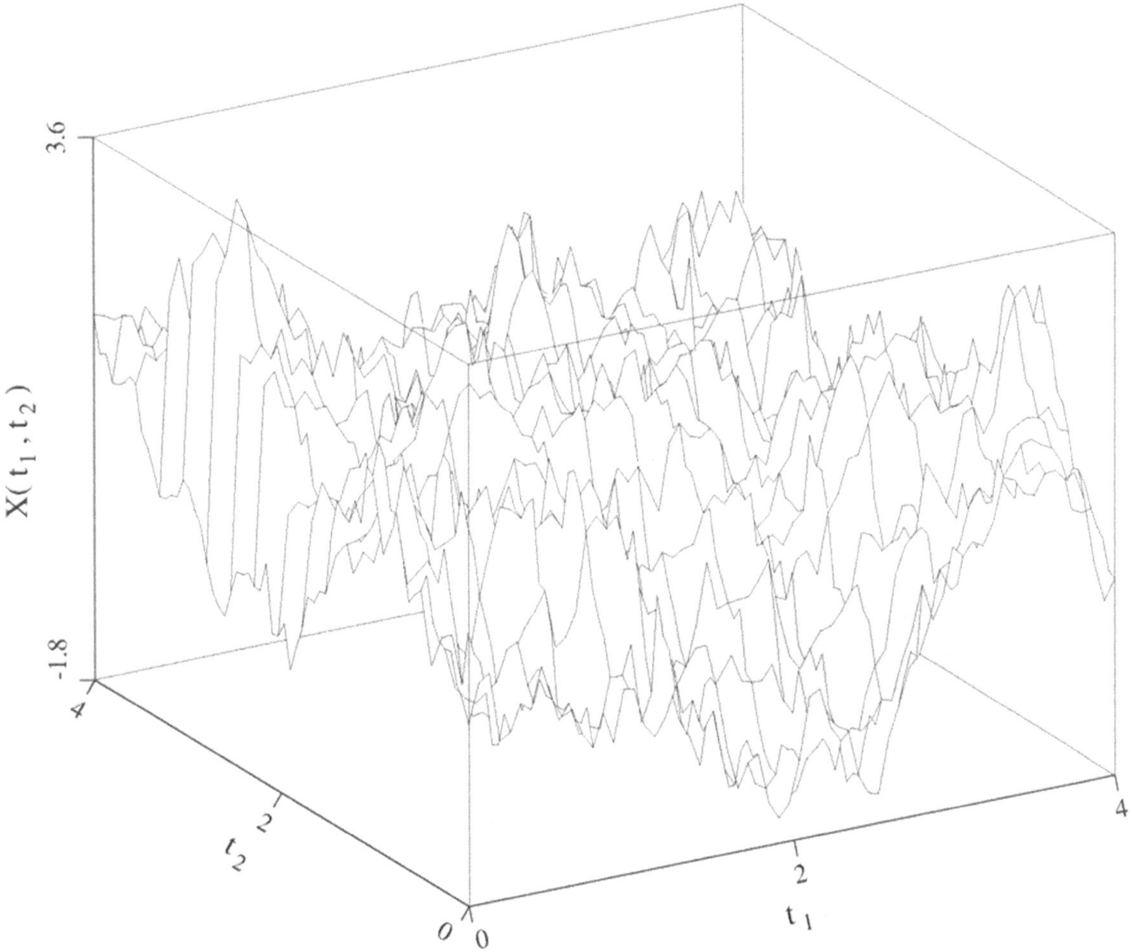

Figure 24. A realization of a two-dimensional random field.

In this section, we will concentrate predominately on two-dimensional random fields, the three-dimensional case generally just involving adding another coordinate. As in the one-dimensional case, a random field is characterized by

1) its first moment, or mean, $\mu(t_1, t_2)$, which may vary in space. If the random field is *stationary*, then the mean does not change with position; $\mu(t_1, t_2) = \mu$.

2) its second moment, or covariance structure, $C(t_1', t_1^*, t_2', t_2^*)$, which gives the covariance between two points in the field, $X(t_1', t_2')$ and $X(t_1^*, t_2^*)$. If the field is *stationary*, then the covariance structure remains the same regardless of where the axes origin is located, that is, the covariance function becomes a function of just the difference, $(\underset{\sim}{t}' - \underset{\sim}{t}^*)$, i.e. $C(t_1' - t_1^*, t_2' - t_2^*)$.
3) its higher-order moments. If the field is Gaussian, it is completely characterized by its first two moments.

We will restrict our attention to just the first two moments of a random field. For simplicity, we will sometimes concentrate on stationary random fields since any random field, X', can be converted to a random field which is stationary in its mean and variance, X, (with zero mean and unit variance) through the transformation

$$X(\underset{\sim}{t}) = \frac{X'(\underset{\sim}{t}) - \mu'(\underset{\sim}{t})}{\sigma'(\underset{\sim}{t})} \tag{147}$$

where $\underset{\sim}{t}$ is a vector denoting spatial position (in two-dimensions, $\underset{\sim}{t}$ has components t_1 and t_2) and $\mu'(\underset{\sim}{t})$ and $\sigma'(\underset{\sim}{t})$ are the mean and standard deviation of X' at the spatial location $\underset{\sim}{t}$.

In the following subsections we investigate various ways that the second moment characteristics of a random field can be expressed.

12.1 The Covariance Function in Higher Dimensions

The covariance function gives the covariance between two points in the field, $X' = X(\underset{\sim}{t}')$ and $X^* = X(\underset{\sim}{t}^*)$. Since the covariance between X' and X^* is the same as the covariance between X^* and X' (that is, it doesn't matter which way you look at the pair) then $C(t_1', t_1^*, t_2', t_2^*) = C(t_2', t_2^*, t_1', t_1^*)$. If the random field is stationary, this translates into the requirement that $C(\underset{\sim}{\tau}) = C(-\underset{\sim}{\tau})$, where $\underset{\sim}{\tau} = \underset{\sim}{t}' - \underset{\sim}{t}^*$ is the spatial *lag* vector having components $\tau_1 = t_1' - t_1^*$, $\tau_2 = t_2' - t_2^*$. For example, for a two-dimensional stationary random field $C(t_1' - t_1^*, t_2' - t_2^*) = C(t_1^* - t_1', t_2^* - t_2')$, or $C(\tau_1, \tau_2) = C(-\tau_1, -\tau_2)$.

In two dimensions, the correlation function is defined as

$$\rho(\tau_1, \tau_2) = \frac{\operatorname{Cov}[X', X^*]}{\sigma'\sigma^*} = \frac{C(\tau_1, \tau_2)}{\sigma'\sigma^*} \tag{148}$$

where σ' and σ^* are the standard deviations of $X' = X(\underset{\sim}{t}')$ and $X^* = X(\underset{\sim}{t}^*)$, respectively. Since we are assuming the random field is stationary, then $\sigma' = \sigma^* = \sigma$, and the correlation function becomes

$$\rho(\tau_1, \tau_2) = \frac{C(\tau_1, \tau_2)}{\sigma^2} \tag{149}$$

An example two-dimensional correlation function is the separable model

$$\begin{aligned}\rho(\tau_1, \tau_2) &= \exp\left\{-\frac{2}{\theta}(|\tau_1| + |\tau_2|)\right\} \\ &= \exp\left\{-\frac{2|\tau_1|}{\theta_1}\right\} \exp\left\{-\frac{2|\tau_2|}{\theta_2}\right\}\end{aligned} \tag{150}$$

which is Markovian in each coordinate direction. Note that even if the directional correlation lengths, θ_1 and θ_2, are equal, this function is not isotropic.

12.2 The Spectral Density Function in Higher Dimensions

In two dimensions, the spectral representation of a stationary random field, $X(t_1, t_2)$, is the double sum

$$X(t_1, t_2) = \mu_X + \sum_{i=-N_1}^{N_1} \sum_{j=-N_2}^{N_2} C_{ij} \cos(\omega_{1_i} t_1 + \omega_{2_j} t_2 + \Phi_{ij}) \tag{151}$$

where, as in the one-dimensional case, C_{ij} is a random amplitude and Φ_{ij} a random phase angle. The variance of $X(t_1, t_2)$ is obtained by assuming the random variables C_{ij} and Φ_{ij} are all mutually independent

$$\sigma_X^2 = \mathrm{E}\left[(X(t_1, t_2) - \mu_X)^2\right] = \sum_{i=-N_1}^{N_1} \sum_{j=-N_2}^{N_2} \tfrac{1}{2}\mathrm{E}\left[C_{ij}^2\right] \tag{152}$$

We define the two-dimensional spectral density function, $S(\omega_1, \omega_2)$, such that

$$S(\omega_{1_i}, \omega_{2_j}) \Delta\omega_1 \Delta\omega_2 = \tfrac{1}{2}\mathrm{E}\left[C_{ij}^2\right] \tag{153}$$

Note that if the correlation function is separable, as is Eq. 150, then both the spectral density and the variance functions will also be of separable form (although in the case of the spectral density function the variance does not appear more than once in the product). The spectral density function corresponding to Eq. 150 is

$$G(\omega_1, \omega_2) = \frac{\sigma^2 \theta_1 \theta_2}{\pi \left[1 + \left(\frac{\theta_1 \omega_1}{2}\right)^2\right] \left[1 + \left(\frac{\theta_2 \omega_2}{2}\right)^2\right]} \tag{154}$$

In the limit as both $\Delta\omega_1$ and $\Delta\omega_2$ go to zero, we can express the variance of X as the volume under the spectral density function

$$\sigma_X^2 = \int_{-\infty}^{\infty} \int_{-\infty}^{\infty} S(\omega_1, \omega_2)\, d\omega_1\, d\omega_2 \tag{155}$$

In the two-dimensional case, the Wiener-Khinchine relationships become

$$C(\tau_1, \tau_2) = \int_{-\infty}^{\infty} \int_{-\infty}^{\infty} S(\omega_1, \omega_2) \cos(\omega_1 \tau_1 + \omega_2 \tau_2)\, d\omega_1\, d\omega_2 \tag{156a}$$

$$S(\omega_1, \omega_2) = \frac{1}{(2\pi)^2} \int_{-\infty}^{\infty} \int_{-\infty}^{\infty} C(\tau_1, \tau_2) \cos(\omega_1 \tau_1 + \omega_2 \tau_2)\, d\tau_1\, d\tau_2 \tag{156b}$$

If we express the components of spatial lag and frequency using vectors, $\underset{\sim}{\tau} = \{\tau_1, \tau_2\}^T$ and $\underset{\sim}{\omega} = \{\omega_1, \omega_2\}^T$, where superscript T denotes transpose, then the Wiener-Khinchine relationships can be written for n dimensions succinctly as

$$C(\underset{\sim}{\tau}) = \int_{-\infty}^{\infty} S(\underset{\sim}{\omega}) \cos(\underset{\sim}{\omega} \cdot \underset{\sim}{\tau})\, d\underset{\sim}{\omega} \tag{157a}$$

$$S(\underset{\sim}{\omega}) = \frac{1}{(2\pi)^n} \int_{-\infty}^{\infty} C(\underset{\sim}{\tau}) \cos(\underset{\sim}{\omega} \cdot \underset{\sim}{\tau}) \, d\underset{\sim}{\tau} \tag{157b}$$

where it is understood that we have a double integral for two dimensions, a triple integral for three dimensions, and so forth. The centered dot denotes the *vector dot product*, for example $\underset{\sim}{\omega} \cdot \underset{\sim}{\tau} = \omega_1 \tau_1 + \omega_2 \tau_2$.

12.3 The Variance Function in Higher Dimensions

In two dimensions, we can define the moving local average of a random field, $X(t_1, t_2)$, over an area of dimension $A = T_1 \times T_2$ to be

$$X_A(t_1, t_2) = \frac{1}{A} \int_{t_1 - T_1/2}^{t_1 + T_1/2} \int_{t_2 - T_2/2}^{t_2 + T_2/2} X(\xi_1, \xi_2) \, d\xi_2 \, d\xi_1 \tag{158}$$

To determine the statistics of X_A, we will first assume that the random field $X(\underset{\sim}{t})$ is stationary, so that we can choose to find the mean and variance of $X_A = X_A(T_1/2, T_2/2)$ as representative,

$$X_A = \frac{1}{A} \int_0^{T_1} \int_0^{T_2} X(t_1, t_2) \, dt_2 \, dt_1 \tag{159}$$

The mean of X_A is

$$\mu_{X_A} = \frac{1}{A} \int_0^{T_1} \int_0^{T_2} \mathrm{E}\left[X(t_1, t_2)\right] dt_2 \, dt_1 = \mu_X$$

Assuming that the random field $X(t_1, t_2)$ has 'point' mean $\mu_X = 0$ and variance σ_X, then the variance of X_A is

$$\begin{aligned}
\mathrm{Var}\left[X_A\right] = \sigma_A^2 = \mathrm{E}\left[X_A^2\right] &=^2 \frac{1}{A^2} \int_0^{T_1} \int_0^{T_1} \int_0^{T_2} \int_0^{T_2} \mathrm{E}\left[X(t_1, t_2) X(\xi_1, \xi_2)\right] d\xi_2 \, dt_2 \, d\xi_1 \, dt_1 \\
&= \frac{1}{A^2} \int_0^{T_1} \int_0^{T_1} \int_0^{T_2} \int_0^{T_2} \mathrm{Cov}\left[X(t_1, t_2), X(\xi_1, \xi_2)\right] d\xi_2 \, dt_2 \, d\xi_1 \, dt_1 \\
&= \frac{\sigma_X^2}{A^2} \int_0^{T_1} \int_0^{T_1} \int_0^{T_2} \int_0^{T_2} \rho(t_1 - \xi_1, t_2 - \xi_2) \, d\xi_2 \, dt_2 \, d\xi_1 \, dt_1
\end{aligned}$$

The same result would have been obtained even if $\mu_X \neq 0$ (at the expense of somewhat more complicated algebra).

Making use of the fact that, for stationary random fields, ρ is constant along diagonal lines where $t_1 - \xi_1$ and $t_2 - \xi_2$ are constant, we can reduce the four-fold integral to a double integral (see Eq. 135), so that

$$\begin{aligned}
\mathrm{Var}\left[X_A\right] &= \frac{\sigma_X^2}{A^2} \int_{-T_1}^{T_1} \int_{-T_2}^{T_2} (|T_1| - |\tau_1|)(|T_2| - |\tau_2|) \rho(\tau_1, \tau_2) \, d\tau_2 \, d\tau_1 \\
&= \sigma_X^2 \gamma(T_1, T_2)
\end{aligned}$$

where, since $A = T_1T_2$, the variance function is defined by

$$\gamma(T_1, T_2) = \frac{1}{T_1^2T_2^2}\int_{-T_1}^{T_1}\int_{-T_2}^{T_2}(|T_1| - |\tau_1|)(|T_2| - |\tau_2|)\rho(\tau_1, \tau_2)\, d\tau_2\, d\tau_1 \tag{160}$$

Some additional simplification is possible if $\rho(\tau_1, \tau_2) = \rho(-\tau_1, \tau_2) = \rho(\tau_1, -\tau_2) = \rho(-\tau_1, -\tau_2)$ (this is called quadrant symmetry, which will be discussed shortly) in which case

$$\gamma(T_1, T_2) = \frac{4}{T_1^2T_2^2}\int_{0}^{T_1}\int_{0}^{T_2}(|T_1| - \tau_1)(|T_2| - \tau_2)\rho(\tau_1, \tau_2)\, d\tau_2\, d\tau_1 \tag{161}$$

Although $\gamma(T_1, T_2)$ is perhaps questionably defined when T_1 or T_2 are negative, we shall assume that an averaging area of size -2×3 is the same as an averaging area of size 2×3, the sign only arising because T_1 is measured in opposite directions. By this assumption, $\gamma(T_1, T_2)$ is automatically *quadrant symmetric.*

The separable two-dimensional variance function corresponding to Eq. 150, is

$$\gamma(T_1, T_2) = \frac{\theta_1^2\theta_2^2}{4T_1^2T_2^2}\left[\frac{2|T_1|}{\theta_1} + \exp\left\{-\frac{2|T_1|}{\theta_1}\right\} - 1\right]\left[\frac{2|T_2|}{\theta_2} + \exp\left\{-\frac{2|T_2|}{\theta_2}\right\} - 1\right] \tag{162}$$

Bibliography

Abramowitz, M. and Stegun, I. eds. (1970). *Handbook of Mathematical Functions*, 10th Ed., Dover, New York.

Devore, J.L. (2003). *Probability and Statistics for Engineering and the Sciences*, 6th Ed., Duxbury Press, New York.

Gradshteyn, I.S. and Ryzhik, I.M. (1980). *Table of Integrals, Series, and Products*, 4th Ed., Academic Press, Toronto.

Journel, A.G. and Huijbregts, Ch.J. (1978). *Mining Geostatistics*, Academic Press, New York, NY.

Law, A.M. and Kelton, W.D. (2000). *Simulation Modeling and Analysis*, 3rd Ed., McGraw-Hill, New York, NY.

Papoulis, A. (1991). *Probability, Random Variables, and Stochastic Processes*, 3rd Ed., McGraw-Hill, Inc., New York, New York.

Priestley, M.B. (1981). *Spectral Analysis and Time Series*, Vol. Volume 1: Univariate Series, Academic Press, New York, NY.

Vanmarcke, E.H. (1984). *Random Fields: Analysis and Synthesis*, The MIT Press, Cambridge, Massachusetts.

Tools and Strategies for Dealing with Uncertainty in Geotechnics

Farrokh Nadim [1]

[1] International Centre for Geohazards / Norwegian Geotechnical Institute, Oslo, Norway

Abstract. Working with uncertainty is an essential aspect of engineering – the larger the uncertainty and the closer to critical, the greater the need for evaluating its effect(s) on the results. To characterize the uncertainties in soil and/or rock properties, the engineer needs to combine, in addition to actual data, knowledge about the quality of the data, knowledge on the geology and, most importantly, engineering judgment. Once the uncertainty in input parameters and model(s) for solving a particular problem are quantified, the engineer has a variety of tools at his disposal to evaluate the uncertainty in the output. The most common practical tools are Monte Carlo simulation techniques, first-order, second moment (FOSM) approach, first- and second-order reliability methods (FORM and SORM), and event tree analysis. Each has its advantages and shortcomings. The more complicated methods often provide more useful information about the possible outcomes of a problem. These methods are described and their applications are demonstrated through example problems. Many geotechnical problems involve several possible failure modes, which may or may not be correlated. These problems should be treated as systems. Component reliability vs. system reliability are discussed, and example calculations are presented.

1 Introduction

Most of the parameters used in geotechnical analyses, in particular the mechanical soil properties, are uncertain. Working with uncertainty is therefore an essential aspect of geotechnical engineering – the larger the uncertainty and the closer to critical, the greater the need for evaluating its effect(s) on the results. The geotechnical engineer tries to deal with the uncertainties by choosing reasonably conservative parameters for the deterministic stability evaluation. This approach, however, fails to address the problem of properly and consistently dealing with uncertainties.

In recent years, new and creative solutions have been developed for geotechnical design, and calculation methods have been improved. Yet the characterization and reduction of uncertainties still is an area where only few researchers were working until a few years ago, even though as early as 1982 Einstein and Baecher stated the following words of wisdom:

> *"In thinking about sources of uncertainty in engineering geology, one is left with the fact that uncertainty is inevitable. One attempts to reduce it as much as possible, but it must ultimately be faced. It is a well recognized part of life for the engineer. The question is not whether to deal with uncertainty, but how?"*

This chapter presents the different methods and approaches for doing probabilistic geotechnical analysis. To characterize the uncertainties in soil and/or rock properties, the engineer needs to combine, in addition to actual data, knowledge about the quality of the data, knowledge on the geology and, most importantly, engineering judgment. Once the uncertainty in input parameters and model(s) for solving a particular problem are quantified, the engineer has a variety of tools at his disposal to evaluate the uncertainty in the output. The most common practical tools are Monte Carlo simulation techniques; first-order, second moment (FOSM) approach; and first- and second-order reliability methods (FORM and SORM). Each has its advantages and shortcomings, and the more complicated methods often provide more useful information about the possible outcomes of a problem. The application of these methods are demonstrated through example problems.

2 Sources and types of uncertainty

Uncertainty modelling of the variables entering an analysis, whether probabilistic or deterministic, requires collection of data, evaluation of the data set(s), selection of a "model" to represent the data, estimation of the uncertainty in the selected model and its significant characteristics, and a verification of the assumptions made. The evaluation of the data set(s) needs recognition of the type of uncertainties, whether the variables are dependent or independent, whether the observations are independent and whether the uncertainties noted are the result of a combination of uncertainties in underlying parameters and processes.

The uncertainties associated with a geotechnical problem can be divided into two categories: aleatory and epistemic. Human errors and omissions, which are not covered herein, would fall into a third category. Within a nominally uniform geological layer, the soil properties can be affected by both aleatory and epistemic uncertainties:

- Aleatory uncertainty represents the natural randomness of a variable. Examples of aleatory uncertainty are the spatial variation of a soil parameter within a nominally uniform geological layer, the variation in the peak acceleration of a design earthquake with a given return period, the variation in the ocean wave height or wind force, and so on. The aleatory uncertainty is also called the inherent uncertainty. Aleatory uncertainty cannot be reduced or eliminated.
- Epistemic uncertainty represents the uncertainty due to lack of knowledge on a variable. Epistemic uncertainty includes measurement uncertainty, statistical uncertainty (due to limited information), and model uncertainty. Statistical uncertainty is due to limited information such as limited number of observations. Measurement uncertainty is due to for example imperfections of an instrument or of a method to register a quantity. Model uncertainty is due to idealizations made in the physical formulation of the problem. Epistemic uncertainty can be reduced, and perhaps eliminated, by collecting more data and information, improving the measurement method(s) or improving the calculation method(s).

Statistical uncertainty is present because the parameters are estimated from a limited set of data, and is affected by the type of estimation technique used. Measurement uncertainty is described in terms of accuracy and is affected by bias (systematic error) and by precision (random error). It can be evaluated from data provided by the manufacturer, laboratory tests and/or scaled tests. Model uncertainty is defined as the ratio of the actual quantity to the quantity predicted by a model. A mean value different from 1.0 expresses a bias in the model, while the standard deviation expresses the variability in the predictions by the model.

Uncertain soil properties and model uncertainty are best defined as random variables described by their mean, standard deviation (or coefficient of variation) and probability distribution function. Figure 1a compares a soil property described deterministically and a soil property described with its uncertainty. In practice, no experienced geotechnical engineer ever determines a deterministic (punctual) value for a soil property. Armed with engineering judgment, we select an appropriate characteristic value, on the basis of the available data, the expected range of values for this property, the type of problem to be analyzed and our experience. Mentally we establish a possible range of values, and select either a most probable value or a somewhat conservative value.

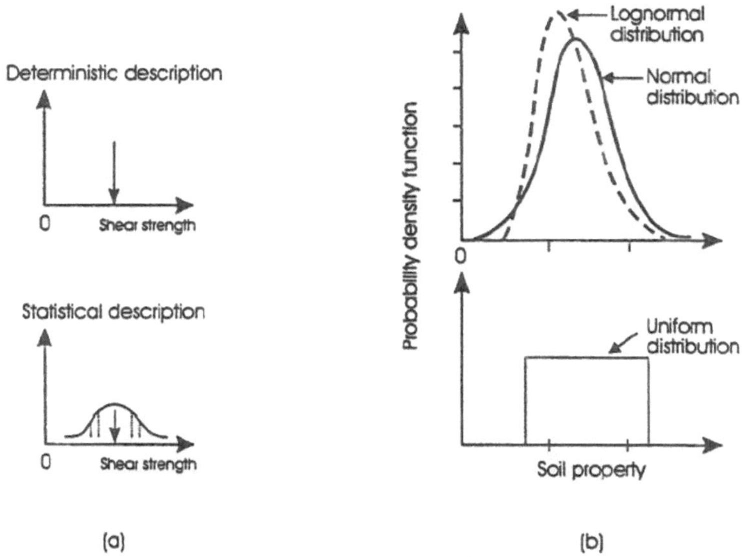

Figure 1. Deterministic and statistical description of soil property.

Figure 1b shows typical probability distribution functions used in geotechnical problems. The normal and lognormal are the most common; the lognormal is often used to characterize variables that do not take negative values. A uniform distribution may also be adequate for an equally likely range of values. These distributions are simple and require little work except the use of standard statistical tables.

Reliability analyses (e.g. Lacasse and Nadim, 1996) show that uncertainties on different soil properties affect differently the reliability of geotechnical analyses. It is therefore important that the uncertainties in analysis parameters be adequately quantified and their effect carefully evaluated. Unfortunately it is not possible to establish a set of guidelines for the evaluation of the uncertainty in soil and rock properties that are valid for all sites. The first rule, when determining the uncertainties related to a soil property and using statistical methods, is to ensure that consistent data populations are used. Major uncertainties have been introduced in the past because of inconsistent data sets (Lacasse and Nadim, 1996). The inconsistency can originate from different soils, different stress conditions, different test methods, stress history, different codes of practice, testing errors or imprecision that are not reported, different interpretations of the data, sampling disturbance, etc.

It can be useful to establish data banks for different types of parameters or geographical locations, or to review the literature and compare one's values to values used by others. These estimates can be biased by the beliefs of the designer. The probabilistic analysis will, however, single out the importance of the hypotheses on the results.

A review was made on test results in the files of Norwegian Geotechnical Institute and data available from the literature. Suspicious data were eliminated. The variability, in terms of coefficient of variation (CoV) and the probability distribution functions are listed in Table 1 (Lacasse and Nadim, 1996).

Table 1. Typical coefficient of variation and distribution of soil properties.

Soil property	Soil type	Prob. distr. function	CoV
Cone resistance	Sand Clay	LN N/LN	Varies greatly from site to site
Undrained shear strength, s_u^*	Clay (triax) Clay (index s_u) Clayey silt	LN LN N	5 - 20% 10 - 35% 10 - 30%
Ratio s_u/σ'_{vo}	Clay	N/LN	5 - 15%
Plastic limit	Clay	N	3 - 20%
Liquid limit	Clay	N	3 - 20%
Submerged unit weight	All soils	N	0 - 10%
Friction angle	Sand	N	2 - 5%
Void ratio, porosity, initial void ratio	All soils	N	7 - 30%
Overconsolidation ratio	Clay	N/LN	10 - 35%

N/LN: Normal and lognormal distributions
* s_u estimated from index tests

3 Model uncertainty

One of the main reasons to place focus on model uncertainty is that it is generally large, and in some cases it can be reduced. Model uncertainty is due to errors introduced by mathematical approximations and simplifications. In probabilistic analysis, model uncertainty is often represented by a parameter (error function) with a normal or lognormal distribution. Model uncertainty is difficult to assess and should be evaluated on the basis of:

- Comparisons of relevant model tests with deterministic calculations
- Expert opinions

- Relevant case studies of "prototypes"
- Information from the literature

To make a reliable estimate of model uncertainty, all relevant mechanisms should be identified and included in the probabilistic models. An important aspect of model uncertainty is the form it takes in the equilibrium function. Model uncertainty is best included in one of three ways:

- Factor on each random variable in the analysis
- Factor on specific components
- Global factor on the limit state function

A general discussion on model uncertainty in geotechnical engineering is provided in Phoon & Kulhawy (2003).

Uncertainty that can arise from the choice of a calculation model is illustrated in Figure 2. Current axial capacity calculation methods used for offshore piles have been derived predominantly from load tests on small piles. Penetration depth, pile length, pile diameter and ultimate load for the largest piles in the reference database are much smaller for the test piles than for those currently used. The uncertainty due to the calculation model is therefore large because the reference database of pile load tests applies to different pile and load conditions than used in design. The linear extrapolation implied when using the calculation models has by no means been verified. The uncertainty due to this extrapolation needs to be included in the estimation of the possibility of a failure.

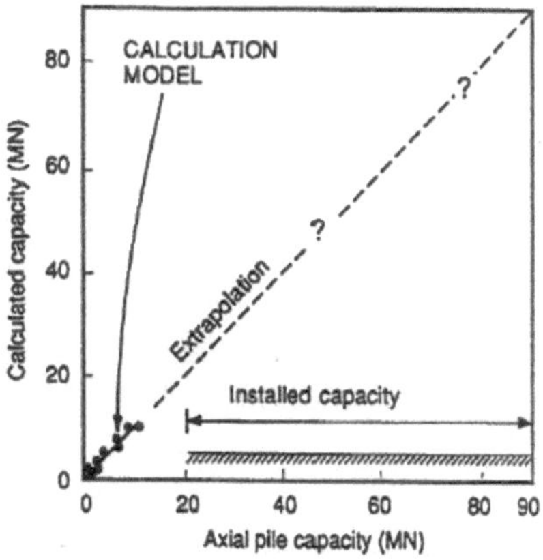

Figure 2. Extrapolation implied by calculation model for pile capacity

Model uncertainty is defined with a mean and a coefficient of variation, and usually a normal or log-normal distribution. Model uncertainty is best included in one of three ways: (1) factor on each random variable in the analysis, (2) factor on friction (each layer) and end bearing components and (3) global factor on the equation describing failure.

Model uncertainty can be evaluated from comparisons between model tests and deterministic calculations, pooling of expert opinions, case studies of prototypes or other model tests, results from the literature, and naturally engineering judgment. To estimate model uncertainty, the relevant mechanisms should first be identified. For example, pooling of 30 international experts on

Table 2. Comparisons of calculated and measured bearing capacities*

Structure	Type of loading	Bias: calculated/measured failure loads
Shallow foundation	Static failure, test 1	0.98-1.01
	Cyclic failure, test 2	0.99-1.15
	Cyclic failure, test 3	1.16-1.17
	Cyclic failure, test 4	1.06-1.23
Tension leg platform	Static failure, test 1	1.00
	Cyclic failure, test 2	1.06
	Cyclic failure, test 3	1.06
	Cyclic failure, test 4	1.02

* (From Dyvik et al, 1989; Andersen et al, 1989; Andersen et al, 1992; Dyvik et al, 1993; Andersen et al, 1993; summarized in Lacasse and Nadim, 1996)

Table 3. Factors affecting model uncertainty

Property/calculations	**Factors of influence**
Undrained shear strength (clay)	• sampling disturbance • test method and scale of laboratory/in situ tests • spatial variability, anisotropy • rate of loading
Friction angle (sand)	• reconstitution of test specimen • density, test method and scale of laboratory test
Pile capacity	• skin friction assumption • limiting values for skin friction and end bearing • subdivision in soil layers • pile installation, residual stresses and plug condition • reconsolidation, rate of loading, cyclic loading, scour • stiffness of pile, pile length, single pile vs pile group • extrapolation from reference database to prototype
Shallow foundations	• position of critical slip surface • modelling of static and cyclic load history • strain-softening and/or progressive failure • testing procedures in reference tests • scale effect, rate of shear and stress conditions • redistribution of stresses and anisotropy • plane strain versus 3-D model, stiffness of structure • model of soil profile and drainage assumptions

pile design gave the consensus that the currently most used pile design method [API RP2A method (API, 1993)] is conservative in medium dense to very dense sand (Lacasse and Goulois, 1989). In dense sands, uncertainties (coefficients of variation of 25% or more) were associated with most of the empirical design factors entering the calculation formulas.

On the other hand, when calculation models have been checked with repeated model tests under different modes of failure and give bias as those shown in Table 2, the model uncertainty is quite small. The key factors affecting the model uncertainty are listed in Table 3.

4 Probabilistic analysis tools

Probability theory and reliability analyses provide a rational framework for dealing with uncertainties and decision making under uncertainty. Depending on the level of sophistication, the analyses provide one or more of the following outputs:

- Probability of failure (or probability of unsatisfactory performance)
- Reliability index
- The most probable combination of parameters leading to failure
- Sensitivity of result to any change in parameters

In his review of slope stability assessment methods, Duncan (2000) pointed out that:

> *"Through regulation or tradition, the same value of safety factor is often applied to conditions that involve widely varying degrees of uncertainty. This is not logical."*

A low safety factor does not necessarily correspond to a high probability of failure and vice versa. The relationship between the factor of safety and probability of failure depends on the uncertainties in load and resistance. Christian et al. (1994) illustrated comparative probabilities for three heights of dikes for the James Bay project: 6, 12, and 23 m. The first two were single stage dikes; the last, a composite dike built in stages. Table 4 gives the estimated factors of safety and probabilities of failure for the three designs. Although the factors of safety are similar (F = 1.5), the probabilities of failure are quite different. The 23-m high dike has significantly lower probability of failure, which is not reflected in the factor of safety.

Table 4. Probabilities of failure for James Bay dikes (Christian et al., 1994)

Case	Factor of safety, F	Probability of failure, P_f
H = 6 m, single stage	1.58	$2.5 \cdot 10^{-2}$
H = 12 m, single stage	1.53	$4.7 \cdot 10^{-3}$
H = 23 m, stage-construction	1.50	$7.1 \cdot 10^{-4}$

Christian (2004) discussed the pros and cons of deterministic and probabilistic safety formats for the evaluation of the stability of existing slopes. The deterministic approach requires that a number of issues be addressed, including what is meant by the conventional factor of safety and how close is the slope to failure. Idealized examples and actual case studies show that slopes with high calculated factors of safety are not necessarily the safest.

Well-established reliability methods, such as FOSM, FORM/SORM and Monte Carlo simulation, which are discussed below, are useful techniques for determining the reliability of geotechnical designs for estimating the probability of failure. The reliability methods also reveal which parameters contribute most to the uncertainty and probability of failure. A simple prescription of a factor of safety to be achieved in all instances is not realistic and may lead to over-design or unsafe situations.

4.1 First-Order, Second-moment Approach

The First-Order, Second-Moment (FOSM) approach (Ang & Tang 1984) provides analytical approximations for the mean and standard deviation of a parameter of interest as a function of the mean and standard deviations of the various input factors, and their correlations.

Consider Y to be a function of random variables $x_1, x_2, \ldots, x_n$; that is

$$Y = f(x_1, x_2, \ldots, x_n) \tag{1}$$

In the general case, $x_1, x_2, \ldots, x_n$ are correlated with covariance matrix [C], i.e. [C] = [σ][R][σ], where [σ] is a diagonal matrix of standard deviations and [R] is the (positive-definite and symmetric) correlation matrix with diagonal elements $R_{ii} = 1$ and non-diagonal elements $R_{ij} = \rho_{ij}$ (ρ_{ij} is the correlation coefficient between variables i and j). In scalar notation, $C_{ij} = \sigma_i\sigma_j R_{ij}$.

Obviously to evaluate the mean and standard deviation of Y, the joint probability density function of $x_1, x_2, \ldots, x_n$ is needed. However, in many practical applications the available information about the random variables is limited to their mean and variance. The approximate mean and variance of the function Y may still be estimated by a Taylor series expansion of the function about the mean values of the random variables and neglecting the higher order terms (Ang and Tang, 1984). If the Taylor series is truncated at its linear terms, the following first-order estimates of mean and variance are obtained:

$$\mu_Y \approx f(\mu_{x_1}, \mu_{x_2}, \ldots\ldots, \mu_{x_n}) \tag{2}$$

$$\sigma_Y^2 \approx \{b\}^T [C]\{b\} \tag{3}$$

where the vector $\{b\}$ denotes $\partial Y / \partial x_i$ evaluated at the mean values of x_i, i.e.:

$$\{b\}^T = \{\partial Y / \partial x_1, \partial Y / \partial x_2, \ldots\ldots\ldots, \partial Y / \partial x_n\}\big|_{\mu_x}$$

If there is no correlation among the variables, Equation 3 can be written as:

$$\sigma_Y^2 \approx \sum_{i=1}^{n} \left(\left. \frac{\partial Y}{\partial x_i} \right|_{\mu_{x_i}} \right)^2 \cdot \sigma_{x_i}^2 \tag{4}$$

Equations 2 and 3 or 4 are referred to as the first-order, second-moment (FOSM) approximations of mean and variance of Y.

The FOSM approximation only provides estimates of the mean and standard deviation, which are not sufficient by themselves for evaluating the failure probability. To estimate the failure probability, one must assume the distribution function for the safety margin or the safety factor beforehand. The first step in estimation of failure probability using any probabilistic method is to decide on what constitutes unsatisfactory performance or failure. Mathematically, this is achieved by defining a performance function G(X), such that G(X) ≥ 0 means satisfactory performance and G(X) < 0 means unsatisfactory performance or "failure". X is a vector of basic random variables including resistance parameters, load effects, geometry parameters and model uncertainty.

Example 1

Consider 3 springs in series with the following stiffness parameters:

Parameter	Mean value	Standard deviation
K_1 (kN/m)	20	2.5
K_2 (kN/m)	15	3
K_3 (kN/m)	30	4

Estimate the mean value of the equivalent stiffness K_{eq}, and its standard deviation.

$$\frac{1}{K_{eq}} = \frac{1}{K_1} + \frac{1}{K_2} + \frac{1}{K_3} \quad \rightarrow \quad K_{eq} = \frac{K_1 K_2 K_3}{K_1 K_2 + K_1 K_3 + K_2 K_3}$$

$\mu_{Keq} \approx (15\times20\times30) / (15\times20 + 20\times30 + 15\times30) = 6.67$ kN/m

x_i	$\partial K_{eq} / \partial x_i$	$\partial K_{eq} / \partial x_i \big\vert_{\mu_{x_i}}$	$(\partial K_{eq} / \partial x_i)^2 \cdot \sigma_{x_i}^2$
K_1	$\left(\frac{K_2 K_3}{K_1 K_2 + K_1 K_3 + K_2 K_3}\right)^2$	0.1111	0.0772
K_2	$\left(\frac{K_1 K_3}{K_1 K_2 + K_1 K_3 + K_2 K_3}\right)^2$	0.1975	0.3512
K_3	$\left(\frac{K_1 K_2}{K_1 K_2 + K_1 K_3 + K_2 K_3}\right)^2$	0.0494	0.0390
$\sigma_{K_{eq}}^2 = \sum (\partial K_{eq} / \partial x_i)^2 \cdot \sigma_{x_i}^2$			0.4674

$$\sigma_{K_{eq}} = \sqrt{0.4674} \approx 0.68$$

Example 2

Consider a structural element with resistance R, subjected to dead load D and live load L. The safety margin (performance function) for this element is defined as:

$$G = R - D - L$$

Given the information below, estimate the mean and coefficient of variation (CoV = σ/μ) of G with and without correlation among the parameters.

Mean values: $\mu_R = 2.8 \quad \mu_D = 1 \quad \mu_L = 0.75$

Standard deviations: $\sigma_R = 0.3 \quad \sigma_D = 0.1 \quad \sigma_L = 0.2$

Correlation coefficients: $\rho_{R,D} = 0.8$ $\rho_{D,L} = 0.3$

FOSM approximation:

$\mu_G = 2.8 - 1 - 0.75 = 1.05$

$\{b\}^T = \{\partial G/\partial R, \partial G/\partial D, \partial G/\partial L\} = \{1\ \ -1\ \ -1\}$

No correlation:

$$[C] = \begin{bmatrix} 0.3 & & \\ & 0.1 & \\ & & 0.2 \end{bmatrix} \begin{bmatrix} 1 & & \\ & 1 & \\ & & 1 \end{bmatrix} \begin{bmatrix} 0.3 & & \\ & 0.1 & \\ & & 0.2 \end{bmatrix} = \begin{bmatrix} 0.09 & & \\ & 0.01 & \\ & & 0.04 \end{bmatrix}$$

$$\sigma_G^2 = \{1\ \ -1\ \ -1\} \begin{bmatrix} 0.09 & & \\ & 0.01 & \\ & & 0.04 \end{bmatrix} \begin{Bmatrix} 1 \\ -1 \\ -1 \end{Bmatrix} = 0.14$$

$\Rightarrow \sigma_G = 0.374$, CoV = 0.374/1.05 = 35.6 %

With correlation:

$$[C] = \begin{bmatrix} 0.3 & & \\ & 0.1 & \\ & & 0.2 \end{bmatrix} \begin{bmatrix} 1 & 0.8 & \\ 0.8 & 1 & 0.3 \\ & 0.3 & 1 \end{bmatrix} \begin{bmatrix} 0.3 & & \\ & 0.1 & \\ & & 0.2 \end{bmatrix} = \begin{bmatrix} 0.09 & 0.024 & 0 \\ 0.024 & 0.01 & 0.006 \\ 0 & 0.006 & 0.04 \end{bmatrix}$$

$$\sigma_G^2 = \{1\ \ -1\ \ -1\} \begin{bmatrix} 0.09 & 0.024 & 0 \\ 0.024 & 0.01 & 0.006 \\ 0 & 0.006 & 0.04 \end{bmatrix} \begin{Bmatrix} 1 \\ -1 \\ -1 \end{Bmatrix} = 0.104$$

$\Rightarrow \sigma_G = 0.323$, CoV = 0.323/1.05 = 30.7 %

The "reliability index", defined as

$$\beta = \frac{\mu_G}{\sigma_G} \tag{5}$$

in which μ_G and σ_G are respectively the mean and standard deviation of the performance function, is often used as an alternative performance measure to the factor of safety (Li & Lumb 1987, Christian *et al.* 1994, Duncan 2000).

The reliability index provides more information about the reliability of a geotechnical design or a geotechnical structure than is obtained from the factor of safety alone. It is directly related to the probability of failure and the computational procedures used to evaluate the reliability index reveal which parameters contribute most to the uncertainty in the factor of safety. This is useful information that can guide the engineer in further investigations. However, the reliability index estimated using the FOSM approach is not "invariant". Table 5 shows the reliability indices for different formats of the performance function using the FOSM method. *R* and *S* in the table represent respectively the total resisting force and the driving force acting on the slope. CoV_R and CoV_S in the table denote the coefficients of variation of the resisting and the loading forces respectively and $F = \mu_R / \mu_S$.

Table 5. Performance function format and FOSM reliability index β (Li & Lumb 1987).

G(X)	β
$R - S$	$\dfrac{F-1}{\sqrt{F^2 CoV_R^2 + CoV_S^2}}$
$\dfrac{R}{S} - 1$	$\dfrac{F-1}{F\sqrt{CoV_R^2 + CoV_S^2}}$
$\ln\dfrac{R}{S}$	$\dfrac{\ln F}{\sqrt{CoV_R^2 + CoV_S^2}}$

4.2 First- and second-order reliability methods (FORM and SORM)

Hasofer & Lind (1974) proposed an invariant definition for the reliability index. The approach is referred to as the first-order reliability method (FORM). As mentioned earlier, the starting point for FORM is the definition of the performance function G(X), where X is the vector of basic random variables. If the joint probability density function of all random variables $F_x(X)$ is known, then the probability of failure P_f is given by

$$P_f = \int_L F_x(X)\, dX \tag{6}$$

where L is the domain of X where G(X) < 0.

In general, the above integral cannot be solved analytically. In the FORM approximation, the vector of random variables **X** is transformed to the standard normal space U, where U is a vector of independent Gaussian variables with zero mean and unit standard deviation, and where G(U) is a linear function. The probability of failure P_f is then (P[…] means probability that …):

$$P_f = P\,[G(U) < 0] \approx P\,[\sum_{i=1}^{n} \alpha_i U_i - \beta < 0] = \Phi\,(-\beta) \tag{7}$$

where α_i is the direction cosine of random variable U_i, β is the distance between the origin and the hyperplane G(U) = 0, n is the number of basic random variables X, and Φ is the standard normal distribution function.

The vector of the direction cosines of the random variables (α_i) is called the vector of sensitivity factors, and the distance β is the reliability index. The probability of failure (P_f) can be estimated from the reliability index β using the established equation $P_f = 1 - \Phi(\beta) = \Phi(-\beta)$, where Φ is the cumulative distribution (CDF) of the standard normal variate. The relationship is exact when the limit state surface is planar and the parameters follow normal distributions, and approximate otherwise. The relationship between the reliability index and probability of failure defined by Equation (7) is shown in Figure 3.

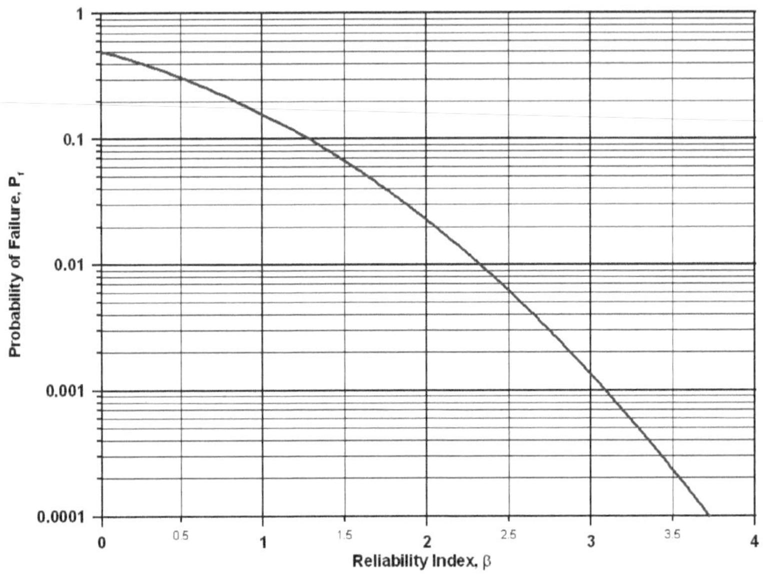

Figure 3. Relationship between reliability index β, and probability of failure P_f.

The square of the direction cosines or sensitivity factors (α_i^2), whose sum is equal to unity, quantifies in a relative manner the contribution of the uncertainty in each random variable X_i to the total uncertainty.

Example 3

Consider an infinite frictional soil slope with thickness H in the vertical direction, soil friction angle φ', slope angle θ, unit weight γ, and pore pressure u at depth H. With the following parameters and probability distribution functions, evaluate the probability of slope failure and its reliability index.

Parameter	μ	σ	**Probability distribution**
H (m)	10.0	1.0	Lognormal
φ' (degrees)	35.0	2.0	Normal
θ (degrees)	20.0	1.5	Lognormal
γ (kN/m^3)	18.0	0.5	Normal
u (kPa)	20.0	7.0	Beta, between 0 and 40

The equation for the safety factor of the slope is:

$$F = \frac{\tan(\phi')}{\tan(\theta)} \cdot \left(1 - \frac{u}{\gamma \cdot H \cdot \cos^2(\theta)}\right)$$

A simple limit state function for the performance of the slope is:

$$G = F - 1$$

i.e. the slope is stable when the factor of safety is greater than 1, and it fails when the factor of safety is less than 1.

Using the software STRUREL (RCP, 1999), the following results are obtained:

Probability of failure: $P_f = 3.9 \cdot 10^{-5}$
FORM reliability index: $\beta = 3.95$

The sensitivity factors for the variables are shown on the figure below. The pie chart shows the squares of the sensitivity factors α_i^2.

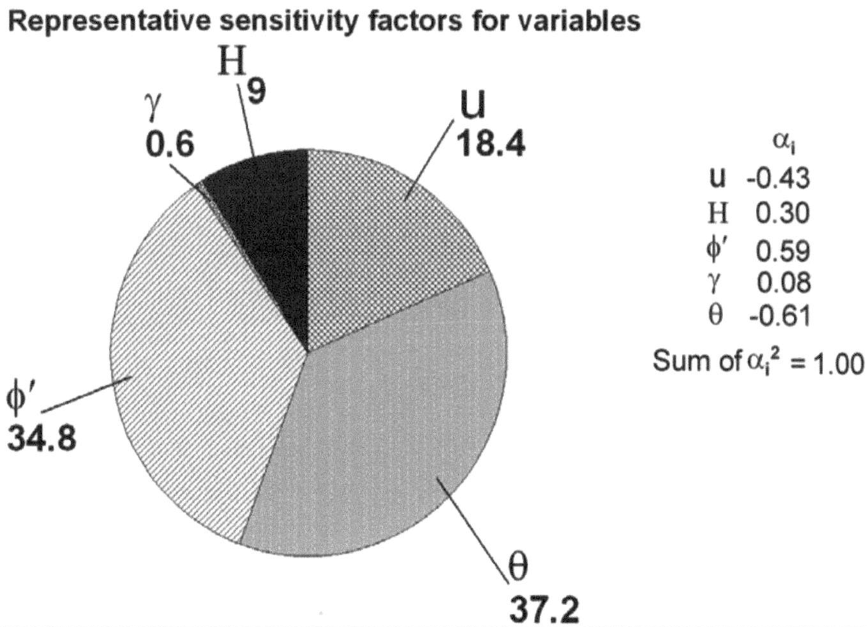

In summary the FORM approximation involves:

1. transforming a general random vector into a standard Gaussian vector,
2. locating the point of maximum probability density (most likely failure point, design point, or simply β-point) within the failure domain, and
3. estimating the probability of failure as $P_f \approx \Phi(-\beta)$, in which $\Phi(\cdot)$ is the standard Gaussian cumulative distribution function.

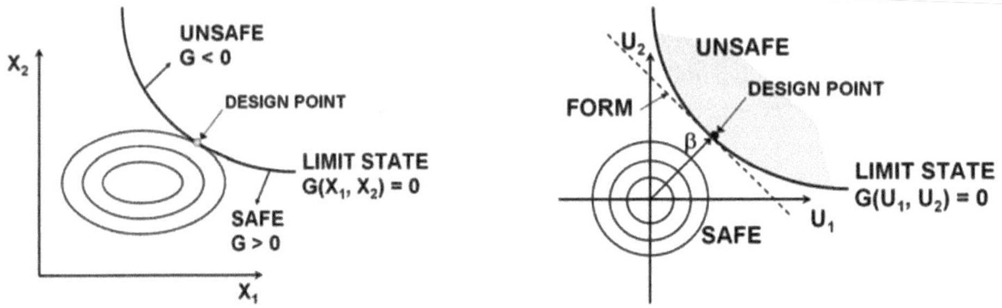

Figure 4. The FORM approximation (right) and definition of β and design point.

An illustration of the design point and graphical representation of β is given in Figure 4.

Low (2003) presented a method for finding the reliability index in the original space. His approach is based on the matrix formulation of the Hasofer-Lind reliability index β (Veneziano, 1974; Ditlevsen, 1981):

$$\beta = \min\sqrt{(X-\mu)^T C^{-1}(X-\mu)} \quad \text{for } \{X : G(X) = 0\} \tag{8}$$

or, equivalently:

$$\beta = \min\sqrt{\left[\frac{x_i-\mu_i}{\sigma_i}\right]^T [R]^{-1}\left[\frac{x_i-\mu_i}{\sigma_i}\right]} \quad \text{for } \{X : G(X) = 0\} \tag{9}$$

in which $X = (x_1, x_2, ..., x_n)$, μ = mean vector of X, C = covariance matrix of X, and R = correlation matrix of X.

Low and Tang (1997) used Equation 9 in preference to Equation 8 because the correlation matrix R is easier to set up, and conveys the correlation structure more explicitly than the covariance matrix C. Geometrically, for a two-variable problem, Equations 8 and 9 can be interpreted as finding the smallest ellipsoid (of the probability distribution of the variables) tangent to the limit state surface, see Figure 5. The key advantage of this formulation is that it can be implemented using built-in functions in EXCEL without programming and EXCEL is widely available on PCs (Phoon and Nadim, 2004).

In the second-order reliability method (SORM), the limit state function is defined as in FORM, but the resulting limit state function is approximated by a second order function (Breitung 1984). However, for geo-problems the probabilities of failure obtained with SORM analyses have been very close to the values obtained with FORM (Lacasse and Nadim, 1999).

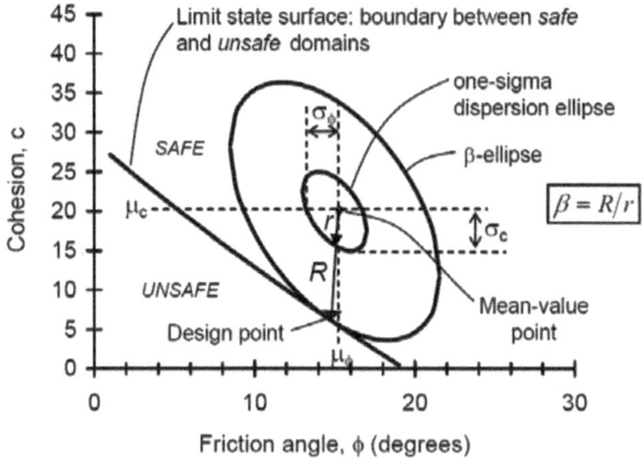

Figure 5. Illustration of β in the plane of original variables (Low, 2003).

4.3 Monte-Carlo simulation

A Monte-Carlo simulation is a procedure, which seeks to simulate stochastic processes by random selection of input values to an analysis model in proportion to their joint probability density function. It is a powerful technique that is applicable to both linear and non-linear problems, but can require a large number of simulations to provide a reliable distribution of the response.

The Monte Carlo simulation technique is implemented in some commercial slope stability analysis packages (e.g. Geo-Slope, 2003). However, when the probability of failure is very small, the number of simulations required to obtain an accurate result directly is so large that, except for very simple (or simplified) problems, it renders the application impractical. In these situations the conditional probability of failure can be determined for various low probability scenarios, and then combined, considering the scenario probabilities. Monte Carlo simulation can be optimized by stratified sampling techniques, for example Latin Hypercube sampling (Iman & Conover 1982). These "organized" sampling techniques considerably reduce the number of simulations required for a reliable distribution of the response.

Example 4

El-Ramly et al. (2003) performed probabilistic slope analysis using Monte Carlo simulation to evaluate the stability of a section of the Syncrude Tailings Dyke in Fort McMurray, Canada, Figure 4.1. The dyke was approximately 44m high. The performance of the dyke was governed by uncertainties about material properties and pore-water pressures, in particular in the disturbed clay-shale (Kca) layer, and the sandy till at toe of the dyke.

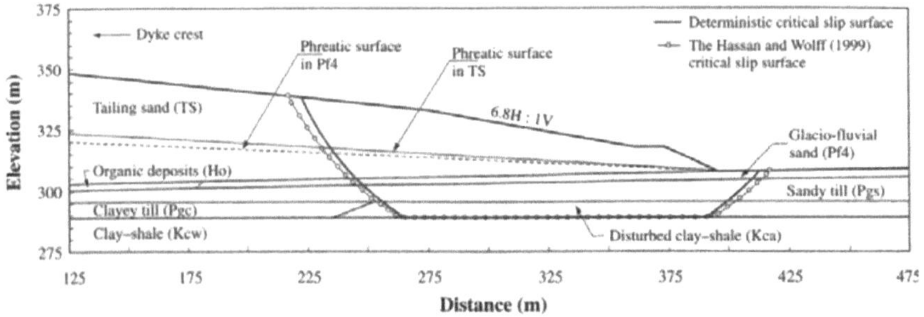

Figure 4.1. Dyke profile and stratigraphy of the Syncrude Tailings Dyke.

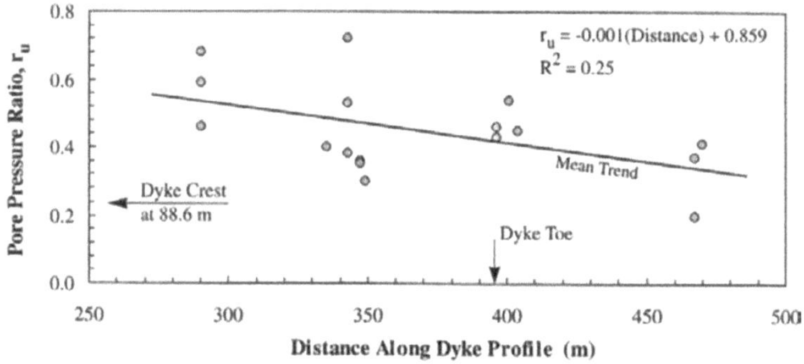

Figure 4.2. Profile of pore pressure ratio in the Kca layer along dyke cross-section, March 1994.

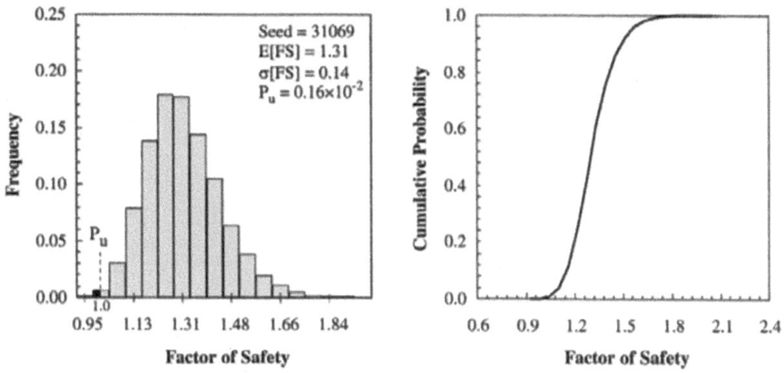

Figure 4.3. Histogram and probability distribution function of the factor of safety - number of simulation iterations equals 34000 (El-Ramly et al. 2003).

The residual shear strength of this material was evaluated from shear box tests on 80 specimens. Based on the histogram of the results obtained from the tests, and spatial averaging to the appropriate scale, the residual friction angle was assumed to have a lognormal probability density function with a mean of 7.5° and a standard deviation of 2.1° in the Monte Carlo simulations.

Substantial amounts of pore pressure data were available from numerous pneumatic and standpipe piezometers along the dyke. Figure 4.2 shows a plot of the pore pressure ratio r_u, defined as the ratio of excess pore pressure to effective vertical stress for hydrostatic conditions, at a section along the dyke profile in March of 1994. The measurements are scattered. It seems that the pore pressure ratio tends to decrease towards the dyke toe. A linear trend fitted to the data using the method of least squares is shown on the plot. The standard deviation of the pore pressure ratio around the mean trend is calculated to be 0.12.

The measured peak friction angles of the sandy till layer belonged to different statistical populations and grouping them together increased the estimate of uncertainty. When all measurements were combined for an approximate statistical assessment, the measured values ranged between 33.3° and 39.2°. The mean and standard deviation of the peak friction angle were calculated to be 35.7° and 2°, respectively.

The pore water pressure in the sandy till layer was assessed from data from 14 piezometers at different depths and locations along Section 53+000E. The pore pressure ratio varied between 0.1 and 0.46 with a mean of 0.30. Due to a large increase in pore pressure ratio towards the dyke toe, the data were divided into two subgroups and the pore pressure ratio modeled by two random variables representing the middle portion of the slope and the toe area respectively.

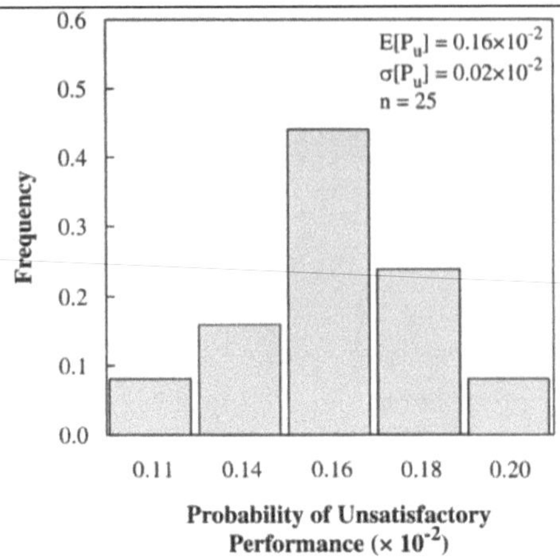

Figure 4.4. Histogram of the probability of unsatisfactory performance (El-Ramly et al. 2003).

A model of dyke geometry, soil stratigraphy and properties, and pore water pressures was developed in an Excel spreadsheet. The Bishop method of slices was used in the model with the limit equilibrium equations rearranged to account for the noncircular portion of the slip surface. Five input parameters are considered variables: the residual friction angle of Kca clay-shale, the peak friction angle of the sandy till, the residual pore pressure ratio in the Kca clay-shale and the pore pressure ratios at the middle and at the toe of the slope in the sandy till. Monte Carlo simulation was performed using @Risk and the prepared spreadsheet model. Figure 4.3 shows the histogram of results and the inferred probability distribution function of the factor of safety. Figure 4.4 shows the probability of unsatisfactory performance, based on the results of 25 repetitions of the 34,000 realizations (e.g. total of 25 × 34,000 realizations). The mean probability of unsatisfactory performance, P_u, is estimated to be $1.6\cdot10^{-3}$, with the 95% confidence interval around the mean ranging between $1.5\cdot10^{-3}$ and $1.7\cdot10^{-3}$.

In fact, the standard deviation of the estimated probability of failure can be computed analytically as $\sigma_{Pu} = \{P_u(1 - P_u)/34000\}^{0.5} = 0.0002$, which is what is seen in Figure 4.4.

4.4 Event tree analysis

For a complex system, the conditions that could lead to any of the potential modes of failure may be quite involved and an event tree analysis is often the optimum way to quantify hazard and risk. Given a number of possible consequences resulting from an initiating event, the sequence of following events need to be identified and their probability of occurrence needs to be quantified. This can be done systematically and effectively through the use of an event tree diagram. The approach is widely used for dams, but is also useful for slopes with complex history, e.g. a rock slope with possibly different volumes sliding over time followed by a tsunami. Ang and Tang (1984) and Whitman (1984) presented several application examples for the method. Figure 6 illustrates event tree analysis.

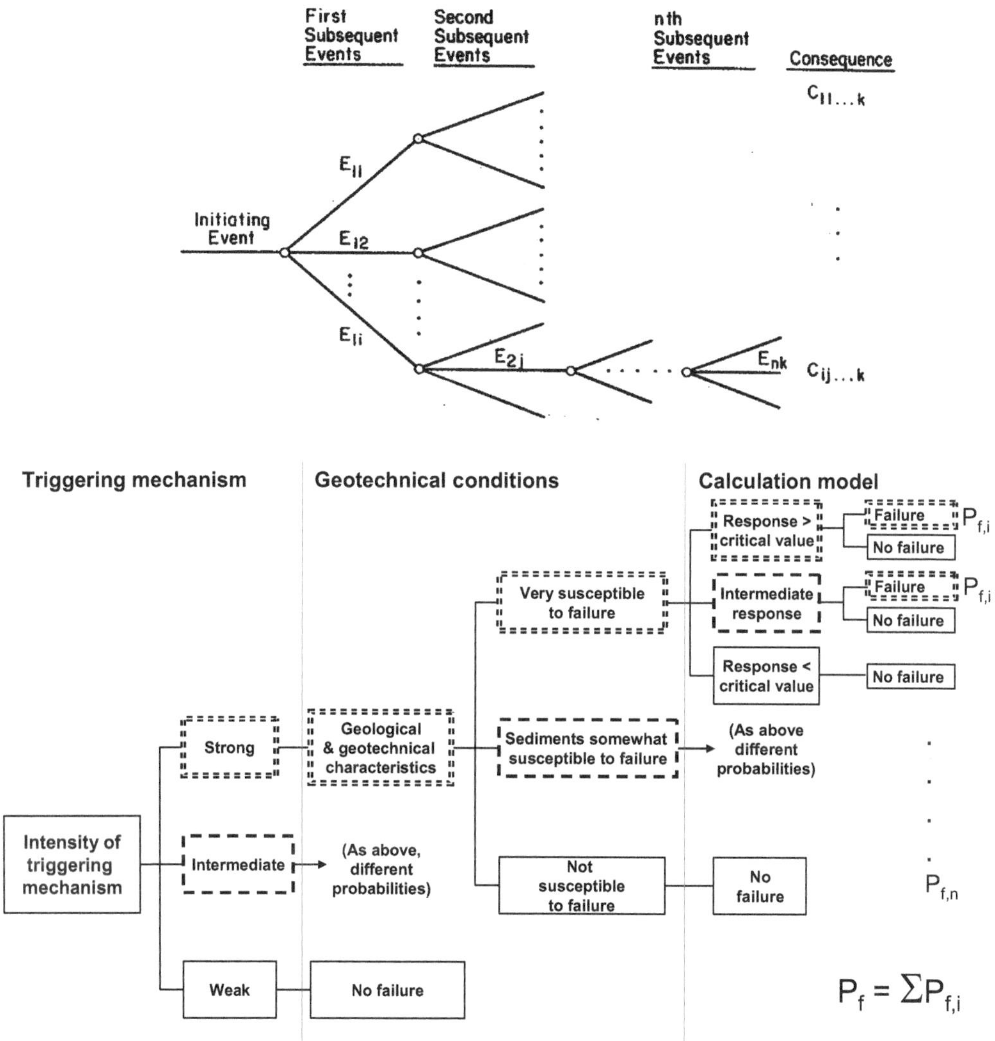

Figure 6. Event tree model and example for the analysis of a slope

A general event tree is shown in Figure 6 with an initiating event, E, and a number of possible consequences, $C_{ij \ldots k}$. It can be observed that a particular consequence depends on the subsequent events following the initiating event; in other word, for a given consequence to occur, a sequence of subsequent events, or *path* in the event tree, must occur. Given an initiating event, there may be several "first subsequent events" that will follow. Obviously, these subsequent events are mutually exclusive. If we assume a particular first subsequent event, a mutually exclusive set of "second subsequent events" is possible. Each path in the event tree, therefore, represents a specific sequence of (subsequent) events, resulting in a particular consequence. The probability associated with the occurrence of a specific path is simply the product of the (conditional) probabilities of all the events on that path.

Each event in the event tree is associated with a probability of occurrence. The probabilities can be obtained by first assigning a verbal descriptor as given below. The sum of the probabilities at any node is always unity, if all possible events have been included. The estimates rely heavily on engineering judgment. Observations are also very useful in assisting judgment. Each outcome in the event tree ends up as failure or no failure. The total probability of failure is the summation of the probabilities of each outcome leading to failure. If data are available, component events should be treated statistically, for example the 100-year and 1000-year rainfall or flood, based on historic data, the earthquake frequency and response spectrum. In practice, the results of an event tree analysis prove even more useful when done for several slopes and compared.

To achieve consistency in the evaluation of the probabilities from one dam to another, conventions have been established to anchor the probabilities. An example of descriptors of uncertainty used in the dam profession goes as follows:

Verbal description of uncertainty	Event probability
Virtually impossible	0.001
Very unlikely	0.01
Unlikely	0.10
Completely uncertain	0.50
Likely	0.90
Very likely	0.99
Virtually certain	0.999

Virtually impossible: event due to known physical conditions or processes that can be described and specified with almost complete confidence.
Very unlikely: the possibility cannot be ruled out on the basis of physical or other reasons.
Unlikely: event is unlikely, but it could happen
Completely uncertain: there is no reason to believe that one outcome is any more or less likely than the other to occur.
Likely: event is likely, but it may not happen
Very likely: event that is not completely certain.
Virtually certain: event due to known physical conditions or processes that can be described and specified with almost complete confidence.

5 System reliability

A system, for example a complex geotechnical structure, consists of many components or elements, whose individual or combined failure can lead to collapse. A simple gravity retaining wall would fail if the lateral forces on the wall exceed the base shear resistance (sliding mode of failure), if the weight of the wall and vertical forces acting on the wall exceed the bearing capacity at the base (bearing capacity mode of failure) or if the driving moment from the external loads exceeds the resisting moment from the weight of the wall (rotational mode of failure). The wall could therefore be thought of as a system that comprises three components whose individual failure would constitute the failure of the wall.

The methods and examples discussed in the previous section generally characterize the performance and reliability of a single component of a complete system. The failure event of a system, in a reliability sense, is defined as the union, or intersection, or combinations thereof, of component failure events. In a graphical representation of a system, the components are represented by boxes that are connected together by lines to form the system. Input and output are marked by arrows (see Figure 7).

It is useful to distinguish between two basic types of systems depending on the logical structure of the components, namely series and parallel systems. In a series system the individual components are connected in series with regard to their function (Figure 7a). A series system will fail if any of its components fail, i.e. the system failure event is the union of all the component failure events. As a simple example, consider a chain consisting of many links. If the weakest link breaks, the chain fails. That is, the least reliable link determines the reliability of the system. If a series system is composed on "n" statistically independent components, then the probability of system failure can be computed from the probability of failure of individual components by the following equation:

$$\mathrm{P}_{\mathrm{f,system}} = 1 - \prod_{i=1}^{n}(1 - P_{f,i}) \approx \sum_{i=1}^{n} P_{f,i} \tag{10}$$

The summation approximation is valid for very small probabilities of failure $\mathrm{P}_{\mathrm{f,i}}$.

Obviously the probability of failure of a series system increases with the number of elements and is largely governed by the probability of failure of its most unreliable element. If all elements of a series system are perfectly correlated, then:

$$\mathrm{P}_{\mathrm{f,system}} = \max[P_{f,i}] \tag{11}$$

Thus the probability of failure of a series system lies within the following bounds:

$$\max[P_{f,i}] \leq \mathrm{P}_{\mathrm{f,system}} \leq 1 - \prod_{i=1}^{n}(1 - P_{f,i}) \tag{12}$$

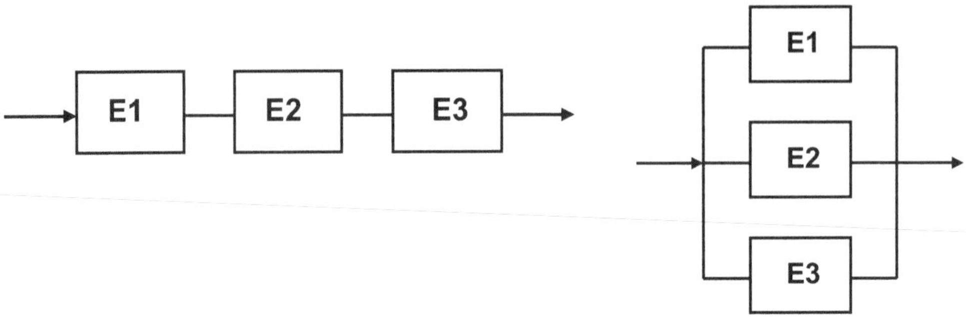

(a) Series system (b) Parallel system

Figure 7. Schematic representation of series and parallel systems.

In a parallel system, the elements of the system are connected in parallel with regard to their function (Figure 7b). This means that a parallel system will fail if all its components fail, i.e. the system failure event the intersection of the component failure events.

If a parallel system is composed on "n" statistically independent components, then the probability of system failure can be computed from the probability of failure of individual components by the following equation:

$$P_{f,system} = P_{f,1} \cdot P_{f,2} \cdot \ldots \cdot P_{f,n} = \prod_{i=1}^{n} P_{f,i} \quad (13)$$

If all elements of a parallel system are perfectly correlated, then:

$$P_{f,system} = \min[P_{f,i}] \quad (14)$$

That is, the probability of failure of a parallel system is never greater than probability of failure of its most unreliable element. The probability of failure of a parallel system, therefore, lies within the following bounds:

$$\prod_{i=1}^{n} P_{f,i} \le P_{f,system} \le \min[P_{f,i}] \quad (15)$$

In constructed facilities, true parallel systems are rare. Consider, for example, a foundation slab supported by six piles. This foundation system, on the first sight, might well be considered a parallel system consisting of six components, as all six piles must fail in order for the foundation to fail. However, the carrying capacities of the piles are strongly correlated. Furthermore, the question of ductile versus brittle failure of the components in the system is of prime importance. While a ductile component may continue to carry load until the other elements of the system yield, a brittle component stops carrying its share of load, leaving the remaining components with even more load.

Example 5

A system, represented by the minimal cut-set below, operates only if there is a path of functional units from A to B.

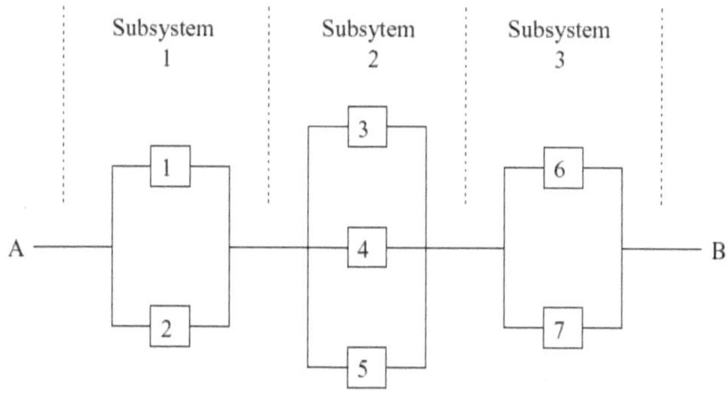

Under a severe event, the following failure probabilities for different components are estimated:

$P_{f,1} = 0.1$
$P_{f,2} = 0.05$
$P_{f,3} = 0.2$
$P_{f,4} = 0.15$
$P_{f,5} = 0.12$
$P_{f,6} = 0.06$
$P_{f,7} = 0.12$

- Evaluate the failure probability and reliability index of each of the 3 subsystems assuming that all units operate and fail independently.

  Subsystem 1: $P_{f,S1} = P_{f,1} \cdot P_{f,2} = 0.1 \cdot 0.05 = 0.005$

  $\beta_{S1} = \Phi^{-1}(1 - P_{f,S1}) = \Phi^{-1}(0.995) = 2.58$

  Subsystem 2: $P_{f,S2} = P_{f,3} \cdot P_{f,4} \cdot P_{f,5} = 0.2 \cdot 0.15 \cdot 0.12 = 0.0036$

  $\beta_{S2} = \Phi^{-1}(1 - P_{f,S2}) = \Phi^{-1}(0.9964) = 2.69$

  Subsystem 3: $P_{f,S3} = P_{f,6} \cdot P_{f,7} = 0.06 \cdot 0.12 = 0.0072$

  $\beta_{S3} = \Phi^{-1}(1 - P_{f,S3}) = \Phi^{-1}(0.9928) = 2.44$

- Evaluate the system reliability index and the system probability of failure.

  $P_{f,\,System} = 1 - (1 - P_{f,S1}) \cdot (1 - P_{f,S2}) \cdot (1 - P_{f,S3}) = 0.0157$

  (Approximate estimate for small probabilities: $P_{f,\,System} \approx P_{f,S1} + P_{f,S2} + P_{f,S3} = 0.0158$)

  $\beta_{System} = \Phi^{-1}(1 - P_{f,System}) = \Phi^{-1}(0.9843) = 2.15$

Most real life systems are mixed systems that could be represented as a series connection of subsystems, where each subsystem comprises parallel components. Some commercial software for computation of system reliability (e.g. STRUREL) require that the system is represented is terms of minimal unions of intersections, also denoted as minimal cut-set.

References

Andersen, K.H., Dyvik, R., Lauritzsen, R, Heien, D., Hårvik, L. and, Amundsen, T. (1989). Model Tests of Gravity Platforms. II: Interpretation. *ASCE Journal of Geotechnical Engineering*, 115(11): 1550-1568.

Andersen, K.H., Dyvik, R., and Schrøder, K. (1992). Pull-Out Capacity Analyses of Suction Anchors for Tension Leg Platforms. In *Proceedings of BOSS'92 Conference,* London, U.K., 2, 1311-1322.

Andersen, K.H., Dyvik, R., Schrøder, K., Hansteen, O.E., and Bysveen, S. (1993). Field Tests of Anchors in clay II: Predictions and Interpretation. *ASCE Journal of Geotechnical Engineering,* 119(10): 1532-1549.

Ang, A.H.S., and Tang, W. (1984). *Probability Concepts in Enginering Planning and Design. Volume I Basic Principles.* John Wiley and Sons, New York, USA.

API (American Petroleum Institute) (1993). *Recommended Practice for Planning, Designing and Constructiong Fixed Offshore Platforms - Load and resistance Factor Design.* API RP2A-LRFD 1st ed. Dallas, Texas, USA.

Breitung, K. (1984). Asymptotic approximations for multinormal integrals. *ASCE Journal of Engineering Mechanics*, 110(3): 357-366.

Christian, J.T. (2004). Geotechnical Engineering reliability: How well do we know what we are doing? *ASCE Journal of Geotechnical and Geoenvironmental Engineering*, 130(10): 985-1003.

Christian, J.T., Ladd, C.C., and Baecher, G.B. (1994). Reliability Applied to Slope Stability Analysis. *ASCE Journal of Geotechnical Engineering* , 120(12): 2180-2207.

Ditlevsen, O. (1981). *Uncertainty modeling with applications to multidimensional civil engineering systems*. McGraw-Hill, New York.

Duncan, J.M. (2000). Factors of safety and reliability in geotechnical engineering. *ASCE Journal of Geotechnical and Geoenvironmental Engineering*, 126(4): 307-316.

Dyvik, R., Andersen, K.H., Madshus, C., and Amundsen, T.(1989). Model Tests of Gravity Platforms I: Description. *ASCE Journal of Geotechnical Engineering*, 115(11): 1532-1549.

Dyvik, R., Andersen, K.H., Hansen, S.B., and Christophersen, H.P. (1993). Field Tests of Anchors in Clay I: Description. *ASCE Journal of Geotechnical Engineering*. 119(10): 1515-1531.

Einstein, H.H., and Baecher, G.B. (1982). Probabilistic and statistical methods in engineering geology. I. Problem statement and introduction to solution. *Rock Mechanics*, Supp. 12: 47-61.

El-Ramly, H., Morgenstern, N.R., and Cruden, D.M. (2003). Probabilistic stability analysis of a tailings dyke on presheared clay-shale. *Canadian Geotechnical Journal*, 40: 192-208.

Geo-Slope International (2003). *SLOPE/W for slope stability analysis, User's guide*, Version 4.23, Calgary, Alberta, Canada.

Hasofer, A.M., and Lind, N.C. (1974). An exact and invariant first order reliability format. *ASCE Journal of Engineering Mechanics Division*, 100(EM1): 111–121.

Iman, R.L., and Conover, W.J. (1982). A distribution-free approach to inducing rank correlation among input variables. *Communications in Statistics*, 11: 311-334.

Lacasse, S., and Goulois, A. (1989). Uncertainty in API Parameters for Predictions of Axial Capacity of Driven Pile in Sand. In *Proceedings of 21st OTC*, Houston, Texas, USA. 353-358.

Lacasse, S. and Nadim, F. (1996). Uncertainties in Characterizing Soil Properties – Plenary paper. In *Proceedings of ASCE Special Technical Publication No. 58: Uncertainty in the Geologic Environment – From Theory to Practice*. Madison, Wisconsin, USA, 1: 49–75.

Lacasse, S., and Nadim, F. (1999). Risk analysis in geo-engineering. *Proceedings of Rocksite 1999 – International Conference on Rock Engineering Techniques for Site Characterization*, Bangalore, India. Also NGI Report 594000-7.

Low, B. K. (2003). Practical probabilistic slope stability analysis. *Proceedings of Soil and Rock America Conference*, MIT, Cambridge, Massachusetts, USA, Verlag Glückauf GmbH Essen, 2: 2777-2784.

Low, B. K., and Tang, W. H. (1997). Reliability analysis of reinforced embankments on soft ground. *Canadian Geotechnical Journal*, 34(5): 672-685.

Li, K.S., and Lumb, P. (1987). Probabilistic design of slopes. *Canadian Geotechnical Journal*. 24: 520-535.

Phoon, K.K., and Kulhawy, F.H. 2003. Evaluation of Model Uncertainties for Reliability-based Foundation Design, In *Proceedings of 9th International Conference on Applications of Statistics and Probability in Civil Engineering*, San Francisco, USA, July 6-9, 2: 1351-135.

Phoon, K.K., and Nadim, F. (2004). Modeling non-Gaussian random vectors for FORM: State-of-the-Art Review. *Workshop on Risk assessment and Geohazards*, Indian Institute of Science, Bangalore, India, 26 November.

RCP (Reliability Consulting Programs) (1999). *STRUREL - A Structural Reliability Analysis Program System*. RCP GmbH, Munich, Germany.

Veneziano, D. (1974). *Contributions to second moment reliability*. Research Report No. R74-33. Department of Civil Engineering, MIT, Cambridge, Massachusetts, USA.

Whitman, R.V. (1984). Evaluating calculated risk in geotechnical engineering. *ASCE Journal of Geotechnical Engineering*, 110(2): 145-188.

Basic Concepts and Applications of Point Estimate Methods in Geotechnical Engineering

Helmut F. Schweiger[1] and Robert Thurner[2]

[1] Computational Geotechnics Group, Institute for Soil Mechanics and Foundation Engineering, Graz University of Technology, Graz, Austria
[2] Keller Grundbau GesmbH, Söding, Austria

Abstract. A procedure for combining a point estimate method (PEM) and deterministic finite element methods for geotechnical analysis is presented. A brief description of the theoretical background is given and an illustrative example analysing a sheet pile wall supporting an excavation is presented to demonstrate the general applicability of the proposed approach. Finally the application to numerical analyses of tunnel excavation is discussed in some detail. A NATM tunnel construction is considered and the results obtained from the suggested approach are compared to the ones obtained from conventional finite element analyses. The results clearly reveal the applicability of the proposed concept for solving practical problems and it is argued that the validity of finite element analyses is extended by introducing stochastic properties of input parameters.

1 Introduction

When using finite element codes in reliability analysis there are some advantages as compared to limit equilibrium methods or other similar methods because with one calculation more than one system-parameter can be obtained. These parameters are the basis for the performance function which will be evaluated with the numerical procedure described in this paper. On the other hand, some difficulties have to be overcome by using finite elements for reliability calculations. The accuracy of the results obtained by finite element codes may vary significantly depending on the solution procedure and the convergence criteria adopted and therefore very tight specifications have to be enforced.

Due to the arguments given above attempts to combine probabilistic concepts and numerical modelling published in the literature are somewhat limited. The approach suggested in this paper involves the application of point estimate methods (PEM) in combination with deterministic finite elements.

First a brief description of the theoretical background of the probabilistic concepts employed will be given before the suggested procedure will be demonstrated by solving a simplified geotechnical problem. The developed methodology will be applied for the analysis of a tunnel excavation according to the principles of the New Austrian Tunnelling Method (NATM) describing the determination of variables, their influence on a specific result and the evaluation of the chosen limit state function. A general discussion on the merits of probabilistic analysis is not given here at it is presented in other chapters of this book.

2 Point Estimate Methods

It is well established that input parameters for geotechnical calculations are associated with uncertainties. This holds for material properties as well as for model parameters which have to be introduced when building a geomechanical model, which itself represents only an approximation to the actual situation in situ.

In order to arrive at a probability of "failure", whereas the term "failure" has a very general meaning here as it may indicate collapse of a structure or in a very general form define the loss of serviceability, a limit state function or performance function *G(X)* of the following form can be defined

$$G(\boldsymbol{X}) = R(\boldsymbol{X}) - S(\boldsymbol{X}) \tag{1}$$

R(X) is the "resistance", *S(X)* is the "action", and ***X*** is the collection of random input parameters. For $G(\boldsymbol{X})<0$ failure is implied, while $G(\boldsymbol{X})>0$ means stable behaviour. The boundary defined by $G(\boldsymbol{X})=0$ separating the stable and unstable state is called the limit state boundary. The probability of failure p_f is defined as:

$$p_f = P[G(\boldsymbol{X}) \le 0] = \int_{G(\boldsymbol{X})\le 0} f(\boldsymbol{X})dx \tag{2}$$

where *f(X)* is the common probability density function of the vector formed by the variables ***X***.

A number of different approaches have been suggested in the literature to integrate Eq. 2 (see, e.g., Benjamin and Cornell, 1970, Li, 1992). In this work the approach proposed by Zhou and Nowak (1988) is used. It is a numerical procedure for computing the statistical parameters of a function *G(X)* of multiple random variables. The sample of basic variables is obtained by transforming a priori the selected points from the standard normal space to the basic variable space. The method is similar to other point estimates, except that points and weights are predetermined in the standard normal space. If the basic variables are standard normal distributed the performance function in Eq.1 can be written as:

$$G(\boldsymbol{Z}) = G(Z_1, Z_2, ..., Z_n) \tag{3}$$

The exact k[th] moment of *G*, $E[G^k(\boldsymbol{Z})]$, may be obtained by evaluating the integral:

$$E[G^k(\boldsymbol{Z})] = \int_{-\infty}^{+\infty} \Phi(z) G^k(z) d_z \tag{4}$$

where $\Phi(z)$ is the cumulative distribution function of standard normal variable *Z*. The integration is performed numerically (Eq.5)

$$E[G^k(\boldsymbol{Z})] \cong \sum_{j=1}^{m} w_j G^k(z_j) \tag{5}$$

where m is the number of points considered, w_j are the weights and z_j are the typical normal points. The numerical operations required are not repeated here in detail but can be summarized as follows:

- Definition of correlation matrix $\boldsymbol{C}$.
- Transformation of $\boldsymbol{C}$ into $\boldsymbol{C_0}$, which represents the correlation matrix in the correlated normal space $\boldsymbol{Y}$ using a Nataf-transformation. $\boldsymbol{C_0}$ contains the correlation coefficients $\rho_{0,ij}$.
- By means of a Cholesky decomposition of $\boldsymbol{C}_0$ one obtains the lower triangular matrix $\boldsymbol{L}_0$.
- $\boldsymbol{L}_0$ enables a mapping of the uncorrelated standard normal space $\boldsymbol{Z}$, where the integration points are predefined by the integration rule, to the correlated normal space $\boldsymbol{Y}$.

$$\boldsymbol{Y} = \boldsymbol{L_0 Z} \tag{6}$$

- Mapping of the correlated standard normal space $\boldsymbol{Y}$ to the correlated non-normal space $\boldsymbol{X}$ by using

$$\boldsymbol{X} = \boldsymbol{F}^{-1}[\Phi(\boldsymbol{L_0 Z})] \tag{7}$$

Thus above operation denotes an approximate method of relating correlated variables from a normal space to an uncorrelated standard normal space, which is necessary in order to be able to perform the required mathematical operations.

With the operations above Eq.5 can be written as

$$E[G^k(\boldsymbol{X})] \cong \sum_{j=1}^{m} w_j G^k(x_{1j}, x_{1j}, \ldots, x_{nj}) \tag{8}$$

whereas $(x_{1j}, x_{2j}, \ldots\ldots, x_{nj})$ is obtained using Eq.6 and Eq.7

$$\begin{pmatrix} y_{1j} \\ . \\ . \\ . \\ y_{nj} \end{pmatrix} = \begin{pmatrix} L_{11} & . & \cdots & . & 0 \\ . & . & . & . & . \\ . & . & L_{ii} & . & . \\ . & . & . & . & . \\ 0 & . & \cdots & . & L_{nn} \end{pmatrix} \begin{pmatrix} z_{1j} \\ . \\ . \\ . \\ z_{nj} \end{pmatrix} \tag{9}$$

$$\begin{pmatrix} x_{1j} \\ . \\ . \\ . \\ x_{nj} \end{pmatrix} = \begin{pmatrix} F_{x_1}^{-1}[\Phi(y_{1j})] \\ . \\ . \\ . \\ F_{x_{n1}}^{-1}[\Phi(y_{nj})] \end{pmatrix} \tag{10}$$

The first subscript i denotes the ith variable and the second subscript j denotes the jth point; w_j and $(z_{1j},.....,z_{nj})$ are the predetermined values given by Zhou and Nowak (1988). The integration rule adopted in this work is $2n^2+1$, which means in term of finite elements that $2n^2+1$ deterministic calculations have to be performed. Other integration rules have been investigated but the above proved to be an optimum compromise between accuracy and computational effort (Thurner, 2000).

3 General Description of Procedure for Practical Application

In the previous section the theoretical basis for performing a probability analysis in geotechnics utilizing a particular point estimate method has been briefly described. So far no assumptions on the calculation model have been made, i.e. the method is very general and can be applied in the context of e.g. simple limit equilibrium analyses as well as in combination with more advanced computational models. In this paper finite element models are chosen as calculation models because the authors believe that numerical modelling is the only feasible way of getting a deeper insight into the behaviour of complex geotechnical structures. In order to achieve a user friendly environment for introducing probabilistic concepts into finite element modelling pre- and post-processing routines have been developed in order to facilitate the additional input and output operations. These routines have been connected to the finite element code PLAXIS (Brinkgreve, 2000), which is used for all calculations presented in the following, but can be easily modified to accommodate other codes. By doing so, the required parameter sets, which are determined by the Point Estimate Method (PEM) chosen, and the subsequent generation of the data sets in a form suitable for input into the PLAXIS code are automatically generated (Thurner, 2000).

In summary the following steps have to be performed:

Step 1: Choice of calculation model (limit equilibrium method, numerical model, ..). In the following a finite model is employed.

Step 2: Decision which of the input parameters are to be taken into account with stochastic properties (number n) in the analysis.

Step 3: Determination of mean value and standard deviation for all variables and definition of the statistic distribution of the parameters.

Step 4: Sensitivity study by performing a Taylor-Series-Finite-Difference-Analysis (TSFD) to identify these parameters which have the most significant influence on certain results. Only these are treated as stochastic parameters for further calculations. This step is optional.

Step 5: Calculation of integration points; i.e. the mathematically defined parameter combinations depending on the chosen integration method and preparation of the relevant data files for the finite element calculations. This step is performed with the program PEM_1 (Thurner, 2000) and additional preprocessing tools developed for this purpose. $2n^2+1$ input files for the finite element code PLAXIS are automatically prepared.

Step 6: Finite element calculations and determination of results such as displacements, stresses, strains and internal forces in structural elements.

Step 7: Definition of suitable performance functions. Critical deformations (e.g. surface settlements) and/or maximum stresses (e.g. in the shotcrete or concrete lining of a tunnel) can be defined.

Step 8: With the results from the finite element calculations the first two moments of the evaluated system parameters can be calculated. For these variables a statistical distribution has to be assumed. In the last step the performance function can be evaluated. The mean value (μ_G) and standard deviation (σ_G) as well as the safety index β_{HL} (Hasofer and Lind, 1974) can be calculated. The integral of the curve G(**X**) < 0 indicates the probability of "failure". The evaluation of the performance function can be done by means of Monte-Carlo simulations or a "FORM analysis" with appropriate computer programmes.

4 Example Sheet Pile Wall

As a first example, an excavation supported by a cantilever sheet pile wall will be considered. Two aspects will be discussed: firstly, the probability of failure and secondly the probability of exceeding a predefined horizontal displacement of the top of the wall. The example is taken from Gutjahr (2000), who analysed the problem using a hypoplastic constitutive model. The calculations are performed using the so-called Hardening-Soil Model implemented into PLAXIS. It accounts for shear and volumetric hardening, includes a stress dependent stiffness distinguishing between primary loading (governed by the parameter E_{50}^{ref}) and unloading/reloading (governed by the parameter E_{ur}^{ref}) and incorporates the Mohr Coulomb failure criterion. The geometry, the finite element mesh and the boundary conditions follow from Figure 1. The material parameters are summarized in Table 1 and it can be seen that only the angle of internal friction (φ) and the reference stiffness (E_{50}^{ref}) are treated as stochastic variables (with a lognormal distribution) for this illustrative case. This leads to 9 calculations to be performed according to the point estimate method with a $2n^2+1$ integration rule (Zhou and Nowak, 1988). The resulting input values for the finite element calculations are given in Table 2.

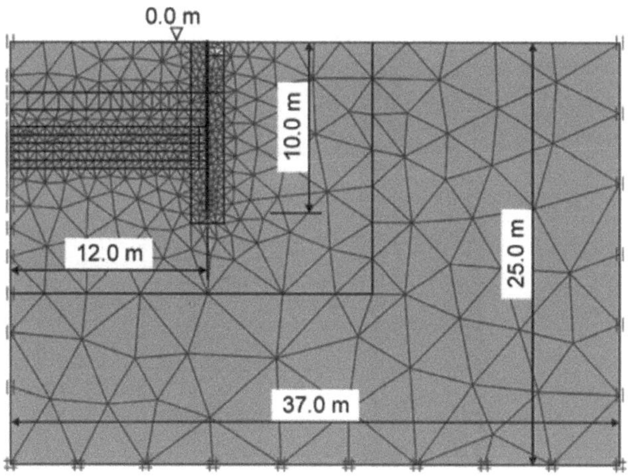

Figure 1. Geometry and finite element mesh for excavation problem.

The analysis has been performed as follows: excavation is simulated in 0.5 m intervals, and for each excavation depth, calculations have been made with the parameters given in Table 2. After reaching equilibrium, a so-called φ-c-reduction has been executed whereas the strength parameters are reduced until no equilibrium can be found. By comparing these strength parameters with the actual ones a factor of safety (termed Msf) is obtained. Thus the limit state function can be written as

$$G(X) = Msf_{FE} - 1 \tag{11}$$

where Msf_{FE} are the values obtained from the finite element analyses as described above.

Table 1. Material parameters for excavation problem.

Material Parameter	μ	υ	σ	5%-Fractile
γ [kN/m³]	17	-	-	17.9
φ [°]	42.1	0.04	1.68	39.4
c [kN/m²]	0	-	-	0
ν [-]	0.2	-	-	0.226
$E_{50}^{ref}=E_{oed}^{ref}$ [kN/m²]	60000	0.2	12000	42480
$E_{ur}^{ref}=3.E_{50}^{ref}$				

Table 2. Input values for calculations with $2n^2+1$ integration rule for excavation problem.

Parameter set	E_{50}^{ref} [kN/m²]	φ [°]
1	58209	42.07
2	95247	42.07
3	35573	42.07
4	58209	45.57
5	58209	38.83
6	82454	44.51
7	82454	39.75
8	41092	44.51
9	41092	39.75

Following the procedure described in section 3 a mean and standard deviation of the Msf-value is obtained for each excavation depth and by assumption of a lognormal distribution of Msf the probability of failure or the reliability index can be obtained. Figure 2 reveals that the variation of the probability index with excavation depth is roughly linear which in turn means a progressive increase of probability of failure (e.g. from $p_f = 9.3E\text{-}09$ for an excavation depth of 6.5 m to $p_f = 2.6E\text{-}04$ for an excavation depth of 7.0 m).

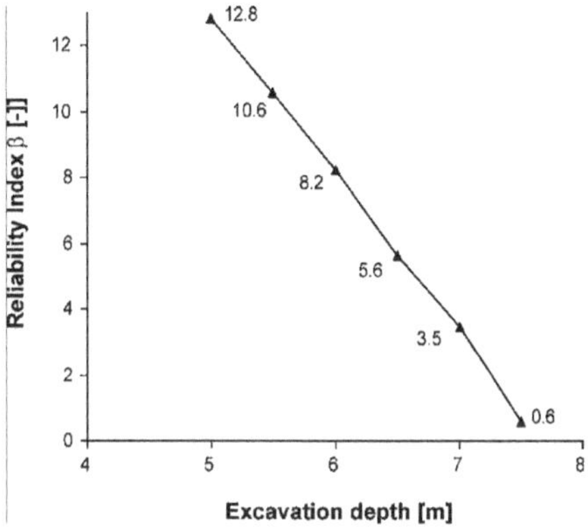

Figure 2. Reliability index vs excavation depth.

In a similar way, a limit state function can be specified with respect to the displacement of the top of the wall thus representing a criterion for the serviceability of the structure. If e.g. 3.5cm are defined as limiting displacement (here chosen arbitrarily) the performance function is

$$G(X) = 0.035 - u_{x,FE} \tag{12}$$

$u_{x,FE}$ being the displacement obtained from the finite element analyses. Figure 3 compares u_x obtained from various analyses: mean value and standard deviation from the probability analysis and results with different deterministic parameters. The probability of exceeding 3.5cm is approx. 0.25 % at a depth of 5.6m and increases rapidly to approx. 55% at 6.1m excavation depth. Of course the 5%-fractile value yields the lowest depth where the displacement of u_x = 3.5cm is reached (5.6m), the design value after Schneider (1999) results in a depth of 6.1m.

It should be mentioned at this point that the results, displacements as well as bending moments, compare well with the ones obtained by Gutjahr (2000). This agreement shows that the proposed approach yields results which are consistent with other methods and experience; it is argued however that the stochastic nature of soils is accounted for in more rational way within the framework presented here.

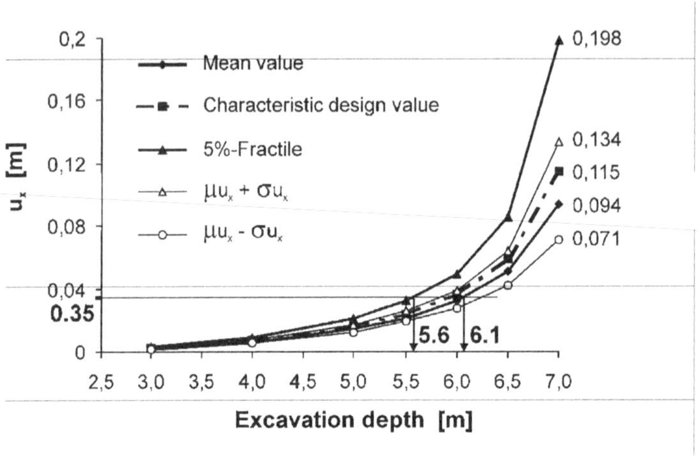

Figure 3. Horizontal displacements vs excavation depth.

5 Application to Numerical Simulation of Tunnel Excavation

In this section, the application of the developed framework to the analysis of tunnel excavation will be shown. The problem considered is closely related to an actual tunnel construction. However, some minor simplifications have been introduced, as it is the intention here to demonstrate the potential of the proposed approach and not to give a full analysis of a real case history. At first, mean and standard deviation of input variables are determined, using different sources of information employing Bayes' theorem. In the second step, the influence of each individual parameter on a particular result is evaluated by means of the TSFD-Method. Finally, the probability of exceeding the admissible stress in the lining is evaluated by the point estimate method described in section 2.

5.1 Geometry and subsoil conditions

The geometry of the shallow tunnel analysed is given in Figure 4 including the assumed, simplified, ground conditions. The tunnel is situated in a limestone formation ("rock"), with a soil layer of an average thickness of approximately 5m on top ("soil").

The tunnel is excavated according to the principles of the New Austrian Tunnelling Method (NATM) thus a shotcrete lining with rock bolts provides the support during construction stages whereas a cast concrete shell serves as final lining. The final lining however will not be dealt with in this paper. Excavation modelling is made in sections and prerelaxation factors of 25 % (for each section) according to the load reduction method have been assumed to account for 3D effects in the 2D plane strain analysis (Schweiger et al., 1997). The shotcrete lining is modelled with elastic beam elements. It is distinguished between the "young" and "old" shotcrete to incorporate the aging effect for the stiffness in an approximate way. The goal of this analysis is a comparison of results obtained by using design values and from probability analysis. In addition the probability of failure during construction is determined.

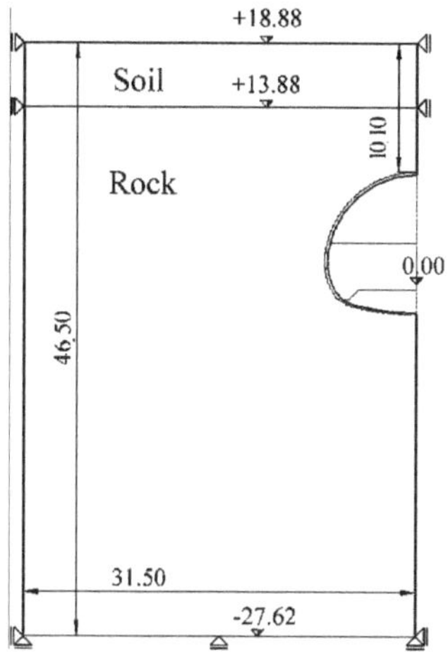

Figure 4. Geometry for tunnel example.

5.2 Determination of input variables

Unfortunately it is a fact that even for large projects site investigation and laboratory data available are rarely sufficient for stochastic modelling. However by utilizing tools such as the Bayes' theorem information obtained from different sources, with different degrees of confidence, can be combined to arrive at the required input parameters. As an example the determination of mean and standard deviation of the cohesion of the rock layer is shown in the following. The following sources of information are utilized:

Laboratory test data
Results from a limited number of uniaxial compression, triaxial and direct shear tests have been available. These yielded a mean value of 90.8 kN/m² and a standard deviation of 30.2 kN/m².

Experience from projects under similar geological conditions
Employing the "3σ-rule" a mean and standard deviation of 80 and 20 kN/m² are obtained respectively.

Rock Mass Rating System (Bieniawski, 1989)
Based on the geological report the mean is 110 kN/m² with a standard deviation of 25 kN/m².

Hoek-Brown Criterion (Hoek, 2000)

The Hoek-Brown criterion is well established in rock engineering and a number of extensions have been proposed since the publication of the basic model including the introduction of a so-called Geological Strength Index (GSI). If GSI, the uniaxial compressive strength and the material parameter m_i are defined with mean values and standard deviation the Mohr Coulomb strength parameters φ and c can be determined (Thurner, 2000). Although this procedure is not recommended as a standard way of determining strength parameters it is included here for demonstration purposes. Doing so the mean value for the cohesion is obtained as 150 kN/m² and the standard deviation as 50 kN/m².

If these sources of information are combined utilizing an appropriate program whereas a weighting is introduced taking into account the degree of confidence one associates to each source, the following is obtained:

mean value for cohesion of rock layer:	100 kN/m²
standard deviation:	30 kN/m²

Obviously an equivalent procedure can be adopted for all other parameters and, if available, values taken from the literature may also be included in the same way.

5.3 Taylor Series Finite Difference-Method and Matrix of Influence

For the given geology, 12 significant parameters were identified from experience and treated as basic variables as shown in Table 3. For soil and rock, the angle of internal friction, the cohesion, the coefficient of lateral earth pressure and the elastic modulus at a reference stress level are variables. ΣMStage1 is an expression used in PLAXIS defining the prerelaxation factor for the top heading and ΣMstage2 for the bench respectively. Uncertainties in thickness and normal stiffness of the shotcrete lining are included in the values for the shear moduli (G_{young} and G_{old}), which are treated as variables.

Using the TSFD-Method (U.S. Army Corps of Engineers, 1992) the influence of each variable on a certain limit state function can be quantified. For the 12 variables shown in Table 3, 25 calculations (2n+1) are required to get an estimation of mean and standard deviation for each limit state function. In this case, the settlement of the crown (u_y) and the relation between admissible and calculated normal force (N_{all}) in the shotcrete lining at the end of construction is evaluated. For the admissible normal force a definition after Schikora and Ostermeier (1988) is used. To include some measure for the mobilisation of the shear strength of the rock mass, the maximum shear strain (gam_s) after the excavation of the top heading is considered. To quantify this effect for each variable separately, a relation between the obtained standard deviation based on all calculations and the standard deviation caused by a single variable is used. This is similar to the sensitivity index α, which can be obtained using a FORM-approach. Doing this with all variables the influence of each variable on the respective result can be assessed and shown in a matrix of influence (Table 4). It can be seen that roughly half of the variables have a significant influence. At this stage a decision can be made, which variables should be used in further calculations and which one can be treated as deterministic values as their influence on the result is not significant. The defined threshold value depends on engineering judgment and should be used carefully (Einstein, 2000). It has been chosen here as approximately 4 %.

Table 3. Basic variables for design value calculations and Taylor Series analysis for tunnel example.

	Parameter	Distribution	μ	σ	Unit
Soil	φ	lognormal	21	2	[°]
	c	lognormal	24	8	[kN/m²]
	E	lognormal	24000	8000	[kN/m²]
	Ko	normal	0.65	0.1	[-]
Rock	φ	lognormal	37	2.5	[°]
	c	lognormal	100	18.75	[kN/m²]
	E	lognormal	200000	31250	[kN/m²]
	Ko	normal	0.7	0.1	[-]
ΣMStage	ΣMstage1	normal	0.25	0.03	[-]
	ΣMstage2	normal	0.25	0.02	[-]
Shotcrete	G_{young}	lognormal	2000000	500000	[kN/m²]
	G_{old}	lognormal	6000000	1500000	[kN/m²]

Table 4. Matrix of influence for tunnel example.

	N_{all}	u_y	gam_s	Total sum
φ-Soil	0	0	1	1
c-Soil	0	0	1	2
E-Soil	1	2	2	5
Ko-Soil	0	0	1	1
φ-Rock	3	4	9	16
c-Rock	4	3	6	13
E-Rock	1	5	4	10
Ko-Rock	4	9	6	19
ΣMStage 1	2	2	4	7
ΣMStage 2	0	0	1	1
G_{young}	12	9	2	22
G_{old}	1	0	0	2

5.4 Evaluation of Limit State Function

Based on the results of the previous section 7 parameters, which are above the defined threshold value (see Table 4), are treated as variables for further investigations with the PEM after Zhou and Nowak (1988), i.e. 99 calculations are required ($2n^2+1$). As an example the allowable normal force for the shotcrete lining will be assessed. According to the results obtained from the design value calculation 15 cm shotcrete thickness is required. The deterministic factor of safety for this case would be FOS = 1.6. Following the formula of Schikora & Ostermeier (1988) the limit state function reads

$$N_{zul} = -\frac{\sigma_R}{F_s} \cdot d \cdot (1 - 2 \cdot \frac{\left(e_a + ABS\left(\frac{M}{N}\right)\right)}{2.1 \cdot d} \tag{13}$$

with

σ_R .... uniaxial strength of shotcrete
F_s .... factor of safety
e_a .... eccentricity
d .... thickness of lining
M .... bending moment
N .... normal force
N_{zul} .... admissible normal force

Figure 5 shows the evaluation of Eq. 13 by means of a Monte-Carlo simulation. The probability of exceeding the admissible force in the lining is $p_f = 1.5E\text{-}6$. If a thickness of 20 cm is chosen it results in $p_f = 1.3E\text{-}7$ and a FOS = 2.1. For a thickness of 10 cm the FOS is below 1.0 and the obtained $p_f = 2E\text{-}4$.

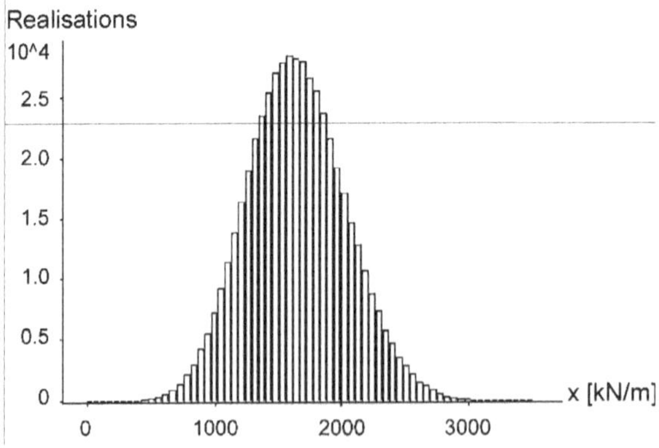

Figure 5. Evaluation of limit state function for normal force in shotcrete lining.

5.5 Comparison for Results Obtained from Probabilistic Analysis and Design Values

In this section the results of the PEM-calculations will be compared with deterministic analyses making different assumptions for design values. The choice of design parameters is often based on personal experience rather than on a sound theoretical basis. In the following some possible assumptions for deterministic analysis will be investigated here, namely

- mean values
- characteristic design values: the mean value reduced by half of the standard deviation is used (approx. 35%-fractile; Schneider, 1999)
- 5%-fractile: for the rock and soil parameters 5%-fractile value are used; for the shotcrete stiffness the 95%-fractile is used > very conservative
- 95%-fractile: for the rock and soil parameters 95%-fractile value are used; for the shotcrete stiffness the 5%-fractile is used.

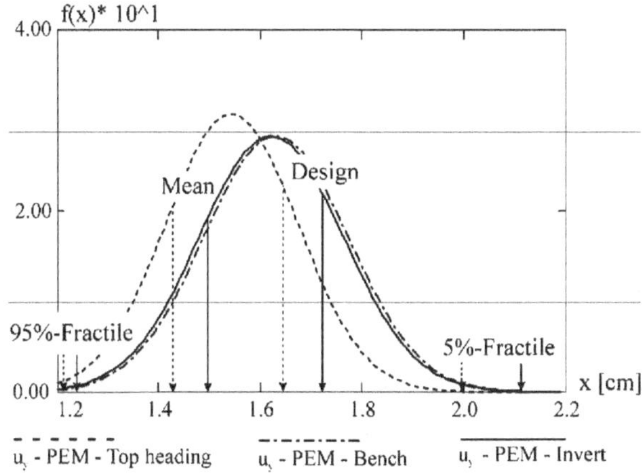

Figure 6. Density distribution for the settlement of the crown for all construction stages.

Table 5. Comparison of input-fractiles of parameters for deterministic analysis and corresponding fractiles from stochastic analysis.

Deterministic calculation value	Fractiles obtained from stochastic model [%]			
	Settlement of crown u_y		Normal force	
	Top heading	Bench	Top heading	Bench
5%-Fractile	0.02	0.001	0.001	0.0001
Design value	22	18.6	23.4	23.6
Mean	81.2	87.1	29.6	27.6
95%-Fractile	99.42	99.79	99.999	99.999

From the 99 calculations mean value and standard deviation for a particular result, e.g. the crown settlement, can be determined. By assuming a normal distribution the probability of obtaining a result above or below the corresponding deterministic analysis can be evaluated. A graphical view of this comparison for the settlement of the crown is given in Figure 6 for all construction stages. In Table 5 input-fractiles and result-fractiles from the stochastic model are summarized for all assumptions given above. It follows that using a of 5% fractile for the input parameters results in a calculated settlement which represents only the 0.02% fractile value from the stochastic analysis, i.e. only 0.02% of the stochastic results yield higher settlements and thus the assumption is more conservative than anticipated. The difference is even higher for the normal force in the lining. If the design value is assumed to be approximately the 35%-fractile, which is sometimes assumed in practice, the result is an approximately 20%-fractile. With an input value representing a 95%-fractile the result is a fractile-value of over 99% for displacements and 99.9% for the normal forces. The latter assumption however is of no practical significance, but it illustrates nicely the fact that a certain input fractile may result in a significantly different fractile for the output, a result which is of course not surprising from a theoretical point of view but often not appreciated in practice.

It follows from this exercise that probabilistic concepts provide a more rational way of introducing the uncertainties in material parameters into numerical calculations than assuming a single design value which is based on assumptions without a sound theoretical background.

6 Conclusion

A framework for introducing probabilistic concepts employing point estimate methods into finite element calculations has been proposed. The theoretical background has been briefly described but more emphasize has been put in this paper in illustrating how the proposed approach can be used in practice. This has been done by means of examples, namely an excavation supported by a sheet pile wall and a tunnel excavation. In the latter example the Bayes' theorem has been employed to arrive at input parameters based on information from different sources, the TSFD Method has been used to identify the parameters having the most pronounced influence on the results and finally a point estimate method has been utilized to evaluate the probability of failure. Comparison with deterministic analyses has been made.

It can be concluded that due to the achievements made in the development of hardware and software it has become feasible to perform probabilistic analyses with high level numerical modelling. It is argued that the uncertainties associated with material and model parameters inevitably inherent in any analysis in geotechnics are covered in a rational and theoretically sound way by a probabilistic approach thus increasing the validity of numerical calculations. The additional complexity introduced seems to be acceptable also from a practical point of view because only a few additional steps (see Figure 7) are required provided the appropriate computational tools are readily available.

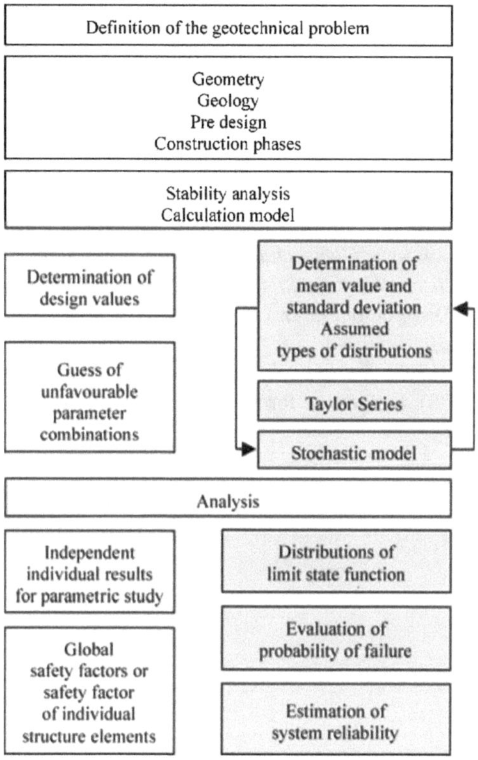

Figure 7. Scheme of typical deterministic analysis with probabilistic extensions.

References

Benjamin, J. and Cornell, C.A. (1970). *Probability, Statistics and Decision for Civil Engineers*, McGraw Hill: New York.

Bieniawski, Z.T. (1989). *Engineering rock mass classifications: a complete manual for engineers and geologists in mining, civil, and petroleum engineering*. Wiley: New York.

Brinkgreve, R. B. J. (2000). *PLAXIS, Finite element code for soil and rock analyses*. Users manual, Rotterdam: Balkema.

Einstein, H.H. (2000). Risikoanalyse in der Geotechnik. *Lecture Notes Short Course*. Graz University of Technology.

Gutjahr, S. (2000). Tragverhalten von nichtgestützten Wänden in Sand. *Workshop Verformungsprognose für tiefe Baugruben*, DGGT:Stuttgart, 181-198.

Hasofer, A.M. and Lind, N.C. (1974). Exact and Invariant Second-Moment Code Format. *Engineering Mechanics Division* 100: 111-121.

Hoek, E. (2000). *Rock Engineering, Course Notes*. http://www.rocscience.com/Hoekcorner.html.

Li, K.S. (1992). Point-Estimate Method for Calculating statistical Moments. *Engineering Mechanics* 118:1506-1511.

Schikora, K. and Ostermeier, B. (1988). Temporäre Sicherung von Tunneln mit Spritzbeton. *Bauingenieur* 63: 399-403.

Schneider, H.R. (1999). Determination of characteristic soil properties. In Barends et al., eds., *Proc. 12th Europ. Conf. Soil Mechanics and Geotechnical Engineering*. Amsterdam. 273-281.

Schweiger, H.F., Schuller, H. and Pöttler, R. (1997). Some remarks on 2-D-models for numerical simulations of underground constructions with complex cross sections. In Yuan, ed. *Proc. 9th Int. Conf. Computer Methods and Advances in Geomechanics*. Wuhan. 1303-1308.

Thurner, R. (2000). *Probabilistische Untersuchungen in der Geotechnik mittels deterministischer Finite Elemente-Methode*. Ph.D. Dissertation, Graz University of Technology. Institute for Soil Mechanics and Foundation Engineering.

U.S. Army Corps of Engineers (1992). *Reliability Assessment of Navigation Structures*. ETL 1110-2-532.

Zhou, J. and Nowak, A.S. (1988). Integration formulas to evaluate functions of random variables. *Structural safety* 5: 267-284.

Basic Concepts and Applications of Random Sets in Geotechnical Engineering

Helmut F. Schweiger[1] and Gerd Peschl[2]

[1] Computational Geotechnics Group, Institute for Soil Mechanics and Foundation Engineering, Graz University of Technology, Graz, Austria
[2] INSITU Geotechnik ZT GmbH, Graz, Austria

Abstract. This paper presents a possible framework for dealing with uncertainties in geotechnical engineering based on more recently introduced mathematical theories. Random sets are employed to describe the variability of material parameters and geometrical data. After a brief introduction of the basics of the proposed approach application to a boundary value problem, namely a deep excavation analysis, is presented. Comparison with in situ measurements is provided and it is shown that the calculated most likely range of displacements compares well with measurements. Furthermore an assessment of the probability of damage of a building, situated adjacent to the excavation, is in line with observed behaviour.

1 Introduction

Uncertainties in determination of the in situ soil profile and material parameters for individual soil layers are one of the important problems geotechnical engineers have to cope with. It is important to realise that different sources of uncertainty exist, material parameters varying in a certain - but known - range may be one of them but simply the lack of knowledge may be the more pronounced one. A rigorous mathematical treatment of all aspects of uncertainties is not straightforward and thus is commonly replaced in practice by engineering judgement. Recent theoretical developments and advances made in computational modelling allow for a more formal consideration of uncertainties and it can be expected that theories and models taking uncertainty into account in the design of geotechnical structures will be more appreciated in near future.

The full scope of uncertainty and its dual nature can be described with the following definitions from Helton (1997):

Aleatory Uncertainty – the type of uncertainty which results from the fact that a parameter can behave in random ways (stochastic, objective uncertainty also known as variability).

Epistemic Uncertainty – the type of uncertainty which results from the lack of knowledge about a parameter (state of knowledge, subjective uncertainty or ignorance).

Aleatory uncertainty is associated with variability in known (or observable) populations and is irreducible, whereas epistemic uncertainties change with the state of knowledge and is therefore reducible. Up to now probability theory has been used to characterise both types of uncertainty, although it is well recognised that it is the aleatory uncertainty which is best dealt by classical probability theory but it can be argued that it is not capable of capturing epistemic uncertainty (e.g., Sentz and Ferson, 2002).

Although some general information on probabilistic parameters can be found in the literature (e.g., Rackwitz, 2000), geotechnical parameters for particular soils, used e.g. in the analysis of practical boundary value problems, are given as intervals in most cases, with no information about the probability distribution across the interval and therefore a formal framework is required in order to encode uncertainty in geotechnical systems. In this paper random sets are proposed to represent uncertainty in finite element reliability analysis.

2 Random Set Theory

Random set theory provides a general framework for dealing with set-based information and discrete probability distributions. It yields the same result as interval analysis, when only range information is available and the same result as Monte-Carlo simulations when the information is abundant.

Basic concepts

Let X be a non-empty set containing all the possible values of a variable x. Dubois and Prade (1990, 1991) defined a random set on X as a pair $(\Im,m)$ where $\Im = \{A_i : i = 1,\ldots,n\}$ and m is a mapping, $\Im \to [0,1]$, so that $m(\emptyset) = 0$ and

$$\sum_{A \in \Im} m(A) = 1. \tag{1}$$

$\Im$ is called the support of the random set, the sets A_i are the focal elements ($A_i \subseteq X$) and m is called the basic probability assignment. Each set, $A \in \Im$, contains some possible values of the variable, x, and $m(A)$ can be viewed as the probability that A is the range of x. Because of the imprecise nature of this formulation it is not possible to calculate the 'precise' probability Pro of a generic $x \in X$ or of a generic subset $E \subset X$, but only lower and upper bounds on this probability: $Bel(E) \leq Pro(E) \leq Pl(E)$. Figure 1 shows possible 'precise' probabilities (Pro) bounded by Pl and Bel. In the limiting case, when $\Im$ is composed of single values only (singletons), then $Bel(E) = Pro(E) = Pl(E)$ and m is a probability distribution function.

Following Dempster (1967) and Shafer (1976) the lower bound Bel and the upper bound Pl of its probability measure are defined, for every subset $E \in X$, by (Tonon et al., 2000a):

$$Bel(E) = \sum_{A_i : A_i \subseteq E} m(A_i) \tag{2}$$

$$Pl(E) = \sum_{A_i : A_i \cap E \neq \emptyset} m(A_i) \tag{3}$$

where the belief function, *Bel*, of a subset E is a set-valued function obtained through summation of basic probability assignments of subsets A_i included in E and the plausibility function, *Pl*, of subset E is a set-valued function obtained through summation of basic probability assignments of subsets A_i having a non-zero intersection with E. They are envelopes of all possible cumulative distribution functions compatible with the data.

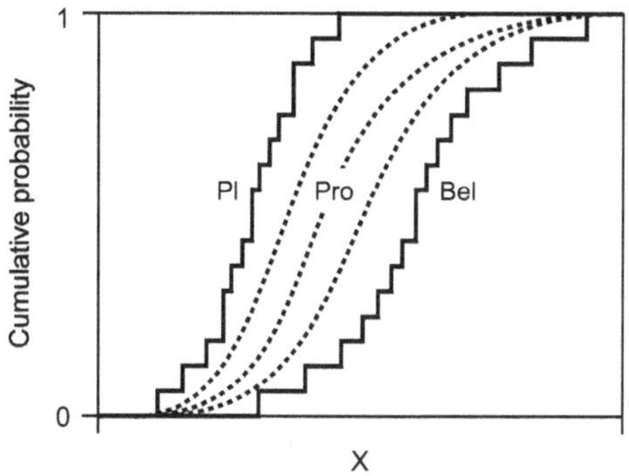

Figure 1. Upper bound (*Pl*) and lower bound (*Bel*) on 'precise' probability (*Pro*).

Finding the bounds on the system response

Random set theory provides an appropriate mathematical framework for combining probabilistic as well as set-based information in which the extension of random sets through a functional relation is straightforward (Tonon et al., 2000a). Let f be a mapping $X_1 \text{x} \ldots \text{x} X_N \rightarrow Y$ and $x_1,\ldots,x_N$ be variables whose values are incompletely known. The incomplete knowledge about $\mathbf{x} = (x_1,\ldots,x_N)$ can be expressed as a random relation R, which is a random set $(\Im,m)$ on the Cartesian product $X_1 \text{x} \ldots \text{x} X_N$. The random set $(\Re,\rho)$, which is the image of $(\Im,m)$ through f is given by (Tonon et al., 2000b):

$$\Re = \{R_j = f(A_i), A_i \in \Im\}; \qquad f(A_i) = \{f(\mathbf{x}), \mathbf{x} \in A_i\} \tag{4}$$

$$\rho(R_j) = \sum_{A_i : R_j = f(A_i)} m(A_i) \tag{5}$$

If $A_1,\ldots,A_n$ are sets on $X_1 \text{x} \ldots \text{x} X_N$ respectively and $x_1,\ldots,x_N$ are random set independent (Ferson et al., 2004; Fetz and Oberguggenberger, 2004), then the joint basic probability assignment is the product measure given by

$$m(A_1 \times ... \times A_n) = \prod_{i=1}^{n} m_i(A_i), \quad A_1 \times ... \times A_n \in \Re \tag{6}$$

If the focal set A_i is a closed interval of real numbers: $A_i = \{x \mid x \in [l_i, u_i]\}$, then the lower and upper cumulative probability distribution functions, $F_*(x)$ and $F^*(x)$ respectively, at some point x can be obtained as follows:

$$F_*(x) = \sum_{i:x \geq u_i} m(A_i) \tag{7}$$

and

$$F^*(x) = \sum_{i:x \geq l_i} m(A_i), \tag{8}$$

Figure 2 illustrates schematically an example of constructing a random set from multiple sources of information given as intervals (focal elements $A_1,...,A_4$ and basic probability assignments $m_1,...,m_4$).

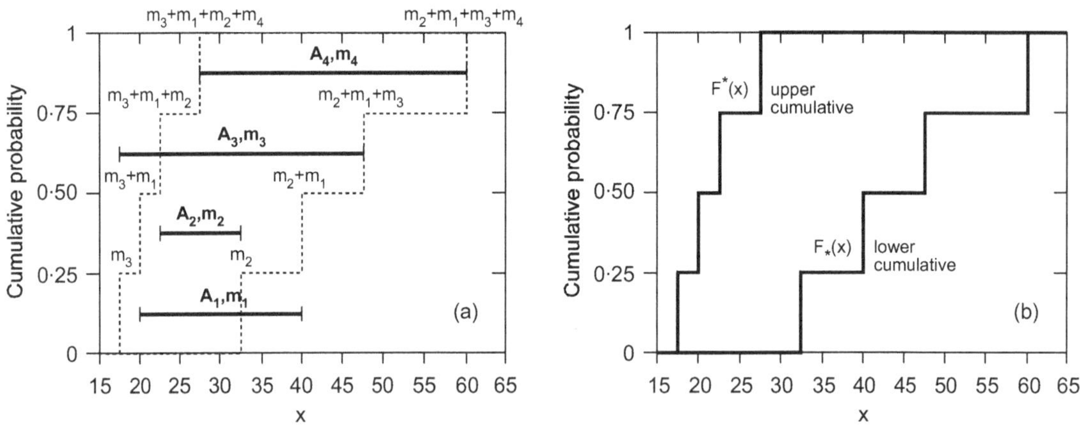

Figure 2. Random set: (a) construction, (b) upper and lower discrete cumulative distribution function.

To obtain the left envelope (Figure 2(a) and 2(b) each contain a left and a right envelope), the distribution of the probability mass of each interval in the calculation matrix is assumed to be concentrated at the lower bound of each source of information given as interval (the low bounds are sorted from smallest to greatest, and the cumulative distribution curve is stepped upward at each value of the horizontal axis representing an interval low bound, by the amount of the basic probability assignment concentrated at that low bound). This envelope is referred to here as the left envelope rather than the upper envelope to reduce ambiguity in terminology. On the other hand, to construct the right envelope, the probability mass for each interval is assumed to be concentrated at the upper bound of the interval.

In the absence of any further information, a random relation, so-called *calculation matrix*, can be constructed by assuming random set independence between marginal random sets (Equation 6). A calculation matrix implies bounds on a corresponding discrete cumulative distribution function (*CDF*). The basic step is the calculation by means of Equation 4 and 5 of the image of a focal element through function f. The requirement for optimisation to locate the extreme elements of each set $R_j \in \Re$ (Equation 4) can be avoided if it can be shown that the function $f(A_i)$ is continuous in all $A_i \in \Im$ and also no extreme points exist in this region, except at the vertices, in which case the Vertex method (Dong and Shah, 1987) applies. This is generally the case for the type of problems considered here. Furthermore, the sensitivity analysis discussed later will reveal if this condition does not hold true. Assume each focal element A_i is a N-dimensional box, whose 2^N vertices are indicated as v_k, $k = 1,\ldots,2^N$. If the vertex method applies then the lower and upper bounds R_{j*} and R_j^* on each element $R_j \in \Re$ will be located at one of the vertices:

$$R_{j*} = \min_k \left\{ f(\mathrm{v}_k) : k = 1,...2^N \right\} \tag{9}$$

$$R_j^* = \max_k \left\{ f(\mathrm{v}_k) : k = 1,...2^N \right\} \tag{10}$$

Thus function $f(A_i)$ which represents in this framework a numerical model has to be evaluated 2^N times for each focal element A_i. The number of all calculations, n_c, required for finding the bounds on the system response is

$$n_c = 2^N \prod_{i=1}^{N} n_i \tag{11}$$

where N is the number of basic variables and n the number of information sources available for each variable. The computational effort involved seems to be very high, but can be reduced if $f(A_i)$ is continuous and a strictly monotonic function with respect to each parameter $x_1,\ldots,x_N$, which is, as mentioned previously, a valid assumption for all analyses discussed in this paper. In this case the vertices where the lower and upper bounds (Equation 9 and 10) on the random set are located can be identified merely by consideration of the direction of increase of $f(A_i)$ which can be done by means of a sensitivity analysis (Peschl, 2004). Thus $f(A_i)$ has to be calculated only twice for each focal element A_i (Tonon et al., 2000b).

Combination of random sets

An appropriate procedure is required if more than one source of information is available for one particular parameter in order to combine these sources. Suppose there are n alternative random sets describing some variable x, each one corresponding to an independent source of information. Then for each focal element $A \in X$

$$m(A) = \frac{1}{n}\sum_{i=1}^{n} m_i(A) \tag{12}$$

Alternative combination procedures have been proposed depending on different beliefs about the truth of the various information sources (e.g., Sentz and Ferson, 2002; Hall and Lawry, 2004) but these will not be discussed here.

Formulation of the reliability problem

Basically, reliability analysis calculates p_f, the probability of failure of a system characterised by a vector $\mathbf{x} = (x_1,\ldots,x_N)$ of basic variables on X. The resistance r of the system can be expressed as $r = g_r(\mathbf{x})$ and the action s as $s = g_s(\mathbf{x})$. The probability of failure p_f is the probability p of $(r \leq s)$ or in general

$$p_f = p(g(\mathbf{x}) \leq 0) = \int\limits_{g(\mathbf{x}) \leq 0} f_X(\mathbf{x})d\mathbf{x} \tag{13}$$

where g is called the 'limit state function', f_X the joint probability density function over the basic variables and p_f is identical to the probability of limit state violation. The resistance r and action s are generally implicit in $\mathbf{x}$. Utilising random set theory the reliability problem is reduced to evaluate the bounds on $p_f = p(g(\mathbf{x}) \leq 0)$ subject to the available knowledge restricting the allowed values of $\mathbf{x}$. If the set of failed states is labelled $F \subseteq X$, the upper and lower bound on the probability of failure are the Plausibility $Pl(F)$ and Belief $Bel(F)$ respectively: $Bel(F) \leq p_f \leq Pl(F)$ where

$$Bel(F) = \sum_{A_i : A_i \subseteq F} m(A_i) \tag{14}$$

$$Pl(F) = \sum_{A_i : A_i \cap F \neq \varnothing} m(A_i) \tag{15}$$

3 Application to Deep Excavation Problem

In this section some results from a back analysis of a practical example, namely a deep excavation (Breymann et al., 2003) on a thick layer of post-glacial soft lacustrine deposit (clayey silt) are presented to demonstrate the applicability of the proposed method. An underground car park has been constructed as open pit excavation with a 24m deep anchored concrete diaphragm wall as retaining construction. Figure 3 plots the cross section of the system and the soil profile. In this analysis particular attention is given to the risk of damage to the existing structures and the reliability of the system by assessing the angular distortion of the building respectively.

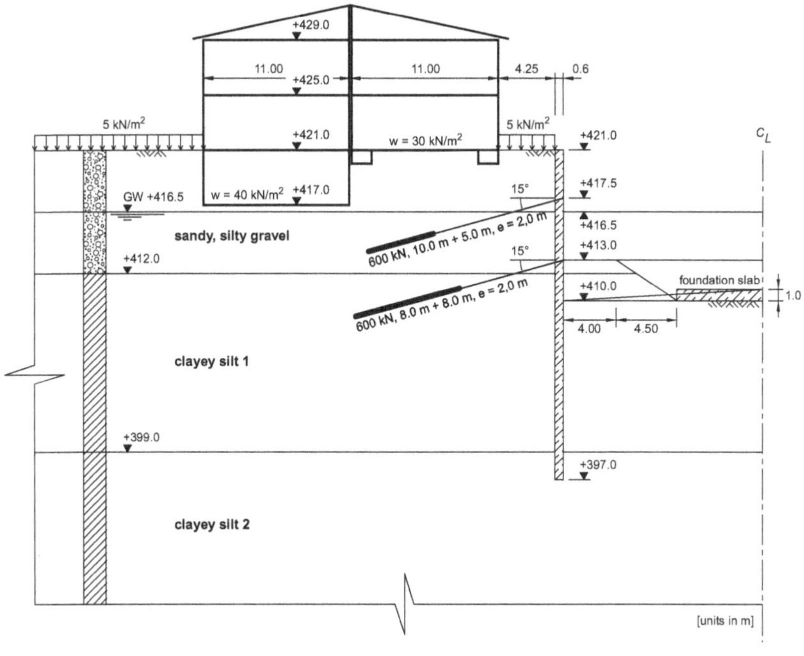

Figure 3. Geometry and subsoil conditions for deep excavation example.

Subsoil conditions and material parameters

The behaviour of the subsoil is characterised by soil parameters established from a number of laboratory and in situ tests. In order to assess the applicability of the proposed approach in practical design situations only data available before the excavation started has been used. Of particular significance for the deformation behaviour of the soft-plastic clayey silt is the deformation modulus E_s (denoted as E_{oed} in the following), gained from one-dimensional compression tests on undisturbed soil samples after pre-loading with the in situ stress of the relevant depth. The full set of parameters used in the analysis performed with the finite element code PLAXIS V8 (Brinkgreve, 2000) and the so-called Hardening-Soil model (HS) is not given here (see Peschl, 2004), only the basic variables used in the random set model are presented in the following.

Basic variables for the random set model

The material parameters for the soil layers which were treated as basic variables are summarised in Table 1. The parameters were established not only from laboratory and in situ tests (geotechnical report) but also from previous experience of finite element analyses under similar conditions (expert knowledge). The table highlights the wide range of certain parameters which in itself to some degree contain engineering judgement of people involved, e.g. a significant capillary cohesion has been assigned to the sandy, silty gravel in the geotechnical report. Due to the fact that the expert knowledge is based on a number of similar projects, the geotechnical report however forms the basis for design, both sources have been equally weighted in this particular case.

Table 1. Basic variables for material parameters (input values).

Soil Information source	c kN/m^2	φ' °	E_s MN/m^2
Sandy, silty gravel			
Geotechnical report	0 - 50·0	35·0 - 37·0	20·0 - 35·0
Expert knowledge	0 - 5·0	34·0 - 36·0	30·0 - 50·0
Clayey silt 1			
Geotechnical report	0 - 20·0	22·0 - 30·0	5·0 - 25·0
Expert knowledge	10·0 - 30·0	24·0 - 28·0	20·0 - 40·0
Clayey silt 2			
Geotechnical report	0 - 20·0	22·0 - 29·0	20·0 - 30·0
Expert knowledge	10·0 - 30·0	24·0 - 28·0	30·0 - 40·0

Two published sources of information were available and these interval estimates were combined using the averaging procedure in Equation 12. As an example the random sets for the effective friction angle φ' and the stiffness E_{oed} of the gravel layer are depicted in Figure 4. Spatial correlation is taken into account in a simplified manner in qualitatively the same way as proposed by Vanmarcke (1983). It reduces the aleatory type of uncertainty of a specific parameter but does not affect the epistemic uncertainty (Peschl, 2004). Most values for the vertical spatial correlation length Θ for sandy gravel materials and clayey silt deposits recorded in the literature are in the range of about 0.5 up to 5.0m. Generally speaking the uncertainty of the spatial correlation length should also be modelled as a random set but this has not been done here for the sake of simplicity and a value of 2.5m has been assumed. The characteristic length, L, has been taken as 55m, which is based on analyses investigating potential failure mechanisms for this problem.

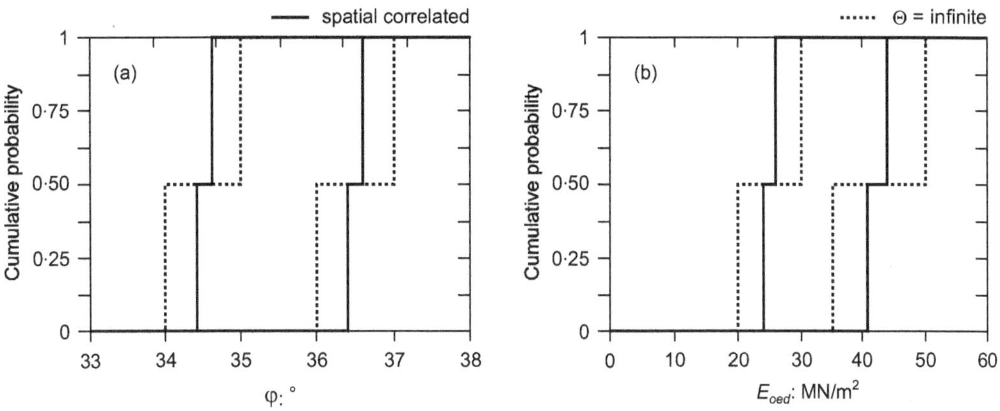

Figure 4. Random sets of input parameters of the gravel layer: (a) friction angle and (b) stiffness parameter for HS model.

Construction steps modelled

The analyses performed were calculated as 2D plane strain problems and do not consider 3D-effects. It could be reasonably assumed that consolidation effects do not play a significant role for the excavation-induced movements and therefore an undrained analysis in terms of effective stresses was performed for the clayey silt layers. The computational steps have been defined as follows (according to the real construction process):

1. Initial stresses
2. Activation of buildings, reset displacements after this step
3. Construction of the diaphragm wall
4. First excavation step to level 417.5m (level of ground surface is 421.0m)
5. Pre-stress of first anchor row with 300kN/m
6. Lowering of the groundwater table to level 413.0m
7. Second excavation step to level 413.0m
8. Pre-stress of second anchor row with 300kN/m
9. Lowering of the groundwater table to level 410.0m
10. Centre excavation to level 410.0m, leaving berm in front of the wall and replace material of the berm
11. Final excavation removing berm and construction of the foundation slab
12. φ/c-reduction in order to obtain a factor of safety for the final excavation

Calculation Results

Before the random set analysis as described previously is performed a sensitivity analysis quantifying the influence of each variable on certain results can be made (Peschl, 2004). For the 9 variables shown in Table 1, 37 calculations are required to obtain a sensitivity score for each variable. In this case the horizontal displacement of the top of the diaphragm wall, u_x, the angular distortion, δ/l, of the adjacent building at each construction step and the safety factor, determined by means of the so-called φ/c-reduction technique (e.g., Griffiths, 1980), at the final construction step is evaluated.

Figure 5 shows the total relative sensitivity and it can be seen that just three of the variables (c and E_s of the gravel layer and E_s of the upper clayey silt layer) have a significant influence on the evaluated results. At this point a decision can be made which variables should be used in further calculations and which can be treated as deterministic values as their influence on the result is not significant. The defined threshold value has been chosen in this case as approximately 5%. Based on the results of the sensitivity analysis the following parameters were considered in the random set model: cohesion for the sandy, silty gravel layer and the stiffness parameters E_{oed}, E_{50} and E_{ur} (but these are correlated) for the sandy, silty gravel and the upper clayey silt layer, i.e. 64 calculations are required (Equation 11). These stiffness parameters are required as input for the Hardening-Soil model. E_{oed} is a tangent modulus from one-dimensional compression tests, E_{50} a secant modulus from triaxial compression and E_{ur} is the unloading/reloading modulus.

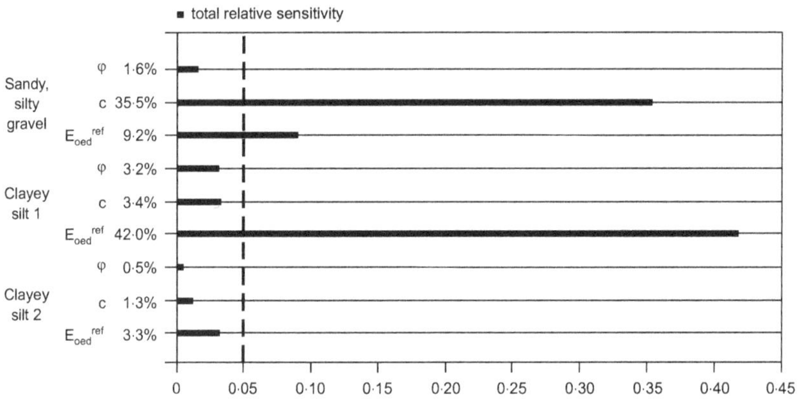

Figure 5. Total relative sensitivity for the deep excavation problem.

Serviceability limit state

The angular distortion δ/l of a structure, with δ being the differential settlement and l the corresponding length, is often used as measure to assess the likelihood of damage. The value tolerated depends on a number of factors such as the structure of the building, the mode of deformation and of course the purpose the building has to serve. A ratio of about 1:600 is used here as a limiting value for the evaluation of the limit state function in order to obtain the reliability in terms of serviceability. Figure 6 depicts the calculated cumulative distribution functions (*CDF*) of the angular distortion δ/l after prestressing of the first anchor row (Fig. 6(a)) and after the second excavation step (Fig. 6(b)). These discrete *CDF*'s were fitted using best-fit methods in order to achieve a continuous function (dotted line in Fig. 6). It is acknowledged that this fitting procedure can be questioned but it is required for the evaluation of the limit state function by means of Monte-Carlo simulations. However, if results are used qualitatively rather than quantitatively valuable information can be extracted despite the simplifications involved. For example the probabilities of exceeding a limiting value of angular distortion are obtained as given in Table 2 for all construction steps. In this particular case the calculated probabilities clearly indicate that damages of the adjacent building can be expected already during the second excavation step by showing a significant jump in the probability of failure (the accuracy of the actual value is of no relevance) and continues to be critical throughout the following construction steps. This is well in line with the observed behaviour where indeed cracks occurred during the second excavation phase.

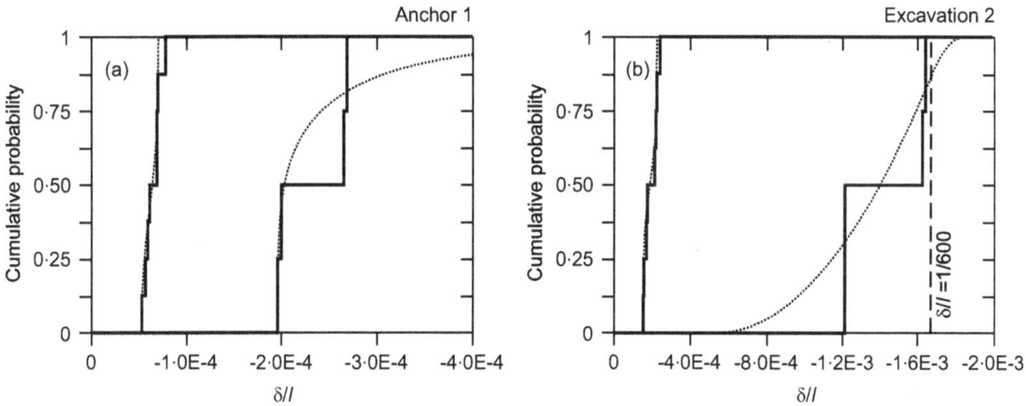

Figure 6. Range of angular distortion δ/l after: (a) first anchor row and (b) second excavation step.

Table 2. Range of probability that $\delta/l \geq 1/600$.

Construction step	**Fitted distribution**		**max pf**	**min pf**
	Upper bound	Lower bound	-	-
Excavation 1	Beta	Gamma	0	0
Anchor 1	Gamma	Beta	3.0E-5	0
Excavation 2	Triangular	Beta	1.3E-1	0
Anchor 2	Gamma	Beta	2.2E-1	0
Excavation 3	Exponential	Beta	7.2E-1	0
Excavation 4	Beta	Beta	9.7E-1	0

In general, the most likely values are defined as values with the highest probability of occurrence, i.e. where the slope of the corresponding cumulative distribution function is steepest. For the purpose of illustration, it is assumed that the most likely values have a probability of 50% as shown in Figure 7, but of course other definitions are possible.

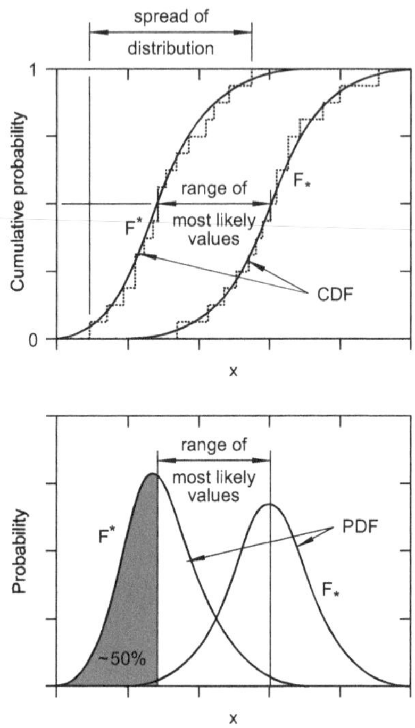

Figure 7. Range of most likely values and spread of distribution.

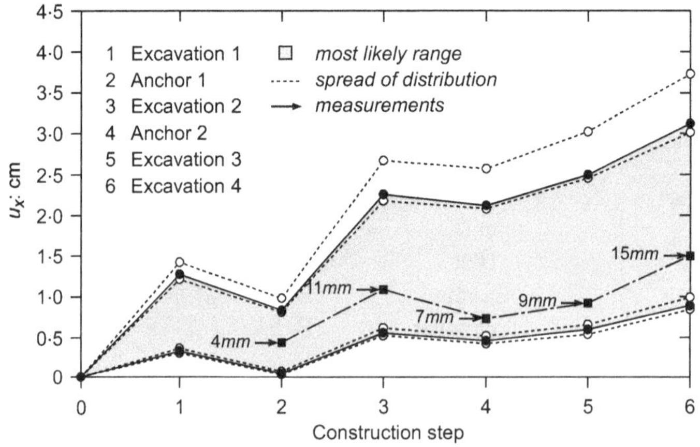

Figure 8. Horizontal displacements of the top of the diaphragm wall (most likely values and spread of distributions).

In situ measurements have been performed in order to control the displacements of the top of the wall. Figure 8 plots these measurements together with calculated values for all construction steps and it can be seen, that measured values compare well to the results of the numerical prediction presented as most likely values including the spread of distributions as defined above. At this stage two conclusions can be made: firstly it follows very clearly from Figure 8 that an analysis with very 'cautious' estimates for soil parameters, represented by the upper limit in Figure 8, grossly overestimates deformations under working load conditions. Secondly, such an analysis can be used to determine the level of soil investigations required in order to arrive at a bandwidth of results which is acceptable. Additional information will usually change the random set, i.e. the *Bel* and *Pl* functions move closer together and therefore the most likely range of results also becomes narrower.

4 Conclusion

Reliability analysis in engineering conventionally represents the uncertainty of the system state variables as precise probability distributions and applies probability theory to generate precise estimates of e.g. the probability of failure or the reliability. However, it has been recognised that traditional probability theory may not capture the full scope of uncertainty (inherent variability and lack of knowledge). Random set theory offers an alternative to traditional probabilistic theory to overcome these shortcomings. The significant innovation of the proposed framework is that it allows for the allocation of a probability mass to sets or intervals and provides a consistent framework for dealing with uncertainties throughout the design and construction of a project, because the model can be refined by adding more information when available depending on the project status (feasibility stage, preliminary design, detailed design, construction) without changing the underlying concept of analysis. It can be used for defining the level of information required on model parameters by specifying acceptable ranges of key results.

As a side effect worst case assumptions in terms of unfavourable parameter combinations have not to be estimated from experience but are automatically generated. The argument that engineering judgement will do the same much faster is not entirely true because in complex non-linear analyses, which become more and more common in practical geotechnical engineering, the parameter set for a highly advanced constitutive model leading to the most unfavourable result in terms of serviceability and ultimate limit state for all construction stages is not easy to define. From a practical point of view the concept of dealing with 'ranges' seems to be more appealing for engineers than working in density distribution functions.

The applicability of the proposed method for solving practical boundary value problems has been shown by analysing the excavation sequence for a deep excavation in soft soil presenting comparison with field measurements.

References

Breymann, H., Moser, M. and Premstaller, M. (2003). Die Gründung des Kongresshauses Salzburg – Konsequenzen aus dem Verformungsverhalten des Salzburger Seetons. *Felsbau* 21 (5). VGE: Essen. 10-17.

Brinkgreve, R. B. J. (2000). *PLAXIS, Finite element code for soil and rock analyses*. Users manual, Rotterdam: Balkema.

Dempster, A. P. (1967). Upper and lower probabilities induced by a multivalued mapping. *Annals of Mathematical Statistics* 38: 325-339.

Dong, W. and Shah, H. C. (1987). Vertex method for computing functions of fuzzy variables. *Fuzzy Sets & Systems* 24: 65-78.
Dubois, D. and Prade, H. (1990). Consonant approximation of belief functions. *Int. J. of Approximate Reasoning 4*: 419-449.
Dubois, D. and Prade, H. (1991). Random sets and fuzzy interval analysis. *Fuzzy Sets and Systems* 42: 87-101.
Ferson, S., Hajagos, J. G., Berleant, D., Zhang, J., Tucker, W. T., Ginzburg, L. and Oberkampf, W. (2004). *Dependence in Dempster-Shafer theory and probability bounds analysis*. Sandia Report SAND2004-3072. Sandia National Laboratories: Albuquerque.
Fetz, T. and Oberguggenberger, M. (2004). Propagation of uncertainty through multivariate functions in the framework of sets of probability measures. *Reliability Engineering and Systems Safety* 85: 73-88.
Griffiths, D. V. (1980). Finite element analyses of walls, footings and slopes. *Proc. Symp. on Comp. Applic. to Geotech. Probs. in Highway Eng.* Pub. PM Geotechnical Analysts Ltd.: Cambridge: U.K. 122-146.
Hall, J. W. and Lawry, J. (2004). Generation, combination and extension of random set approximations to coherent lower and upper probabilities. *Reliability Engineering & System Safety* 85: 89-101.
Helton, J. C. (1997). Uncertainty and Sensitivity Analysis in the Presence of Stochastic and Subjective Uncertainty. *Journal of Statistical Computation and Simulation* 57: 3-76.
Peschl, G. M. (2004). *Reliability Analyses in Geotechnics with the Random Set Finite Element Method.* Ph.D. Dissertation Graz. University of Technology. Institute for Soil Mechanics and Foundation Engineering,
Rackwitz, R. (2000). Reviewing probabilistic soils modelling. *Computers and Geotechnics* 26: 199-223.
Sentz, K. and Ferson, S. (2002). *Combination of evidence in Dempster-Shafer theory*. Report SAND2002-0835. Sandia Nat. Laboratories: Albuquerque.
Shafer, G. (1976). *A Mathematical Theory of Evidence*. Princeton: Princeton University Press.
Tonon, F., Bernardini, A. and Mammino, A. (2000a). Determination of parameters range in rock engineering by means of Random Ret Theory. *Reliability Engineering and System Safety* 70: 241-261.
Tonon, F., Bernardini, A. and Mammino, A. (2000b). Reliability analysis of rock mass response by means of Random Set Theory. *Reliability Engineering and System Safety* 70: 263-282.
Vanmarcke, E. H. (1983). *Random Fields – Analysis and Syntesis*. Cambridge, Massachusetts: MIT-Press.

On Some Aspects of Reliability Computations in Bearing Capacity of Shallow Foundations

Wojciech Puła

Institute of Geotechnics and Hydrotechnics, Wrocław University of Technology, Wrocław, Poland

Abstract. The chapter deals with bearing capacity of spread foundations in the context of reliability computations. In the first part evaluations base on the recommendations given by Polish Standard PN-81/B-03020 (1981). Consequently some most important ideas concerning bearing capacity suggested by this standard are presented and compared with analogical statements of *Eurocode* EC7 (1997). Next some reliability computations are carried out under an assumption that each individual soil property is modelled by a single random variable throughout the earth body considered. But such approach seems to be too simple. In order to evaluate credible reliability indices when bearing capacity of a shallow foundation is considered it is reasonable to describe soil strength properties in terms of random field's theory. As a next step the selected random field can be spatially averaged by means of a procedure introduced by Vanmarcke (1977). Earlier experiences have proved that, without applying spatial averaging procedure, reliability computations carried out in the context of foundation's bearing capacity had given significantly small values of reliability indices (large values of failure's probability) even for foundations which were considered as relatively safe. On the other hand the volume of the area under averaging strongly affects results of reliability computations. Hence the selection of the averaged area constitutes a vital problem and has to be dependent on the failure mechanism under consideration. In the present study local averages associated with kinematically admissible mechanism of failure proposed by Prandtl (1920) are considered. Soil strength parameters are assumed to constitute anisotropic random fields with different values of vertical and horizontal fluctuation scales. These fields are subjected to averaging along potential slip lines within the mechanism under consideration. Next examples of equations for variances of random variables resulting from averaging procedure are shown. By numerical examples it is demonstrated that for reasonable proportions (from practical viewpoint) between horizontal and vertical fluctuation scales the reliability indices resulting in two-dimensional case only slightly differs from resulting that obtained in one-dimensional. This means that the simpler one-dimensional approach can be usually utilised when reliability measures of shallow strip foundation are carried out.

1 Basic Assumptions and Explanations

Within this chapter the focus will be oriented towards spread foundations. In this case the assumption that the ground resistance on the sides of the foundation does not contribute significantly to the bearing capacity resistance will be accepted. Moreover only the case of drained resistance will be considered. On the beginning some examples basing on the Polish Standard PN-81/B-03020. *Foundation bases. Static computations and design* will be presented. Therefore some equations proposed by this Standard are firstly presented. For comparison some analogical equations suggested by *Eurocode* EC7 are shown.

The bearing capacity can be evaluated by means of the following formula, which appears both in the PN-81/B-03020 (1981) and the *Eurocode* EC7 (1997):

$$Q_f = \overline{B}\overline{L}\left[N_c c i_c s_c + N_q q \gamma_q i_q s_q + \frac{1}{2} N_\gamma \gamma_\gamma i_\gamma \overline{B} s_\gamma \right] \tag{1}$$

where N_q, N_c, *and* N_γ are the bearing capacity factors defined by EC and PN as:

$$N_q = \exp(\pi \tan\varphi)\tan^2\left(\frac{\pi}{4} + \frac{\varphi}{2}\right) \tag{2}$$

$$N_c = \left(N_q - 1\right)\cot\varphi \tag{3}$$

$$\text{EC7:} \quad N_\gamma = 2\left(N_q - 1\right)\tan\varphi \qquad \text{PN:} \quad N_\gamma = 1.5\left(N_q - 1\right)\tan\varphi \tag{4}$$

Equation (1) utilizes effective dimensions of the foundation, namely

$$\overline{B} = B - e_B \qquad \text{and} \qquad \overline{L} = L - e_L \tag{5}$$

with e_B and e_L eccentricity with respect the width B and the length L, respectively. The shape coefficients (in the case of rectangular shape of foundation) s_q, s_γ, s_c differs significantly in the EC7 and the PN, namely:

$$\text{EC7:} \qquad s_q = 1 + \left(\frac{\overline{B}}{\overline{L}}\right)\sin\varphi, \quad s_\gamma = 1 - 0.3\left(\frac{\overline{B}}{\overline{L}}\right), \quad s_c = \frac{s_q N_q - 1}{N_q - 1} \tag{6}$$

$$\text{PN:} \qquad s_q = 1 + 1.5\left(\frac{\overline{B}}{\overline{L}}\right), \quad s_\gamma = 1 - 0.25\left(\frac{\overline{B}}{\overline{L}}\right), \quad s_c = 1 + 0.3\left(\frac{\overline{B}}{\overline{L}}\right) \tag{7}$$

Finally the load inclination coefficients i_q, i_γ, i_c are, due to correction suggested by Orr and Farrel (1999), take the form as in the German standard DIN 4017 (1979)

$$i_q = \left[1 - \frac{H}{V + \overline{B}\overline{L} c \cot\varphi}\right]^{m_1}, \qquad i_\gamma = \left[1 - \frac{H}{V + \overline{B}\overline{L} c \cot\varphi}\right]^{m_1 + 1}, \qquad i_c = i_q - \frac{1 - i_q}{N_c \tan\varphi}, \tag{8}$$

where m_1 takes the value

$$m_1 = m_B = \frac{\left[2 + \left(\frac{\overline{B}}{\overline{L}}\right)\right]}{\left[1 + \left(\frac{\overline{B}}{\overline{L}}\right)\right]} \tag{9}$$

when H acts in the direction of B, or takes the value

$$m_1 = m_L = \frac{\left[2 + \left(\frac{\overline{L}}{\overline{B}}\right)\right]}{\left[1 + \left(\frac{\overline{L}}{\overline{B}}\right)\right]} \tag{10}$$

when *H* acts in the direction of *L*. In the case of Polish Standard PN-81/B-03020 (1981) load inclination coefficients are not given in the closed form and they should be determined with the help of nomograms. However, in practical cases resulting values are not far from that given by equations (8).

2 Assumptions Concerning Reliability Computations

Structural reliability problems are usually described by the so-called limit state function $g(\mathbf{x})$. The argument **x** of the function g is a random vector $\boldsymbol{X}=(X_1,X_2,...,X_n)$ consisting of basic random variables defining loads, material properties, geometrical quantities, *etc.* as well as some other properties considered as deterministic. The function $g(\mathbf{x})$ is defined in the following way:

$$g(\mathbf{x}) = \begin{cases} > 0\, for the safe state of the structure \\ < 0\, for the failure state of the structure \end{cases} \tag{11}$$

The hypersurface $g(\mathbf{x}) = 0$ is called the limit state surface. Within this chapter the limit state function will be evaluated in the form of

$$g = mQ_f - N \quad , \tag{12}$$

Where *m* is the model uncertainty factor, which characterized goodness of fit between the model and reality. It can be assumed as a random variable or as a constant parameter.

As a reliability measure the probability of failure is used

$$p_F = \int_{\{g(\mathbf{x})<0\}} f_{\mathbf{X}}(\mathbf{x})d\mathbf{x} \tag{13}$$

As a equivalent measure the reliability index β will be applied. Both measures are related each other by the equation

$$p_F = \Phi_0(-\beta) \quad , \tag{14}$$

provided that $p_F < 0.5$, where Φ_0 is the standard normal one-dimensional distribution function. In the framework of this chapter as a computational tool to evaluate the probability of failure as well as the reliability index the SORM method (see e.g., Hohenbichler et al, 1987 or Ditlevsen and Madsen, 1995).

In order to evaluate the influence of individual parameters on the reliability index β, some sensitivity parameters α_i can be defined as follows:

$$\alpha_i = \frac{1}{\left\| \mathbf{y}^* \right\|} \frac{\partial \beta}{\partial y_i} \bigg|_{\mathbf{y}=\mathbf{y}^*} , \tag{15}$$

where $\mathbf{y}^*$ is so called "design point", which determines the index β in the FORM method (see e.g. Ditlevsen and Madsen, 1995).

3 Examples of Reliability Assessments of Shallow Strip Foundation

Consider a shallow strip foundation as presented in Figure 1. Some soil properties as well as loads have been assumed as random variables. These are specified in the Table 1. The bearing capacity has been evaluated according to equation (1). For the case of strip foundation it can be accepted that the shape coefficients s_q, s_γ, s_c are all equal to one. The inclination coefficients have been assumed in the form given by equation (8). Other coefficient in Equation (1) took the form according to the PN-81/B-03020 (1981) as well as the model uncertainty factor $m = 0.9$.

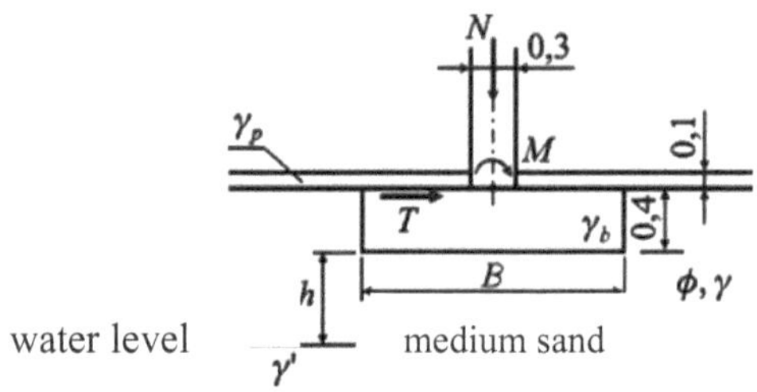

Figure 1. Scheme of shallow strip foundation considered in the example.

Reliability computations were carried out by means of the SORM method. Resulting reliability indices, for several dimensions of the base width, are shown in the Table 2.

It turned out that that minimal width necessary to carry the acting load is $B = 2.3$ m (with the tolerance of 0.05 m). This width corresponds to the value of reliability index $\beta = 1.63$, which seems to be rather small. At same time the ISO 2394 (1998) code suggests beta values equal to $\beta = 3.1$ for small, $\beta = 3.8$ for moderate and $\beta = 4.3$ for large failure consequences. Thus, the value obtained in our example is significantly smaller than any of the recommended values. On the other hand, long practice in applying the PN-81/B-03020 (1981) code shows that that foundations designed according to them can be considered as very safe. Note that in the above example each soil property was modelled by a single random variable without spatial averaging.

Next sensitivity of the reliability index β was examined with respect to soil friction angle coefficient of variation as well as its type of probability distribution. Obtained values of the index β versus the foundation width B for several values of coefficient of variation of the soil friction angle are shown in Figure 2.

Table 1. Probabilistic characteristic of soil and loads parameters.

Soil property	Mean value	Standard Deviation σ_X	Probability distribution	Sensitivity Parameters α
Soil friction angle φ	32°	4.6°	lognormal	0.973
Soil Unit weight γ	18.2 kN/m^3	1.092 kN/m^3	normal	0.008
Concrete floor unit weight γ_p	23.0 kN/m^3	1.38 kN/m^3	normal	0.001
Soil Unit weight under water level γ'	9.8 kN/m^3	0.588	normal	0.024
Unit weight of foundation material γ_b	24.0 kN/m^3	-	nonrandom	
Ground water level h	1.00 m	0.06 m	uniform	0,021
Axial load normal to the base N	300 kN	45.0 kN	lognormal	-0.205
Load tangent to the base T	20.0 kN/m	3.0 kN/m	lognormal	-0.055
Moment M	15,0 kNm/m	2.25 kNm/m	lognormal	-0.012

Table 2. Selected values of reliability measures obtained in the example.

Width of foundation B [m]	Reliability index β	Probability of failure p_F
1.8	1.01	0.15720
2.2	1.51	0.06514
2.3	1.63	0.05147
2.4	1.75	0.04049
2.8	2.17	0.01493
3.2	2.56	0.00523
3.6	2.92	0.00176
4.0	3.25	0.00058

The effect of various probability distributions illustrates Figure 3. The vital role of the friction angle φ variability for β-index values can be recognised by observing the sensitivity parameters in the last column of the Table 1.Figure 1 show that the appropriate selection of the coefficient of variation of φ can be essential for the bearing capacity reliability computations. On the other hand selection of probability distribution type is a less important factor, provided not large foundation width.

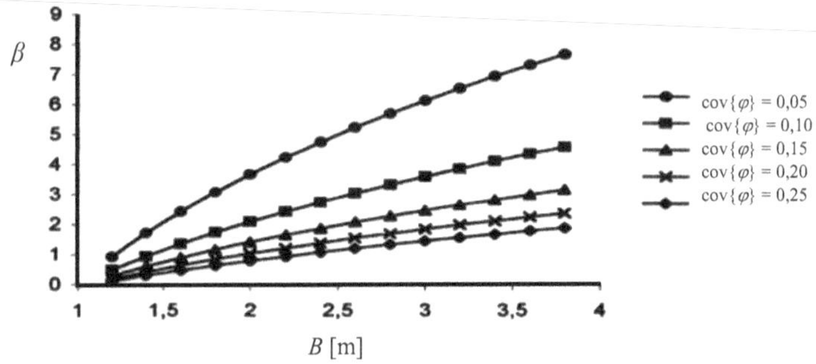

Figure 2. Reliability index β versus width of the foundation B and variation coefficient of φ.

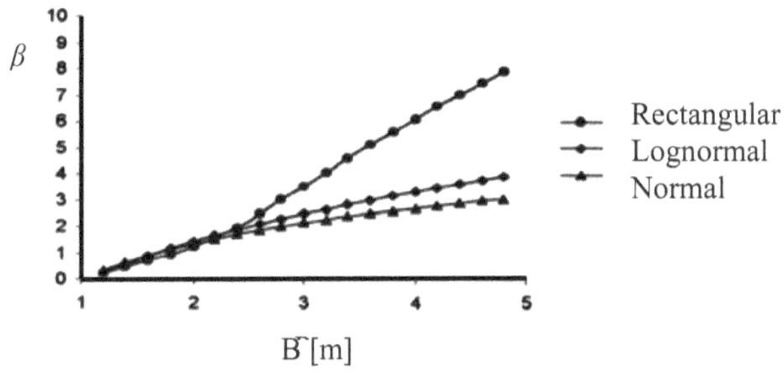

Figure 3. Index β versus width of the foundation B for three different probability distribution of φ.

An effect of mutual correlation between soil properties: the friction angle φ and the unit weight γ is demonstrated in Figure 4. It can be seen that the positive correlation can be almost negligible.

On the other hand a correlation effect can be important when a we deal with a cohesive soil. A cohesive soil strength parameters (friction angle φ and cohesion c) are usually treated as negatively correlated. In this case the value of the correlation coefficient gives remarkable influence on the reliability index β. An example of results obtained for a shallow strip foundation resting on clay is shown in Figure 5.

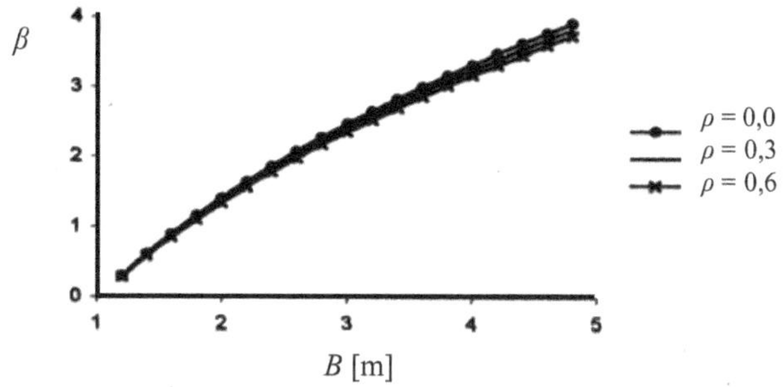

Figure 4. Reliability index β versus width of the foundation B for three different correlation coefficients of φ and γ. Both random variables φ and γ are lognormally distributed with expected values $E\{\varphi\} = 18^{\circ}$ and $E\{c\} = 31$ kPa for φ and γ, respectively, and coefficients of variations 0,15 in both cases.

4 Spatial averaging application

Note that in the above example each soil property was modelled by a single random variable without spatial averaging. Consider now again the above example (with characteristics given in Figure 1 and Table 1) but assume additionally that the random field of the internal friction angle of soil under the base of foundation is subject of one-dimensional (with depth) spatial averaging.

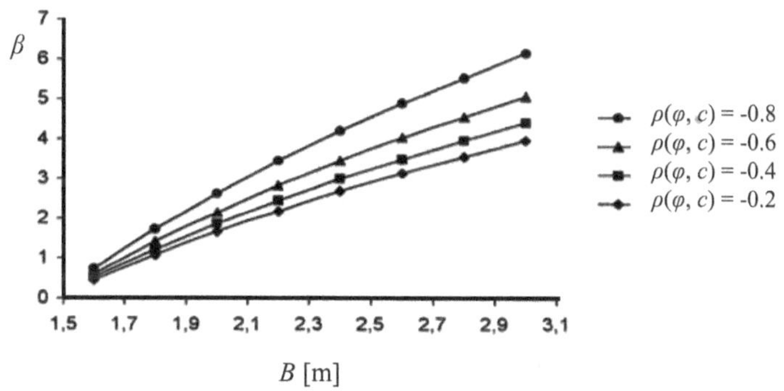

Figure 5. Reliability index β versus width of the foundation B for four different correlation coefficients of φ and c in clay.

The spatial averaging, proposed by Vanmarcke (1977), introduces a new random field (moving average random field) defined by the following equation:

$$\varphi_L = \frac{1}{L} \int_L \varphi(z) dz \tag{16}$$

where $\varphi(z)$ is the random function, which describes random variability of the friction angle φ with the depth z and L is the volume (length) of averaging. The function $\varphi(z)$ is assumed to be stationary with constant mean value m_φ and constant point variance σ_φ^2. The variance of φ_L can be computed as:

$$\mathrm{VAR}[\varphi_L] = \sigma_L^2 = \gamma(L)\sigma_\varphi^2, \tag{17}$$

where $\gamma(L)$ is the variance function. Let us assume now the following form of the variance function:

$$\gamma_3(L) = \frac{\frac{\pi}{\delta} L \cdot \mathrm{erf}\left(\frac{\sqrt{\pi}}{\delta} L\right) - 1 + \exp\left(-\pi\left(\frac{L}{\delta}\right)^2\right)}{\pi\left(\frac{L}{\delta}\right)^2} \tag{18}$$

where

$$\mathrm{erf}(t) = \frac{2}{\sqrt{\pi}} \int_0^t \exp(-x^2)dx . \tag{19}$$

The variance function given by (18) corresponds to the Gaussian correlation function of the random field, namely

$$\rho(\Delta z) = \exp\left(-\pi\left(\frac{\Delta z}{\delta}\right)^2\right), \tag{20}$$

which widely spread in many application of random fields (Rackwitz 2000). The parameter δ is a fluctuation scale (correlation length) and is assumed as $\delta = 0.8$ m. The size of averaging L is now a subject of parameter study and is selected as equal to double width of the foundation ($L = 2B$). Results of computations of reliability indices β are shown in Figure 6.

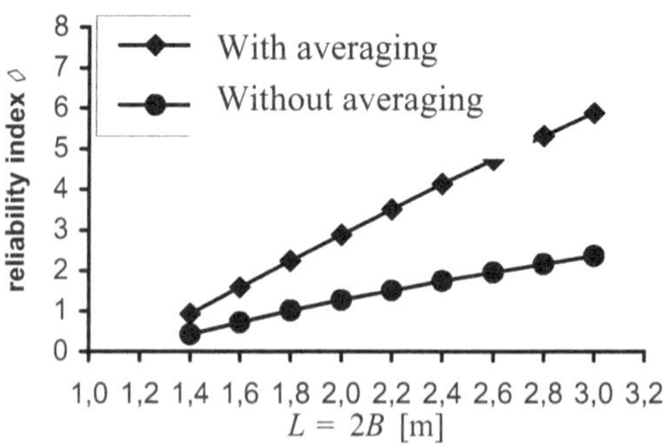

Figure 6. Reliability indices computed without spatial averaging and with the spatial averaging.

Comparing curves in Figure 2 one finds a great influence of the spatial averaging on β index values. For example for the above mentioned width $B = 2.3$ m β index increases from $\beta = 1.63$ to $\beta = 3.83$. The last value satisfies first two of three threshold values specified by ISO 2394 (1998) and mentioned in the section 3. The above example demonstrates that effect of inadequate β index values can be reduced by applying spatial averaging of fields of soil strength parameters.

However, the area of averaging L , which is in fact the argument of the variance function, was selected arbitrarily as equal to the double width of foundation $2B$. The selection could be justified that changes in subsoil leading to the failure are mainly located to the depth $2B$ under the base of foundation. Let us observe now what will happen if we replace the size of averaging $L = 2B$ by $L = B$. The results are demonstrated in Table 3.

It is evident that size decreasing of averaging significantly reduces reliability indices. This is also confirmed by the graphs in Figure 7, presenting values of reliability index as a function of fluctuation scale δ for three types of variance function: $\gamma_3(L)$ – given by eqn. (18), $\gamma_1(L)$

$$\gamma_1(L) = \begin{cases} 1 & for \quad L \le \dfrac{\delta}{2} \\ \dfrac{\delta}{L}\left(1 - \dfrac{\delta}{4L}\right) & for \quad L > \dfrac{\delta}{2} \end{cases} \tag{21}$$

and $\gamma_2(L)$

$$\gamma_2(L) = \begin{cases} 1 - \dfrac{L}{3\delta} & dla \quad L \le \delta \\ \dfrac{\delta}{L}\left(1 - \dfrac{\delta}{3L}\right) & dla \quad L > \delta \end{cases} \tag{22}$$

Table 3. Selected values of reliability indices obtained with the averaging and without averaging for two different sizes of averaging area

Width of foundation b [m]	Reliability index β without averaging	Reliability index β averaging $L = B$	Reliability index β averaging $L = 2B$
1.8	1.01	1.71	2.24
2.2	1.51	2.71	3.51
2.3	1.63	2.96	3.83
2.4	1.75	3.20	4.13
2.8	2.17	4.17	5.31
3.2	2.56	5.11	6.43
3.6	2.92	6.00	7.48
4.0	3.25	6.86	8.47

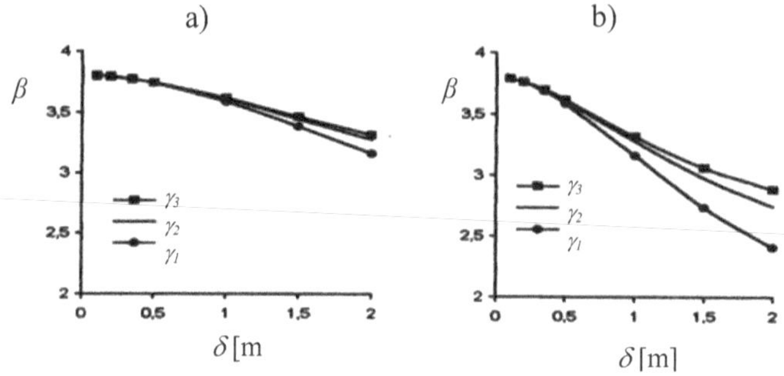

Figure 7. Reliability index β as a function of fluctuation scale values for three different variance functions. Figure a) shows results with spatial averaging of averaging size $L = 2B$. Figure b) shows results with spatial averaging of averaging size $L = B$.

Due to its definition the variance function depends both on scale of fluctuation δ and the averaging area L. The fluctuation scale δ can be considered as soil parameter and can be determined by field and can be determined by field and testing. Opposite to δ, none testing could gives the size of L. Consequently to avoid the loss of uniqueness in reliability computations the L must be carefully selected and precisely defined among assumptions for a problem under consideration.

It seems that to get proper values of variance reduction in bearing capacity problems it is necessary to carry out the spatial averaging along potential slip surfaces associated with a mechanism of failure. Below a kinematically admissible Prandtl's mechanism (Prandtl 1920) for shallow strip foundation is considered.

5 Averaging Along Potential Slip Surfaces

Assume that strength parameters, φ (angle of internal friction) and c (cohesion), constitute two-dimensional log-normal random field with the following covariance function (Gaussian):

$$R(\Delta x, \Delta z) = \sigma_X^2 \exp\left\{-\left[\left(\frac{\Delta z}{\omega_1}\right)^2 + \left(\frac{\Delta x}{\omega_2}\right)^2\right]\right\} , \tag{23}$$

where the parameters ω_1 and ω_2 are proportional to the fluctuation scales, viz.

$$\omega_1 = \frac{\delta_v}{\sqrt{\pi}} \qquad \omega_2 = \frac{\delta_h}{\sqrt{\pi}} , \tag{24}$$

and δ_v, δ_h denote the fluctuation scale in the vertical and horizontal direction, respectively. For distinct values of ω_1 and ω_2 the fields are no longer isotropic. Consider now a shallow strip foundation and a kinematically admissible Prandtl's mechanism (Prandtl, 1920) as shown in Figure 8. In further consideration the foundation width will be denoted by *b*.

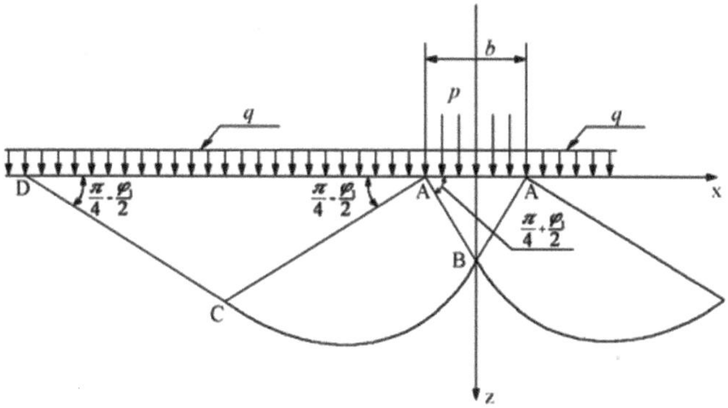

Figure 8. Schematic presentation of the Prandtl mechanism.

In the case discussed here, the averaging has to be carried out along potential slip lines of soil masses. By assuming a parametric representation of the slip line in the form

$$l_i: \quad \begin{matrix} x = x(t_i), \quad z = z(t_i) \\ dla \quad t_i \in [a_i, b_i] \end{matrix} \quad , i = 1,2,3 \tag{25}$$

(index *i* denotes the *i*-th line), the following equation for the covariances is obtained (see Rackwitz 2000, Puła 2002)

$$\mathrm{Cov}(X_{l_i}, X_{l_j}) = \frac{1}{|l_i||l_j|} \int_{l_i} \int_{lj} R(x_1, z_1, x_2, z_2) dl_i dl_j = \frac{1}{|l_i||l_j|} \int_{a_i}^{b_i} \int_{a_i}^{b_j} R(x(t_i), z(t_i), x(t_j), z(t_j)) \times$$
$$\times \sqrt{\left[\frac{dx}{dt_i}\right]^2 + \left[\frac{dz}{dt_i}\right]^2} \sqrt{\left[\frac{dx}{dt_j}\right]^2 + \left[\frac{dz}{dt_j}\right]^2} dt_i dt_j \tag{26}$$

In the particular case, when $i = j$ the representation (26) corresponds to the variance of VAR(X_{l_i}). Referring to the equation (26), there is $l_1 = AB, l_2 = BC, l_3 = CD$.

In order to evaluate the probability of failure, or equivalent reliability index, it is necessary to derive equations for the variances and covariances of random variables which result from the averaging process. The averaging is performed separately along each segment of the potential slip line, namely AB, BC, CD. The final expressions are provided below. Note that all these equations have been derived starting from equation (26), utilising the covariance function (23), and employing parametric equations of each segments AB, BC, CD. In order to avoid complex mathematical transformations and details, only final formulae are presented here (for details see Puła 2004).

The variance of X_{AB}:

$$\mathrm{Var}\{X_{AB}\} = \frac{\sigma_X^2}{h_B} \left(\omega_{0B} \sqrt{\pi}\, \mathrm{erf}\left(\frac{h_B}{\omega_{0B}}\right) + \frac{\omega_{0B}{}^2}{h_B} \exp\left(-\frac{h_B^2}{\omega_{0B}{}^2}\right) - \frac{\omega_{0B}{}^2}{h_B} \right) , \tag{27}$$

where

$$h_B = \frac{b}{2}\operatorname{tg}\left(\frac{\pi}{4}+\frac{\varphi}{2}\right), \quad \omega_{0B} = \sqrt{\frac{\omega_1^2\omega_2^2a^2}{\omega_2^2a^2+\omega_1^2}} \quad \text{and} \quad a = \tan\left(\frac{\pi}{4}+\frac{\varphi}{2}\right). \tag{28}$$

It is easy to see that

$$\lim_{\omega_2\to\infty}\omega_{0B} = \omega_1, \tag{29}$$

in which case eqn. (27) reduces to the solution obtained for one-dimensional case (see Puła 2002).

The variance of X_{CD}. The variance of the random variable X_{CD} can be obtained by means of the equation (10), by substituting h_C for h_B and ω_{0C} for ω_{0B} in accordance to the following equations

$$h_C = \frac{b}{2}\exp\left(\frac{\pi}{2}tg\varphi\right) \text{ and } \omega_{0C} = \sqrt{\frac{\frac{\omega_1^2\omega_2^2}{a^2}}{\frac{\omega_2^2}{a^2}+\omega_1^2}} = \sqrt{\frac{\omega_1^2\omega_2^2}{\omega_2^2+a^2\omega_1^2}} \tag{30}$$

where a is the same as in the eqn. (28). As before, the following asymptotic formula holds

$$\lim_{\omega_2\to\infty}\omega_{0C} = \omega_1 \quad . \tag{31}$$

Again, one can expect now that for large values of ω_2 the reduction of variance could be very similar that that obtained in one-dimensional case.

The variance of X_{BC}. In the case of log-spiral, the following equation is obtained in a polar coordinate system:

$$\operatorname{Var}[X_{BC}] = \sigma_X^2\alpha_0 \int_{\left(\frac{\pi}{4}+\frac{\varphi}{2}\right)}^{\left(\frac{3\pi}{4}+\frac{\varphi}{2}\right)}\int_{\left(\frac{\pi}{4}+\frac{\varphi}{2}\right)}^{\left(\frac{3\pi}{4}+\frac{\varphi}{2}\right)} \left\{\exp\left[-\frac{r_0^2}{\omega^2}\left[(\sin\theta_1)\exp(\theta_1\tan\varphi)-(\sin\theta_2)\exp(\theta_2\tan\varphi)\right]^2\right]\right\}\times \tag{32}$$
$$\times\exp(\theta_1\tan\varphi)\exp(\theta_2\tan\varphi)\,d\theta_1 d\theta_2$$

where

$$\alpha_0 = \operatorname{tg}^2\varphi\frac{\exp\left[-\left(\frac{\pi}{2}+\varphi\right)(\operatorname{tg}\varphi)\right]}{\left[\exp\left(\frac{\pi}{2}\operatorname{tg}\varphi\right)-1\right]^2}, \qquad r_0 = \frac{b}{2\cos\left(\frac{\pi}{4}+\frac{\varphi}{2}\right)}\exp\left[-\left(\frac{\pi}{4}+\frac{\varphi}{2}\right)\operatorname{tg}\varphi\right] \tag{33}$$

The integral in the equation (32) can not be expressed in any closed form, hence it has to be evaluated numerically.

In the same way similar, however more complex, expressions for covariances X_{AB} and X_{BC}, X_{BC} and X_{CD} and X_{AB} and X_{CD} can be established (Puła 2004).

6 The Limit State Function

To carry out reliability computations, it is always necessary to define a limit state function f, which specifies the conditions at failure, as it was indicated in section 2 Here the limit state function is defined as:

$$f = Q_f - P\,, \tag{34}$$

where P denotes the axial force imposed on the foundation and Q_f is the bearing capacity. The bearing capacity is a sum of three components: an effect of weightless cohesive soil (Q_1); an effect of loading in the vicinity of foundation (Q_2); an effect of the self-weight of the soil (Q_3). To make the local averaging procedure possible, it is necessary to consider strength parameters of the soil along each slip line AB, BC and CD (see Figure 1). Denoting by φ_1, c_1,φ_2, c_2,φ_3, c_3 the values of these parameters along AB, BC and CD, respectively, and comparing the work of external load to the total dissipation of internal energy, the value of the limit load (bearing capacity) can be established. Final equations are as follows (Puła 2004):

$$Q = \frac{bQ_0}{\sin\left(\frac{\pi}{4} - \frac{\varphi_1}{2}\right)} \tag{35}$$

where

$$Q_0 = Q_1 + Q_2 + \frac{b}{2}\left(Q_{31} + Q_{32} + Q_{33}\right) \tag{36}$$

and

$$Q_1 = c_1 \frac{\cos\varphi_1}{2\sin\left(\frac{\pi}{4} - \frac{\varphi_1}{2}\right)} + c_2\left[\exp(\pi\tan\varphi_2) - 1\right]\frac{1}{2\sin\left(\frac{\pi}{4} - \frac{\varphi_1}{2}\right)\tan\varphi_2} + c_3\exp(\pi\tan\varphi_2)\frac{\cos\varphi_3}{2\sin\left(\frac{\pi}{4} - \frac{\varphi_1}{2}\right)} \tag{37}$$

$$Q_2 = (q + \gamma D)\exp(\pi\tan\varphi_2)\cos\left(\frac{\pi}{4} - \frac{\varphi_1}{2}\right)\frac{\cos\left(\frac{\pi}{4} + \frac{\varphi_1}{2} - \varphi_3\right)}{\sin\left(\frac{\pi}{4} - \frac{\varphi_1}{2}\right)} \tag{38}$$

$$Q_{31} = -\frac{1}{4}\gamma\cos\left(\frac{\pi}{4} - \frac{\varphi_1}{2}\right) \tag{39}$$

$$Q_{32} = \frac{\gamma}{2\left(1+9\tan^2\varphi_2\right)4\sin^2\left(\frac{\pi}{4}-\frac{\varphi_1}{2}\right)} \times$$

$$\times\left\{\left[3\tan\varphi_2\sin\left(\frac{\pi}{4}+\frac{\varphi_1}{2}\right)-\cos\left(\frac{\pi}{4}+\frac{\varphi_1}{2}\right)\right]\exp\left(\frac{3}{2}\pi\tan\varphi_2\right)+\left[3\tan\varphi_2\sin\left(\frac{\pi}{4}-\frac{\varphi_1}{2}\right)+\cos\left(\frac{\pi}{4}-\frac{\varphi_1}{2}\right)\right]\right\} \tag{40}$$

$$Q_{33} = \frac{\gamma\cos\varphi_3\exp\left(\pi\frac{3}{2}\tan\varphi_2\right)\cos\left(\frac{\pi}{4}+\frac{\varphi_1}{2}-\varphi_3\right)}{8\sin^2\left(\frac{\pi}{4}-\frac{\varphi_1}{2}\right)} \ . \tag{41}$$

In the expressions above, γ denotes unit weight of soil, q is the overburden pressure, b is the foundation width and D is the depth. Further computations of failure probabilities and equivalent reliability indices were carried out by SORM method.

7 Numerical Examples

The numerical analysis pertains to a shallow strip foundation resting on a cohesionless soil. Computations examine the influence of the foundation width b, on the results of reliability study. The simulations were carried out assuming that the vertical scale of fluctuation is constant and equal to $\delta_v = 0.8$ m. . For horizontal fluctuation scale δ_h three different values are established $\delta_h = 3\delta_v\sqrt{\pi}$, $\delta_h = 10\delta_v\sqrt{\pi}$ and $\delta_h = 30\delta_v\sqrt{\pi}$, which corresponds to the following dependences between parameters of the correlation function $\omega_2 = 3\omega_1$, $\omega_2 = 10\omega_1$ and $\omega_2 = 30\omega_1$, respectively. This assumption reflects the phenomenon that the scale of fluctuation in the case of soil properties is significantly larger in the horizontal than in vertical direction, which is reported in the literature (Cherubini 1997, 2000). For comparison a one-dimensional averaging is considered employing the vertical fluctuation scale only. The slip surfaces in the Prandtl's mechanism have been determined for mean value of the friction angle φ_1. They were trrated as non-random during the reliability computations. The main parameters of the example are provided in Table 4, while the results are given in Tables 5 through 7 and Figures 9, 10 below.

Table 4. Probabilistic characteristic of parameters

Soil property	Mean value	Standard Deviation σ_X	Probability distribution
Friction angle φ	32°	4.6°	lognormal
Unit weight of the soil γ	18.2 kN/m^3	1.092 kN/m^3	normal
Axial force P	300 kN	45 kN	lognormal
Unit weight of concrete	24.0 kN/m^3	-	nonrandom
Overburden pressure q	23.0 kN/m^3	-	nonrandom

Table 5. Reliability indices. One-dimensional case

Width of the foundation b [m]	One-dimensional case			
	β_5	β_6	β_7	β_8
1.2	0.91	1.25	1.49	1.43
1.4	1.33	1.81	2.21	2.13
1.6	1.72	2.33	2.91	2.83
1.8	2.09	2.81	3.59	3.50
2.0	2.44	3.27	4.25	4.15
2.2	2.77	3.70	4.88	4.77
2.4	3.09	4.10	5.49	5.38
2.6	3.40	4.49	6.08	5.97
2.8	3.69	4.86	6.65	6.53
3.0	3.97	5.21	7.19	7.08
3.2	4.24	5.55	7.72	7.61
3.4	4.50	5.88	8.23	8.12
3.6	4.75	6.19	8.72	8.61
3.8	5.00	6.49	9.20	9.09

It is noted that the reliability indices, β, given in Tables 5,6,7 as well as Figures 5 and 6, have been computed under a set of distinct assumptions, which are listed below:
β_5: the friction angle is modelled by a single random variable without spatial averaging; β_6: the friction angle is modelled by three independent random variables φ_1, φ_2, φ_3 as described in section 6, but spatial averaging is not incorporated; β_7: three independent random variables involved φ_1, φ_2, φ_3 with one-dimensional spatial averaging; β_8: three correlated random variables φ_1, φ_2, φ_3 incorporated with one-dimensional spatial averaging.

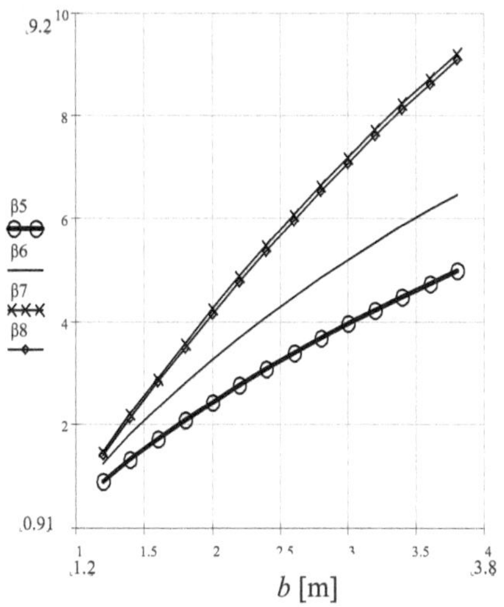

Figure 9. Reliability indices. One-dimensional case. Graphs corresponds to results in Table 5.

Table 6. Reliability indices. two-dimensional case, uncorrelated random variables.

Width of the foundation b [m]	Two-dimensional case			
	β_9	β_{10}	β_{11}	β_{12}
1.2	1.50	1.50	1.49	1.49
1.4	2.24	2.22	2.22	2.21
1.6	2.95	2.92	2.92	2.91
1.8	3.63	3.60	3.60	3.59
2.0	4.29	4.26	4.25	4.25
2.2	4.93	4.89	4.88	4.88
2.4	5.55	5.50	5.49	5.49
2.6	6.14	6.09	6.08	6.08
2.8	6.71	6.66	6.65	6.65
3.0	7.26	7.21	7.20	7.19
3.2	7.80	7.74	7.73	7.72
3.4	8.31	8.25	8.24	8.23
3.6	8.81	8.74	8.73	8.72
3.8	9.29	9.22	9.21	9.20

β_9: three independent random variables φ_1, φ_2, φ_3 employed with two-dimensional spatial averaging, where $\omega_2 = 3\omega_1$; β_{10}: three independent random variables φ_1, φ_2, φ_3 involved, with two-dimensional spatial averaging, where $\omega_2 = 10\omega_1$; β_{11}: three independent random variables φ_1, φ_2, φ_3 with two-dimensional spatial averaging, where $\omega_2 = 30\omega_1$; $\beta_{12} = \beta_7$: three independent random variables φ_1, φ_2, φ_3 with one-dimensional spatial averaging; β_{13}: three correlated random variables φ_1, φ_2, φ_3 with two-dimensional spatial averaging, where $\omega_2 = 3\omega_1$; β_{14}: three correlated random variables φ_1, φ_2, φ_3 with two-dimensional spatial averaging, where $\omega_2 = 10\omega_1$; β_{15}: three correlated random variables φ_1, φ_2, φ_3 with two-dimensional spatial averaging, where $\omega_2 = 30\omega_1$; $\beta_{16} = \beta_8$: three correlated random variables φ_1, φ_2, φ_3 incorporated with one-dimensional spatial averaging.

The results of reliability indices computations for one-dimensional case, as presented in Table 5 and Figure 9, demonstrate the importance of the spatial averaging. There is 60% to 85% increase in the value of reliability indices due to spatial averaging. It is worth mentioning that modelling friction angle by three random variables (each one corresponding to one of segments of the slip line; β_6) instead of one (β_5), increases reliability indices significantly. On the other hand, the effect of correlation between averaged random variables seems to be very small.

In the case of two-dimensional averaging, for all three values of the horizontal fluctuation scale (even for $\omega_2 = 3\omega_1$), reliability indices only slightly differ from those obtained for one-dimensional case – curves in Figure 10 (left hand and right hand part) coincide. For large horizontal fluctuation scale ($\omega_2 = 30\omega_1$) they are almost the same ($\beta_{11} \approx \beta_{12}$ and $\beta_{15} \approx \beta_{16}$). On the other hand, comparing the left hand side graph and the right hand side graph in Figure 10, one finds that the effect of correlation could be neglected.

Some more details as well as many other examples concerning both cohesive and cohesionless soils one finds in the author's book (Puła 2004).

Table 7. Reliability indices; Two-dimensional case. Correlated random variables.

Width of the foundation b [m]	β_{13}	β_{14}	β_{15}	β_{16}
1.2	1.49	1.46	1.44	1.43
1.4	2.22	2.18	2.15	2.13
1.6	2.92	2.89	2.84	2.83
1.8	3.61	3.57	3.52	3.50
2.0	4.27	4.23	4.18	4.15
2.2	4.91	4.86	4.81	4.77
2.4	5.53	5.47	5.42	5.38
2.6	6.13	6.07	6.02	5.97
2.8	6.70	6.64	6.59	6.53
3.0	7.26	7.19	7.14	7.08
3.2	7.79	7.72	7.68	7.61
3.4	8.30	8.23	8.19	8.12
3.6	8.80	8.72	8.69	8.61
3.8	9.28	9.20	9.17	9.09

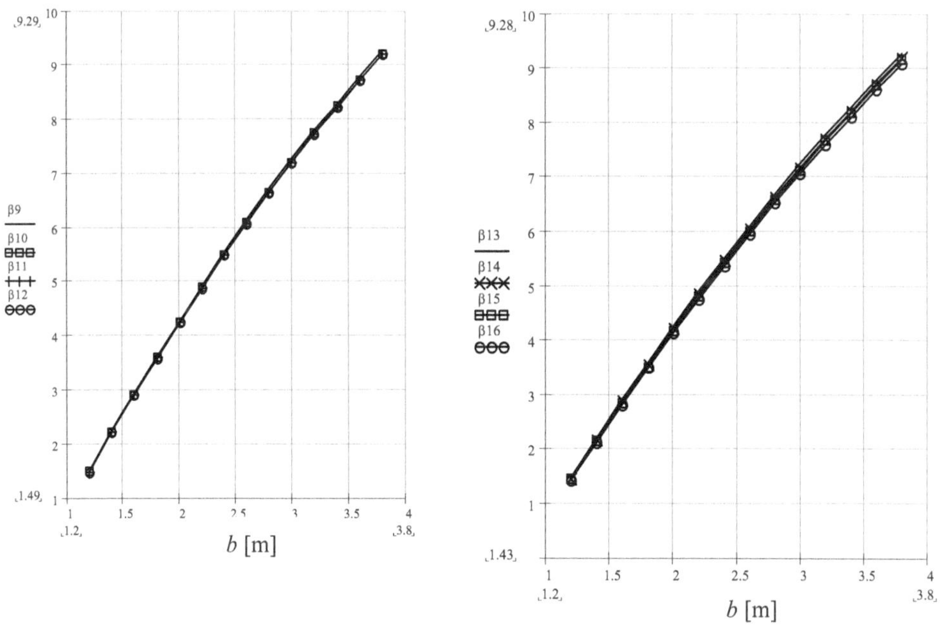

Figure 10. Reliability indices; uncorrelated (left hand side) and correlated (right hand side) case.

8 Closing Remarks

The chapter demonstrates a reliability approach to the bearing capacity computations of shallow strip foundations. The examples on the first part base on approach given by the Polish Standard PN-81/B-03020. However it is demonstrated that the general approach as well as applied equations are not very far from that suggested by *Eurocode* EC7.

It is demonstrated that appropriate selection of the friction angle coefficient of variation plays vital role for values of the safety measures. It is usually more important than the effect of the probability distribution selection for this quantity.

In the case of cohesive soils the effect of mutual correlation between strength parameters φ and c is significant and should be taken into account.

The example considered shows that the stochastic modelling basing on assigning each individual soil property single random variable can be insufficient for obtaining adequate safety measures to well-design foundations. Usually it is necessary to it is reasonable to describe soil strength properties in terms of random field's theory. Next the selected random field should be spatially averaged by means of a procedure introduced by Vanmarcke (1997).

In order to avoid the loss of uniqueness in reliability computations the averaging area must be carefully selected and precisely defined among assumptions for a problem under consideration. It seems that to get proper values of variance reduction in bearing capacity problems it is necessary to carry out the spatial averaging along potential slip surfaces associated with a mechanism of failure. Here a kinematically admissible Prandtl´s mechanism (Prandtl 1920) for shallow strip foundation is considered.

The numerical studies have shown that by incorporating spatial averaging one can significantly reduce standard deviations of soil strength parameters, which leads to a significant increase in reliability indices (decrease in failure probabilities). This is a step forward in making reliability measures more realistic in the context of well-designed (according to standards) foundations.

The results of numerical computations have also demonstrated that for reasonable, from practical point of view, values of horizontal scale of fluctuation (about 10 to 20 times greater than values of vertical fluctuation scale), the reliability measures obtained from two-dimensional averaging are almost the same as those corresponding to one-dimensional averaging. This means that in the case of shallow strip foundations one-dimensional (along the depth) averaging can be sufficient, which simplifies computations and requires a smaller amount of statistical data (vertical fluctuation scale instead of both vertical and horizontal scales).

References

Cherubini, C. (1977). Data and Consideration on the variability of geotechnical properties of soils. In *Proceedings of the ESREL Conference*, Lisboa 1987: 1538-1591.

Cherubini, C. (2000). Probabilistic approach to the design of anchored sheet pile walls. *Computers and Geotechnics,* Vol. 26, No. 3-4: 309-330.

DIN 4017 (1979). *Part 2: Shear failure calculations for shallow foundations with oblique and eccentric loading.* DIN, Berlin.

Ditlevsen, O., Madsen, H.O., *Structural Reliability Methods,* J. Wiley & Sons, Chichester, 1996.

ENV 1997-1 (1997). Eurocode 7: *Geotechnical design - part 1: general rules*. European Committee for Standarisation (CEN), Brussels.

Hohenbichler M., Gollwitzer S., Kruse W., Rackwitz R. (1987). New light on first and second-order reliability methods. *Structural Safety*, 4,1987, 267-284.

ISO 2394 (1998). *General principles on reliability of structures*. International Standard.

Orr, T.L.L., Farrell, E.R. (1999). *Geotechnical design to Eurocode 7*. Springer Verlag, London.

PN-81/B-03020. (1981). Polish Standard. Foundation bases. Static computations and design.

Prandtl, L., (1920). Über die Härte plastischer Köroer, *Nachr. Kgl. Ges. Wiss.* Göttingen, Math. Phys. Kl.:74-85.

Puła, W. (2002). On spatial averaging in reliability computations of shallow foundations. In *Numerical models in Geomechanics. Proceedings of the Eight Int. Symposium on Numerical Methods in Geomechanics, Rome 2002*. Rotterdam: A.A. Balkema.421-426.

Puła W. (2004): *Applications of Structural Reliability Theory to Foundations Safety Evaluation* (in Polish)., Wrocław: Wroclaw University of Technology Press.

Rackwitz, R. (2000). Reviewing probabilistic soils modeling. *Computers & Geotechnics* 25 (3-4): 199-223.

Vanmarcke E.H. (1977). Probabilistic Modeling of Soil Profiles. *Joural. of the Geoechnical Eng. Div.*, ASCE, Vol. 103, No. GT11: 1227-1246.

Application of the response surface method

Wojciech Puła[1] and Jerzy Bauer[2]

[1] Institute of Geotechnics and Hydrotechnics, Wrocław University of Technology, Wrocław, Poland
[2] Institute of Mining Engineering, Wrocław University of Technology, Wrocław, Poland

Abstract. The response surface method can be applied to numerous fields of knowledge. In general, this method consists in approximation an unknown function by known function chosen appropriately. It can be successively utilised in reliability measures computations when the failure criterion does not depend on random variables explicitly. Within this study an application of the response surface method in dealing with random settlement in geotechnical engineering is discussed. In practical applications, an explicit closed form of settlement function $U(\mathbf{X})$ is only rarely known. Most often, if program of the finite element method is available, we are able to determine the function values for assumed material properties, loads and geometrical constants of the FEM computation model. In order to obtain the limit state function in the form appropriate for reliability computations, one can model a closed form of $U(\mathbf{X})$ by means of non-linear regression. To simplify the reliability computations, rather simple functions, e.g. polynomials of the second degree are in use. The approximation in the vicinity of a design point (known from preliminary FORM computations) is especially convenient in computing the probability of failure, because the neighbourhood of this point affects most strongly the value of a failure probability. In order to make computational algorithm more efficient the application of neural network with a hyperbolical activation function is suggested. If $U(\mathbf{X})$ is a continuous function then the three-layered neural network with one hidden layered containing necessary number of neurones could give a satisfactory result. Neural networks allow approximation in cases of large variability intervals of independent variables preserving sufficient accuracy. Furthermore they make possible to overcome some numerical drawbacks of second order approximation known as "false branches problem".

1 Response Surface Method

The response surface method applied to numerous fields of knowledge is exhaustively described in a number of monographs (eg. Myers 1971, Box and Draper 1996). In general, this method consists in approximating an unknown function by the known function chosen appropriately. This approximation can be based on the results of experiments and also on the results of numerical computations, e.g. results obtained by means of the finite element method. In the case of numerical computations, a relationship between the model parameters $x_1, x_2, ...x_n$, which are introduced as input data, and the values obtained as output data $y = f(x_1, x_2, ...x_n)$ is defined. Roughly establishing such a relationship allows us to replace a troublesome numerical procedure with a simple analytical relationship, which helps to forecast a response of the model being analysed in the input set.

The response surface method was adapted to the reliability analyses of engineering structures in eighties (eg. Rackwitz 1982, Wong 1985, Faravelli 1989). However, in order to simplify reliability computations, rather simple functions such as polynomials of the second degree are often used (Bucher and Bourgund 1990). Methods of approximation are diverse, depending on the aim of this operation. In any case algorithms based on regression models may be used (compare next section). The approximation in the vicinity of a design point, which is found in the framework of the FORM method (see e.g. Ditlevsen and Madsen 1996) is especially convenient in computing both the probability of failure or the reliability index, because the neighbourhood of this point affects most strongly the value of a failure probability (Rackwitz 1982, Engelund and Rackwitz 1992).

2 Non-linear Regression Method in Response Surface Modelling for Settlement Analysis Purpose

Examining the serviceability limit state of shallow foundations, one finds it convenient to know the function describing their settlement, depending on such variables as loads, geometric quantities and material constants, which may be considered as random variables. This function expressed as $U(\mathbf{X})$ enables one to introduce the limit state function (c.f. section 3 in chapter 12) for further reliability computations. In practical applications, an explicit closed form of settlement function $U(\mathbf{X})$ is only rarely known. Most often, whenever program of the finite element method is available, we are able to determine the function values for the assumed material properties, loads and geometrical constants of the FEM computation model. In order to obtain the limit state function in the form appropriate for reliability computations, one can model a closed form of $U(\mathbf{X})$ by means of a non-linear regression. Generally speaking, the form of the function U can be described by the following non-linear regression model.

$$\mathrm{E}(U) = f(\mathbf{X}, \mathbf{B}) + err \tag{1}$$

where E (U) is an expected value of U, $\mathbf{X}$ is a vector of m independent variables, $\mathbf{B}$ is a vector of k unknown parameters of regression model f and err is a random variable describing the error of estimation of function U. The components of vector B are determined by the process of minimising the sum of the squares of differences between given quantities U_i and predicted quantities $\hat{U}_i$

$$\Psi = \sum_{i}^{n} \left(U_i - \hat{U}_i\right)^2 \tag{2}$$

for the set of n data: (U_i, X_{il}), $i = 1, 2, \ldots, n$, $l = 1, 2, \ldots, m$. Random variable err of the estimation error takes the zero expected value and its standard deviation is equal to:

$$s_{err} = \sqrt{\frac{\Psi_{\min}}{n-k}} \quad . \tag{3}$$

There exist two classic methods of finding the minimum of function Ψ, namely the linearisation method and the steepest gradient method (Marquardt 1963). However, in computations they exhibit a slow convergence because of the frequently found shape of function Ψ in the space of parameters B_i. This space has the form of an elongated, tortuous canyon with steep sides and a slight slope at its bottom. In such cases, the directions of regression parameter changes obtained using both methods are almost perpendicular to one another. Therefore in our computations we used program NLIN2, which is based on Marquardt's compromise (Marquardt 1966). This allowed select an indirect, optimal direction for the increment of parameter vector, thus rapidly accelerating the iteration convergence.

In this chapter a non-linear model with respect to independent variables and a linear model with respect to regression parameters were applied. The required accuracy of calculating parameters B_i was obtained in five iteration steps.

3 FEM Assessment of Shallow Foundation Settlements

Nowadays, one can find on the market various commercial programs that utilise the finite element method. We have taken advantage of the program PHASES (Hoek et al. 1992), which is a specialised program for plane state of strain. The problem of determining maximum settlements of flexible shallow foundations resolves itself into a boundary problem of plane strain state of elasticity theory. If this problem, however, concerns the medium treated as an infinite or a half-infinite space it cannot be solved by means of the procedure most frequently used. According to this procedure one considers a sufficiently large area so that the effect of the assumed boundary constraints of this area on the parameters of interest are negligible. Unfortunately, such a procedure in the case of determining displacement induced by the foundation load in the elastic half-plane leads to ambiguous results. The value of the determined settlement tends to infinity if the discretised area enlarges to infinity. This is associated with a characteristic singularity of the solutions describing displacement in a plane state of strain induced by loads of non-zero resultants. For all such loads, displacement of each point of a half-plane tends theoretically to infinity. So, it is possible to determine explicitly the displacements only in relation to the chosen reference point. In order to assure uniqueness of the solutions we have assume a recommendation given by the building code PN-81/B-03020 (1981). According to this code, settlement of a foundation is determined by an elastic layer where stresses induced by the foundation exceed normal stresses caused by the dead weight of the soil by 30%. Utilising some results of the elasticity theory the thickness of this layer can be easily determined (Bauer and Puła 2000) as a function of the foundation's width, unit weight of the soil and the load applied. A computational model is presented in Figure 1.

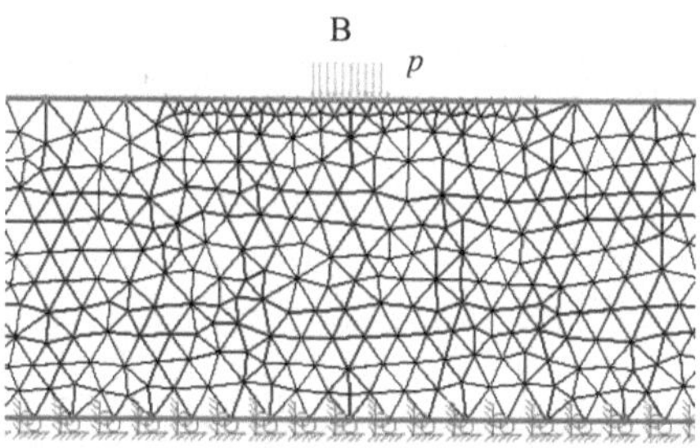

Figure 1. Boundary conditions of the subsoil and the strip foundation FEM model considered.

The free surface is loaded with the load $p = 0.4$ MPa uniformly distributed on the segment whose length equals $B = 2.5$ m. The layer of 10.52 m thickness and 82.5 m width is fixed downward. The sides (edges) of this layer are entirely devoid of the freedom of horizontal displacements. In the neighbourhood of the imposed load, the number of finite elements is increased. The computational model of the FEM method comprises 1941 linear elements of triangular shape and has 2100 degrees of freedom. Taking advantage of this model the maximum displacements for the necessary values of Young's modulus and Poisson's ratio of soil were computed.

4 Response Surface Models

In reliability calculations, it would be best to have an analytical form of the expression defining settlements of shallow foundations as a function of their dimensions, loads, geological subsoil conditions and material constants. Because of the non-existence of such a formula, we have been forced to obtain an approximate form of the necessary function of settlement. This may be done in various ways, e.g. by the use of recommendations given in building codes. However, if one makes use of the finite element method in order to obtain the set of settlement results, he may approximate these results, for example by substituting them with a polynomial of the second degree. The accuracy of computations can be improved by supplementing the polynomial of the second degree equation with an additional term whose form depends on the character of the problem under consideration. In the case of a foundation, it may be the term taking the form of an equation describing a sample settlement in the oedometer.

The forms of settlement function depend on the parameters considered as random variables describing the foundation settlement in reliability calculations. In the present paper, material constants of soil, i.e. Young's modulus and Poisson's ratio, will be considered as random quantities.

In the considerations presented below, we answer the question as to whether or not proposed forms of approximate function describing shallow foundation settlement are suitable for reliability computations.

4.1 Parabolic Response Surface

A polynomial of the second degree function is most commonly used in the response surface method applied to reliability problems. In the case of foundation settlements when only two random variables are assumed, we have:

$$U(X_1, X_2) = B_1 + B_2 X_1 + B_3 X_2 + B_4 X_1^2 + B_5 X_2^2 + B_6 X_1 X_2 + err \tag{4}$$

where $U(X_1, X_2)$ is settlement of the foundation at the point beneath the centre, X_1 and X_2 represent Young's modulus E and Poisson's ratio ν, respectively, and *err* is the random variable of the approximation error (zero-expectation random variable).

The algorithm enabling one to determine the final values of the coefficients B is a modified procedure of non-linear regression which is based on the ideas reported in papers by Engelund and Rackwitz (1992) and El Meligy et al (1997). In the case where a shallow foundation is considered it can be reduced to the following steps:

- given the expected values of variables X_1 and X_2 , one may assume that an approximate interval of their variation is a double standard deviation of length with the expected value in the centre, for each variable
- if a pair of values of material constants is chosen from the interval of their variation, a set of values for foundation settlement could be obtained. It should be stressed that the response surface model (4) comprises a mixed term with coefficient B_6. Because of the above, the rectangular variability domain of E and ν should be covered uniformly in points whose respective values of settlement are determined. Computational points along the diagonal of a rectangle of the variation of material constants are not sufficient. Insertion of the mixed term into Equation (4) may prove to be troublesome or pointless when the number of random variables in the problem exceeds two
- treating Equation (4) as a regression model, the initial values b_i of parameters B_i are determined by means of linear regression (compare Table 1)
- the surface $U(X_1, X_2, ..., X_n)$ obtained in such a way is utilised to create a limit state function $g(\mathbf{x})$ of the form:

$$g(\mathbf{x}) = u_{\max} - U(X_1, X_2, ..., X_n) \tag{5}$$

  where $u_{\max}$ stands for a maximal allowable settlement
- next, the design point on the limit state surface is found using the FORM method
- the above steps are repeated in the neighbourhood of a design point. Intervals of variability are reduced, for example, to approximately half a standard deviation from the values determining a design point. This requires a new set of coefficients $\{b_i\}$
- Most often the design point coordinates defined in such a way are sufficiently accurate, which can be verified by MES computations. However, if this accuracy proves to be inadequate, the procedure described should be repeated with further reduction of variability ranges in neighbourhoods of design point coordinates. As before a new set of coefficients $\{b_i\}$ must be determined by the regression procedure.

Table 1. Parabolic regression coefficients versus Poisson's ratio mean values.

Successive regression coefficients	Poisson's ratio mean values			
	0.1	0.2	0.3	0.4
B_1	0.397839	0.410466	0.407413	0.436019
B_2	-0.030604	-0.032386	-0.031190	-0.043214
B_3	-0.060239	-0.080708	-0.057315	0.149711
B_4	0.000768	0.000832	0.000722	0.001025
B_5	-0.166729	-0.211499	-0.301765	-0.881754
B_6	0.004073	0.006113	0.007455	0.024773
Standard deviation s_{err}	0.000405	0.000458	0.001469	0.001812

For the last design point, respective reliability index β is determined and, next, failure probability according to the SORM procedure is assessed.

The above given approach to reliability computations involving response surface methodology may be hampered by the appearance of false design points. It should be noted that in the case of the limit state surface, whose left side is a polynomial of the second degree, more than one design point could exist. This results from the properties of a second degree polynomial surface. If on this surface a saddle point does not exist, then the intersection of the surface by a plane parallel to the plane of independent variables yields a close surface (line). The changing of physical independent variables into standard variables transforms this line into a closed contour with the origin of coordinates in its interior. This case could be troublesome in FORM and SORM computations.

If a saddle point occurs on the surface, then its intersection by the plane parallel to the surface of independent variables yields two branches along both of which the limit state condition of settlement function (4) is fulfilled ($g(\mathbf{x}) = 0$). Such a phenomenon is shown in Figures 2, 3 and 4.

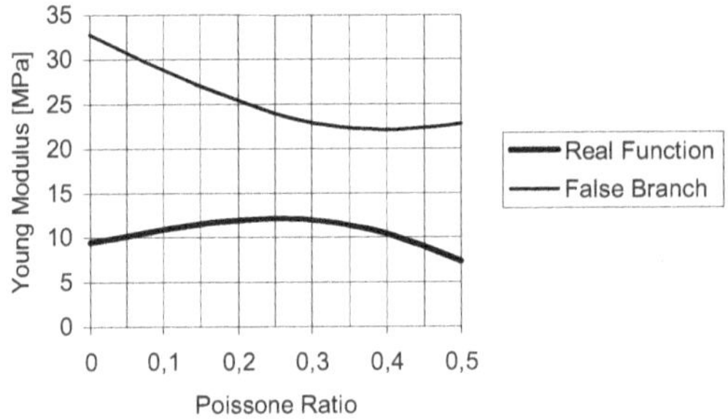

Figure 2. Two branches of the limit state function in the „physical variable space”.

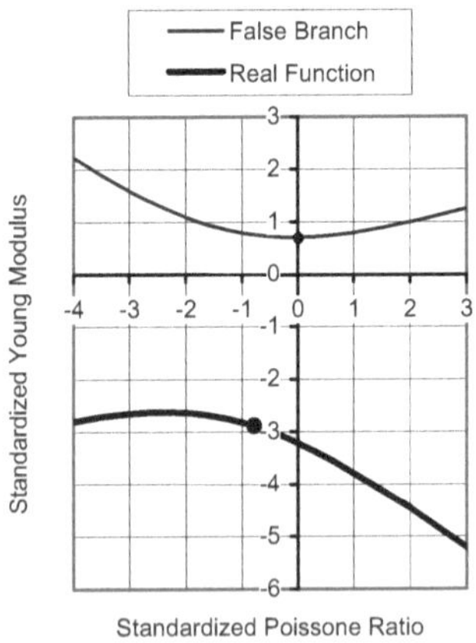

Figure 3. Two branches of the limit state function in the standardised normal space. Random variables of the Young modulus and Poisson's ratio have normal distributions.

Figure 2 presents fragments of the curves $g(\mathbf{x}) = 0$ in the system of „physical coordinates" (X_1, X_2), and Figs. 3 and 4 show these curves in the space being transformed (Y_1, Y_2). On each branch a design point can be found, i.e. the point of shortest distance from the origin of the coordinate system. Both points can yield intuitively satisfactory calculation values of particular physical parameters. In the examples presented, a false design point yields the lowest value of reliability index β.

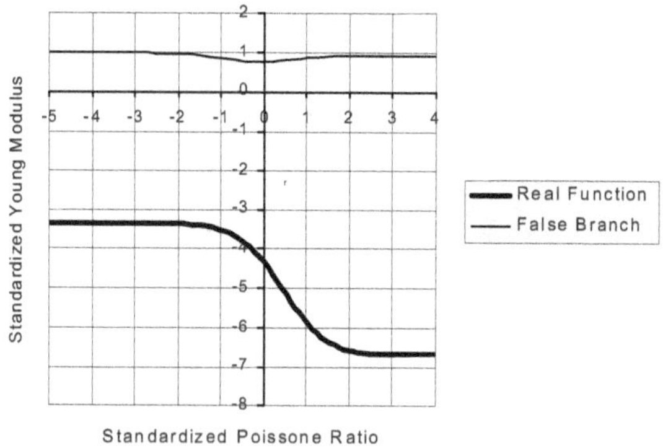

Figure 4. Two branches of the limit state function in the standardized normal space. The random variable of Young's modulus has a log-normal distribution and Poisson's ratio is of a beta distribution.

In order to check which design point represents a false solution, its „physical coordinates" should be applied as material constants to the FEM program, thus enabling an evaluation of the foundation settlement. If the settlement calculated in such a way is close to the limit settlement, the design point is the real one. When the false design point is positioned at the shortest distance from the origin of coordinate system then, defining the limit state function, some bounds for random variables ranges should be introduced to force the programme to find the local minimum distance from the origin. In some examples, in the course of reliability computations, design points with a too high coordinate value of Young's modulus, about 23 MPa, were obtained. However, earlier calculations, done by means a simplified method suggested by the code PN-81/B-03020, displayed a value of about 11 MPa. In order to make the indication of "false design point" impossible, the term $(X_1 - 13)^{10}$ was inserted into the settlement function (4) for values of variable X_1 exceeding 13 MPa. This insertion modified it in the following way:

$$U = \begin{cases} B_1 + B_2 X_1 + B_3 X_2 + B_4 X_1^2 + B_5 X_2^2 + B_6 X_1 X_2 + err & for \quad X_1 < 13 \\ B_1 + B_2 X_1 + B_3 X_2 + B_4 X_1^2 + B_5 X_2^2 + B_6 X_1 X_2 + (X_1 - 13)^{10} + err & for \quad X_1 \geq 13 \end{cases} \tag{6}$$

As a result of this constraint, the settlement function is smooth with its derivatives to the ninth order at line (hypersurface) $X_1 = 13$.

Such corrections can be made even after the preliminary series of computations has been completed. It should be noted that this trouble might appear not only in the case of settlement analysis, but also in each reliability problem, which is solved based on the response surface method involving a polynomial of the second degree as an approximation function.

4.2 Parabolic Approximation with Oedometric Term

Parabolic approximation may appear to be sufficiently accurate only for narrow intervals of variation of Poisson's ratio and Young's modulus. Therefore in Table 1, as much as four approximations for four various expected values of Poisson's ratio are presented. To arrive at sufficiently accurate approximations over a wide interval of random parameters, the second-order polynomial response surface is improved by adding the "oedometric term"of the form

$$U_1(E,\nu) = a\frac{1-\nu-2\nu^2}{E(1-\nu)} \tag{7}$$

This yields the following settlement function:

$$U = B_1 + B_2X_1 + B_3X_2 + B_4X_1^2 + B_5X_2^2 + B_6X_1X_2 + B_7\frac{1-X_2-2X_2^2}{X_1(1-X_2)} + err \tag{8}$$

The results of approximation of the foundation settlements obtained using the above expression are presented in Table 2.

Table 2. Regression coefficients for two ranges of the Young modulus. Parabolic regression models with an additional term of the oedometric modulus.

Successive regression coefficients	Ranges of Young's modulus	
	5 – 30 MPa	5 – 20 Mpa
B_1	0.237703	0.339195
B_2	-0.013756	-0.027901
B_3	-0.005953	-0.067943
B_4	0.000245	0.000700
B_5	-0.050773	-0.106475
B_6	-0.000066	0.004179
B_7	0.390865	0.448634
Standard deviation s_{err}	0.005668	0.003168

The results presented are valid for the entire interval of variability of Poisson's ratio from 0 to 0.5. In two columns, approximations over wide and narrow intervals of Young's modulus are shown. Mean errors of estimation are exposed on the last line. The oedometric term enables a relatively accurate approximation over a wide interval of variability of random parameters of subsoil elasticity.

The effect of two branches of the limit state function also appears in this case, which force us to impose restrictions on parameters variability in the limit state function, for example in the form presented in Equation (6).

5 Examples of Computation of Reliability Indexes

5.1 Probabilistic modeling of subsoil characteristics

The reliability computations were performed for the subsoil of a small elastic modulus (the expected value E(*E*) = 20 MPa), since in such subsoil random variability of compressibility characteristics is considerably greater than in the case of large moduli, the occurrence of which was proved by laboratory tests. Moreover, the settlements of such subsoil are more pronounced compared to the subsoil of a great elastic modulus, which can make the effects associated with random variability unnoticeable. In the examples presented, the compressibility characteristics of subsoil, in other words, Young's modulus *E* and Poisson's ratio ν, treated as an elastic medium, are random variables. The choice of proper probability distributions, and especially of variance (coefficients of variations), is highly important. Unfortunately, knowledge about statistical distributions of the values of parameters *E* and ν for soils is very poor. Because the laboratory determination of those parameters in relation to soils is troublesome, high-number samples are determined very rarely, although they are essential in defining probability distribution or in estimating statistical moments. Literature (Meyerhof 1995, Becker 1996) 14provides only the information that the coefficients of Young's modulus variability may reach great values. In the case of subsoil whose value of Young's modulus *E* is low, the choice of log-normal distribution seems reasonable, since for this probability distribution the value of *E* is not negative even at low expected values and great variance. This distribution was often used to describe the physical characteristics of soil, especially the strength parameters (e.g. El-Meligy 1995, Brząkała et al 1995).

It is worth mentioning that Poisson's ratio changes within a relatively narrow interval. There is no available information about the random variation of this parameter in soils. Some authors accept the opinion that the randomness of Poisson's ratio can be neglected in an analysis of settlements taking place in the case of elastic subsoil (Brząkała and Puła 1996). In most examples analysed in this paper, we assumed for Poisson's ratio a beta distribution symmetrically distributed around the expected value. Beta distributions are of bounded supports. This fact corresponds with relatively narrow variation intervals of Poisson's ratio. Plots representing probability density functions of these distributions are shown in Figure 5. In all cases shown in Figure 5, the coefficient of variation equalled 15%. For the expected value E(ν) = 0.4, the variation interval [0.3, 0.5], and the 15% coefficient of variation, the plot of density changed its shape (see Figure 6). This shape was unacceptable because it provided increased values of probabilities at the ends of the variation interval. A decrease in the coefficient of variation at the same limits of the variation interval made this distribution similar to rectangular distribution (compare Figure 6). Finally, a rectangular distribution within interval [0.3, 0.5] was accepted. The coefficient of variation equals 14.5% in this case. It is evident that in order to keep the random variations of Poisson's ratio in a reasonable interval of values, one has to assume that the coefficient of variation does not exceed 15%.

There is no information available dealing with the correlation between parameters *E* and ν. The results reported by some researchers (Tsytowitch 1973) and the authors of the Polish building code PN-81/B-03020 (1981) may lead us to the conclusion that this correlation is negative; that is, the statistical decrease in the value of Poisson's ratio corresponds to a statistical increase in the value of Young's modulus in soils. In order to make it possible to calculate failure probability when

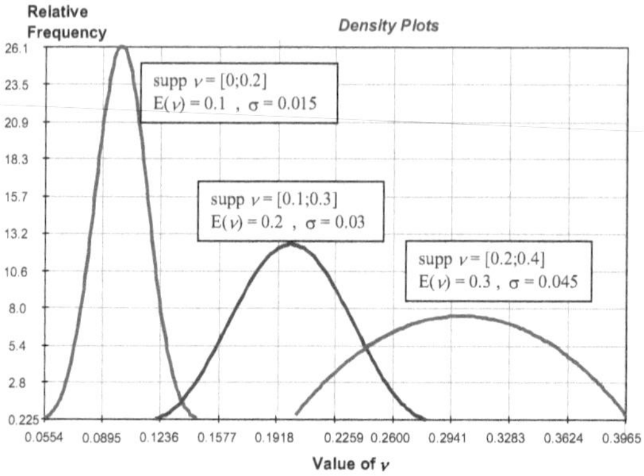

Figure 5. Poisson's ratio probability density functions of a beta type.

the correlation between random variables is taken into account, one has to determine the joint distribution of the random vector. If the random vector is not the normal one, this task can be difficult since it is necessary to have a large sample size for statistical verification in case of two- and multidimensional densities. Thus we made use of the so-called Nataf's distributions (compare Liu and Der Kiureghian 1986). Following the Nataf procedure, one finds multidimensional probability distribution, whose marginal distributions and the coefficients of correlation between them are assumed earlier. This allows us to search for failure probability, provided that one-dimensional distributions of particular random parameters and their mutual correlation relationships are known. Of course, one should be aware of the fact that such a procedure leads to certain approximations.

5.2 Results of Numerical Computations

In numerical examples, exceeding the allowable settlement value of the point beneath the centre of a foundation is assumed as a failure criterion. This leads to the limit state surface defined by (5). Next, reliability index β and the equivalent probability of failure were evaluated according to FORM and SORM procedures. In all examples presented, it was assumed that the distribution of Young's modulus is a log-normal one with expected value E(E) = 20 MPa and standard deviation σ_E = 3 MPa. The statistical moments yielded a 15% coefficient of variation. The threshold value of the settlement was u_{max} = 0.12 m. All reliability computations were done using COMREL system (STRUREL 1995). Presentation of the results comprised only reliability indexes β_{SORM} computed by the SORM method.

In the first series of examples, settlement computations were done using the finite element method and then the results were subjected to approximation by means of the parabolic response surface method described in section 4.1. For each of four expected values of Poisson's ratio under consideration the approximating surface was different. Equations describing these surfaces are presented (as coefficients) in Table 1.

The assumptions accepted for the random variability of parameters were the same as specified above, but with one exception: for the mean value of coefficient E(ν) = 0.4 additional computations were performed. Namely, it was assumed that the probability distribution of ν was normal, and that the coefficient of variation equals 15%. It was assumed that random variables E and ν were stochastically independent. The results are gathered in Table 3.

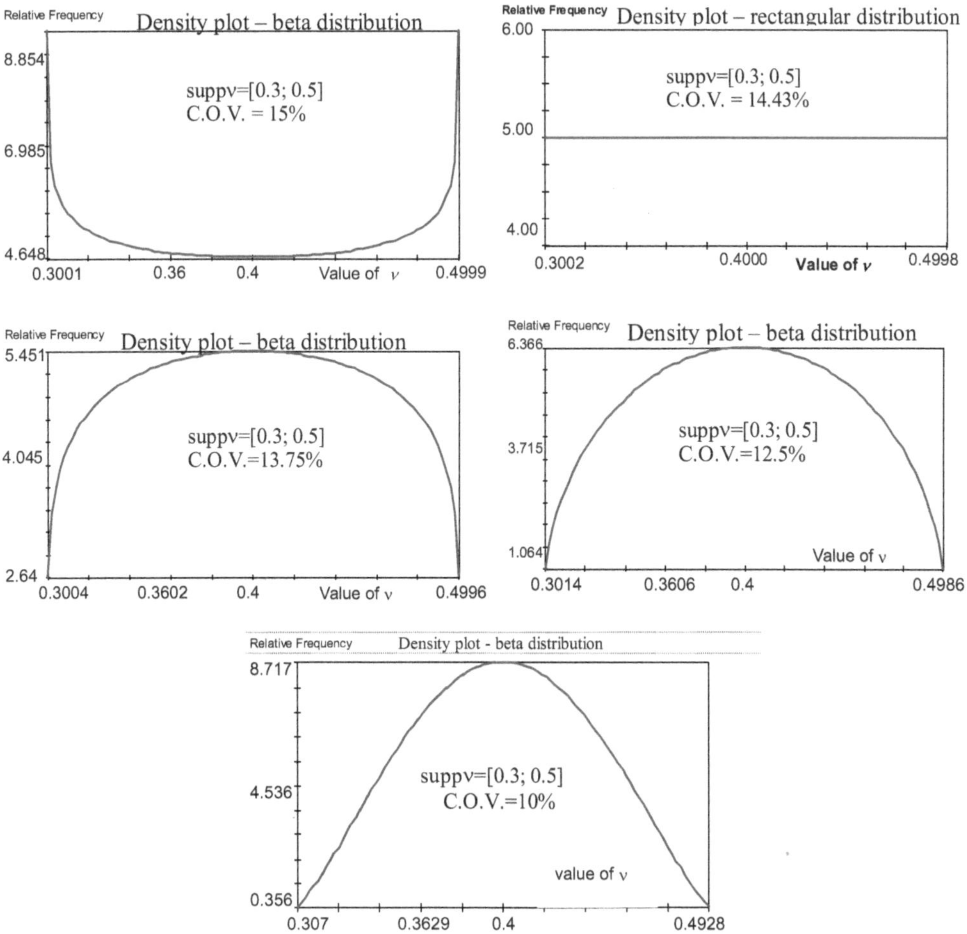

Figure 6. Modelling the p.d.f. for Poisson's ratio (E(v) = 0.4)

Computations were done for four various expected values of Poisson's ratio, whose probability distributions were described in section 6.1. Additionally, in the last column, one can find the values of settlement obtained for the design values of E and ν computed by the finite element method. The aim of such a computation was to check how accurate the approximation was of the limit state surface obtained when using the response surface method. According to assumptions of the SORM method, the design point occurs on the limit state surface; hence the settlement should be exactly 0.12 m at u_{max} = 0.12 m. The results in the last column of Table 3 prove that the accuracy of approximation can be considered satisfactory.

Table 3. Reliability calculations for parabolic response surfaces of four Poisson's ratio mean values. A constraint was introduced in the equations of response surfaces to prevent false design point evaluations. 0.4^*, 0.4^{**} - respectively, normal and uniform distribution of Poisson's ratio.

Mean value of Poisson's ratio ν	Reliability indexes β	Design points		Sensitivity factors		Limit displacements
		E [MPa]	ν	α_E	α_ν	FEM
0.1	2.43	13.78	0.099	0.999	0.030	0.12007
0,2	2.73	13.24	0.188	0.990	0.138	0.12007
0.3	3.20	12.49	0.258	0.964	0.256	0.12074
0.4^*	3.80	11.84	0.321	0.931	0.355	0.11846
0.4^{**}	3.80	11.77	0.325	0.951	0.297	0.11896

E and ν stochastically independent

Reliability index β increases with the increase in the mean value of Poisson's ratio, because the increase of the latter results in the decrease of soil settlements (for given value of the Young modulus E). The design values of Young's modulus decrease with the increase in the mean value of Poisson's ratio. It should be stressed that at small values of Poisson's ratio (0.1 or 0.2) the random variability of this parameter is of minor significance when the reliability index is taken into account. This regularity is proved by small values of sensitivity factors α_ν for mean values 0.1 and 0.2. The effect of randomness ν on the value of index β heightens with the increase of the mean value of ratio ν. At the mean value of ν equal to 0.4, the results obtained are similar both for rectangular and normal distributions, although a stronger effect of the randomness of Poisson's ratio on the reliability index yields a normal one.

In order to have one response surface for the entire range of Poisson's ratio as well as to increase the range of Young's modulus variation, the computations were done by means of a response surface with the oedometric term (section 4.2, Equation (8). The computations comprise two cases, a broad (5-30 MPa) and narrow (5-20 MPa) interval of Young's modulus. The coefficients appropriate for the surface described by Equation (8) are presented in Table 2.

The results of reliability computations are gathered in Tables 4 and 5. The results in Table 5 differ only slightly from those in Table 3 both in the reliability indexes and design values of the parameters. This statement is also valid when the effect of randomness on Poisson's ratio is considered. Taking into account the fact that parabolic fittings were highly accurate (four surfaces were dependent on the mean value of the value of Poisson's ratio), we can recognise the approximation obtained by means of the surface (8) as equally useful. However, the surface (8) is advantageous since it contains practically the whole interval of Poisson's ratio values. Also, the settlement values computed by means of the finite element method (the last column in Table 5) are very close to the threshold assumed ($u_{max} = 0.12$ m).

Table 4. Reliability calculations of parabolic response surfaces with an oedometric term. In the input data Young modulus has a 5-30 MPa range. 0.4*, 0.4** - respectively, normal and uniform distribution of Poisson's ratio.

Mean value of Poisson's ratio ν	Reliability indexes β	Design points		Sensitivity factors		Limit displacements
		E [MPa]	ν	α_E	α_ν	FEM
0.1	2.25	14.35	0.099	0.955	0.024	0.11530
0.2	2.48	13.94	0.193	0.951	0.094	0.11368
0.3	2.84	13.40	0.268	0.930	0.218	0.11148
0.4*	3.33	12.80	0.330	0.889	0.354	0.10866
0.4**	3.35	12.77	0.333	0.906	0.302	0.10846

E and ν stochastically independent

Table 5. Reliability calculations of parabolic response surfaces with an oedometric term. In the input data, the Young modulus has a smaller 5-20 MPa range. To carry out the necessary calculations there was a need to add a constraint to avoid false design points. 0.4*, 0.4** - respectively, normal and uniform distribution of Poisson's ratio.

Mean value of Poisson's ratio ν	Reliability indexes β	Design points		Sensitivity factors		Limit displacements
		E [MPa]	ν	α_E	α_ν	FEM
0.1	2.51	13.65	0.099	0.989	0.033	0.12121
0.2	2.82	13.11	0.188	0.981	0.128	0.12126
0.3	3.29	12.47	0.257	0.956	0.260	0.12104
0.4*	3.83	11.88	0.307	0.902	0.411	0.12056
0.4**	3.89	11.65	0.324	0.941	0.3234	0.12034

$\mathrm{Cor}(E, \nu)=0$

On the other hand, the results in Table 4 differ from those in Tables 5 and 3. This means that when the range of values of Young's modulus is increased, the fitting accuracy decreases despite the repetition of the regression procedure. The results of settlement presented in the last column of Table 4 also confirm the poorer accuracy of the fitting, as they differ more significantly from the threshold set (u_{max} = 0.12 m) than do those in Table 5. On the other hand, if we assume that accuracy tolerance of the settlement computations is closer, being ca 0.5 cm, then the results in Table 4 may also prove to be satisfactory. Moreover, the values of reliability indexes β in Table 4 are smaller than those presented in Table 5, which suggests that computations were done with a certain safety margin. Comparison of the results in Tables 4 and 5 allows one to acknowledge certain facts which are essential when the response method is considered in terms of reliability problems. Although the expected value of Young's modulus E equalled 20 MPa, the surface fitted the case of the modulus interval (5; 30) but fitted worse than it did in the case of the interval (5; 20). This proves that when reliability problems are taken into account, the fitting of the response surface in the vicinity of a design point is much more essential than the same fitting in the neighbourhood of the values of the point composed of the expected values of random parameters.

In order to estimate the effect of correlation between parameters E and ν on the reliability indexes, the computations of various values of the correlation coefficient were done. Some of the results are shown in the Tables below. In Table 6, one finds the results for positive values of correlation coefficient ρ = 0.3 and also the results obtained when it is assumed that Poisson's ratio ν does not change randomly (constant value).

Table 6. The effect of the correlation coefficients between Young modulus and Poisson's ratio on reliability calculation results. The parabolic response surfaces with an oedometric term and Young's modulus in the 5-20 MPa range was applied. 0.4*, 0.4** - respectively, normal and uniform distribution of Poisson's ratio.

Mean value of Poisson's ratio ν	Reliability indexes β	Sensitivity factors		Design points		Reliability indexes β	Design points	
		α_E	α_ν	E [MPa]	ν		E [MPa]	ν
0.1	2.49	0.989	0.027	13.69	0.088	2.51	13.64	0.1
0.2	2.73	0.986	0.096	13.26	0.167	2.83	13.02	0.2
0.3	3.09	0.977	0.164	12.69	0.235	3.40	11.97	0.3
0.4*	3.47	0.940	0.317	12.22	0.279	4.35	10.41	0.4
0.4**	3.61	0.979	0.155	11.81	0.311	ν - deterministic		
	Cor(E, ν)=0.3							

In Table 7, the results for negative values of correlation coefficient ρ equal to –0.2 and –0.6 are shown. Comparison of the results in Tables 5, 6 and 7 leads to the conclusion that the reliability index decreases with the increase of a positive correlation between the parameters E and ν and increases when a negative correlation between E and ν is higher. In the case of small expected values of Poisson's ratio, i.e. E(ν) = 0.1 and E(ν) = 0.2, the results for cases with zero-correlation and non-random (deterministic) parameter ν are very similar, but they may be different when the expected value of Poisson's ratio is greater.

Table 7. The effect of the negative correlation coefficients on reliability calculation results. The parabolic response surfaces with an oedometric term and Young's modulus in the 5-20 MPa range was applied. 0.4*, 0.4** - respectively, normal and uniform distribution of Poisson's ratio.

ν	β	Design points		Sensitivity factors		β	Design points		Sensitivity factors	
		E [MPa]	ν	α_E	α_ν		*E* [Mpa]	ν	α_E	α_ν
0.1	2.53	13.65	0.106	0.989	0.034	2.57	13.54	0.122	0.989	0.031
0.2	2.89	12.97	0.210	0.979	0.143	3.08	12.60	0.246	0.979	0.129
0.3	3.48	12.27	0.276	0.934	0.327	4.08	11.29	0.351	0.906	0.385
0.4*	4.17	11.61	0.328	0.867	0.477	5.47	10.89	0.378	0.741	0.646
0.4**	4.19	11.51	0.334	0.893	0.427	5.51	11.19	0.360	0.710	0.680
	cor(E,ν)=-0.2					cor(E,ν)=-0.6				

If the assumption of a negative correlation in natural soil deposits is accepted and if there is an insufficient sample size to estimate the correlation coefficient statistically, then it would be safest to do computations assuming the stochastic independence of random variables *E* and ν. Should, however, the correlation be positive (as in the case of some kinds of rocks), in the reliability computations the coefficient of correlation would have to be determined exactly in every case. Table 7 shows that the value of reliability index β decreases when the correlation becomes higher, especially at greater values of the ratio ν. Taking into account the effect of probability distribution of Poisson's ratio whose expected value E(ν) equals 0.4 (uniform or Gaussian), we can arrive at the conclusion that this effect is not important. Smaller values of the reliability index and greater values of sensitivity coefficient α_ν are obtained for normal distributions. It seems, however, that a uniform distribution is better, because it guarantees that, from the theoretical viewpoint, any unreal values of Poisson's ratio do not occur.

6 An application of neural network approximators to response surface model

As it has been mentioned in the end of the section 4 the effect of two branches of the limit state function had appeared also in the case of parabolic response surface with the oedometric term. Difficulties caused by a false design point occurrence when the parabolic response surface is used in reliability computations are consequences of properties of the second degree polynomial function as well as the necessity of rectification of variability intervals of independent random variables in the vicinity of investigated design point. To overcome this difficulties an application of the neural network technique is proposed, namely the utilising of neural network with a hyperbolical activation function. If a continuous function is approximated then the three-layered neural network with one hidden layered containing necessary number of neurones could give a satisfactory result. This is a consequence of the Kolmogorov theorem (Hecht-Nielson 1991). Neural networks allow approximation in cases of large variability intervals of independent variables preserving sufficient accuracy. To get required accuracy, according theorem, the number of neurones in the hidden layer is increased. Hence, there is no need to enlarged accuracy of approximation in the vicinity of the design point and therefore avoiding of creation false design points is possible. In the example presented here a three-layered neural network of two input neurones representing two independent variables, Young's modulus E and Poisson's ratio ν, will be used. The hidden layer consists of two neurones with hyperbolic tangent as the activation function. The summation neurone with a bias is at exit. The architecture of the applied network is presented in Figure 7.

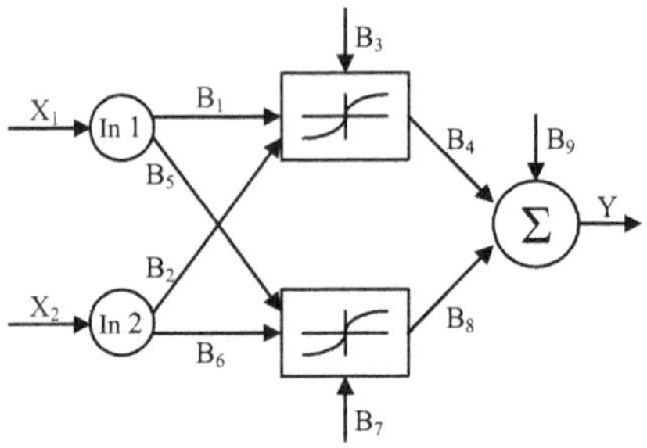

Figure 7. The architecture of the neural network applied. The network consists of three layers. Two neurones are in the entrance layer.

This network includes nine weights, that have to be found in the learning process. As the consequence of the applied network the following response surface will be obtained:

$$Y = B_4 \tanh(B_1 X_1 + B_2 X_2 + B_3) + B_8 \tanh(B_5 X_1 + B_6 X_2 + B_7) + B_9 \tag{9}$$

Instead of classical method of learning a network by the algorithm of backwards error propagation the nine weights $B_1 \ldots B_9$ were evaluated in a one iteration process of non-linear regression method based on Marquarth algorithm (Marquarth 1963). The equation (9) has been applied to create the response surface for settlements in the example described in Section 5.1 (U = Y). Next the standard reliability computations were carried out.

In Table 8 results of these computations are presented. It is worth mentioning that a single surface was able with high accuracy to approximate settlements for entire variability area of Poisson's ratio as well as wider area of Young modulus (5 - 30 Mpa). The obtained results are almost identical with those presented in Table3. This means that the response surface originated from neural networks can improve the fitness of approximation. On the other hand neural network approximations allows to avoid false design points. This way the response surface originated from neural networks, despite of some computational troubles like saturation particular neurones, gives an efficient way for reliability computations concerning settlements of foundations. It seems to be the best one among responses surfaces presented considered above.

Table 8. Results of reliability calculations for response surface originated from neural network. In the input data Young modulus and Poisson's ratio have full their ranges. 0.4*, 0.4** - respectively, normal and uniform distribution of Poisson's ratio.

Poisson ratio mean values	Indexes of reliability	Design points		Sensitivity factors		Limit displacements
ν	β	E [Mpa]	ν	α_E	α_ν	FEM
0.1	2.49	13.66	0.099	0.997	0.039	0.12112
0.2	2.79	13.11	0.190	0.991	0.115	0.12126
0.3	3.22	12.52	0.260	0.968	0.260	0.12056
0.4*	3.77	11.95	0.309	0.911	0.409	0.11985
0.4**	3.82	11.75	0.324	0.947	0.317	0.11932
	Cor(E,ν)=0					

7 Single pile settlement

The purpose of this section is to demonstrate that utilising the neural network based response surface allows to overcome difficulties in finding a design point when the FORM method is applied. The design point can be hardly obtain if the limit state surface is given by a close regular surface as, for example, a circle or an ellipse in two-dimensional case. This may happen when the response surface based on the polynomial of the second degree is applied. An example of single pile settlement analysis will help to illustrate this phenomenon.

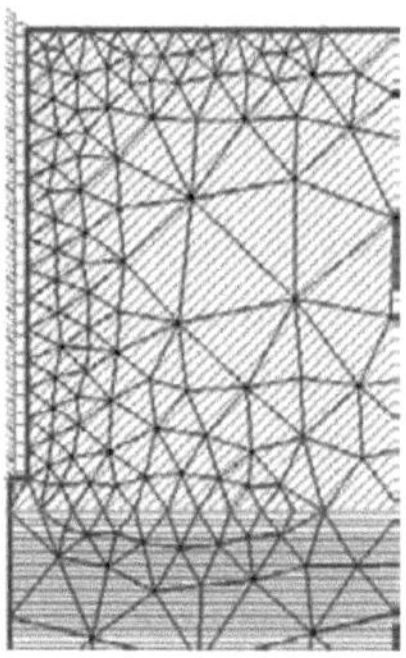

Figure 8. Finite element model for pile settlement computations.Young's moduli of upper and lower layers are lognormally distributed:$(E_1) = 60$ MPa;$\sigma_E = 12$ MPa; $E(E_2) = 80$ MPa, $\sigma_E = 16$ MPa. Poisson's ratio for both layers: $\nu = 0.25$ (non-random).

Consider now a single pile embedded in a two-layered linear-elastic subsoil as it is presented in figure 3. Additionally it is assumed that: The thickness of the upper layer is h_1=7.65, pile's length and diameter are L = 7.15 m and D = 0.6 m, respectively. The pile is axially loaded by a force P = 816 kN. As an allowable level of settlement u_{max} = 0.03 m is assumed.

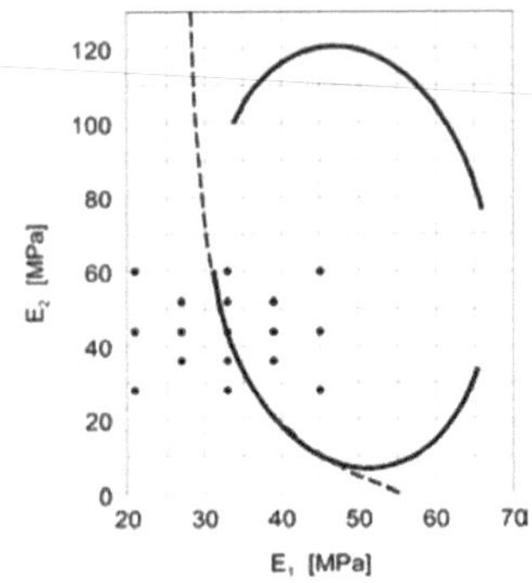

Figure 9. Limit state curves. The ellipse ori-ginates from second degree approximation. Dash line comes from neural network approximation. Dots represent FEM simulations.

Two random variables, Young's moduli E in both layers, were considered as stochastically independent. In order to obtain a response surface approximation a set of finite element simulation, near allowable settlement level u_{max} , were carried out. The mesh considered shown in figure 3. The pairs of moduli applied in finite element computations are marked by dots in figure 4. As a first attempt of finding the limit state surface by response surface method an approximation by a polynomial of the second degree was examined. This was done in accordance with the eqn. (4), where X_1 and X_2 stand for moduli E_1 and E_2, respectively. The line obtained can be regarded as a good approximation of the limit state surface within the area marked by dots (area of FEM simulations). The other part of the curve is obtained automatically as a result of applying equation (4). Checking coefficients (eqn. 4) by well-known mathematical criterion it is easy to demonstrate that the curve is an ellipse in this case. In the figure 4 the gaps in ellipse are due to software drawbacks. Next the reliability computations, by means of the SORM method, in order to find measures p_F and β were carried out. The shape of the curve and especially its image in the standard normal space (see Figures 9 and 10) produces false design point, which is marked as $\mathbf{y}_F^*$ in Figure 10. This is caused by unjustified curvature produced by second degree approximation in the vicinity of the real design point, which is marked as $\mathbf{y}_T^*$ in Figure 10. Finally the global minimum was found at the upper part of the curve instead of the local one. The coordinates of the point $\mathbf{y}_F^*$ (in physical coordinates) were E_1 = 65.66 MPa and E_2 = 79.85 MPa. For these values the settlement was 0.0156 m that is far form the threshold u_{max} = 0.03 m (this part of the curve is far from the approximation region and does not reflect the settlement value). In Section 4.1 a method of inserting some constrains for random variables ranges in the limit state function was proposed. But applying constrains in the presented example did not lead to satisfactory results due to second order approximation drawbacks.

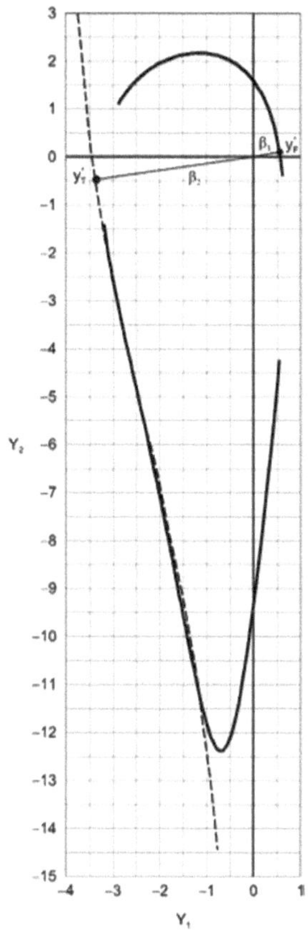

Figure 10. The solid line shows the image of the ellipse in the standard normal space. Only the false design point, denoted by $\mathbf{y}_F^*$, could be foun in this curve. The true design point, $\mathbf{y}_T^*$, was found in neural network curve (dash line).

As a consequence in the next step the second degree approximation was replaced by a surface resulting from neural network approach, accordingly to equation (9). The resulting curve in "physical space" is marked in Figure 9 by a dash line. The field of approximation is the same as before (marked by dots in Figure 10). However, it was checked that the area of well fitness is far greater than in the case of second degree. The image of the curve in standard normal space is illustrated in Figure 10. Both curves in normal space and physical one show the lack of danger of obtaining any false design point. The "real" design point was obtained, which is marked as $\mathbf{y}_T^*$ in Figure 10. This point has the following values of moduli $E_1 = 30.27$ MPa and $E_2 = = 71.07$ MPa and for them the settlement is 0.0307 m, which means that the design point belongs to the limit state surface (approximately).

The above example demonstrates how the neural base response surface could help to overcome difficulties in finding a design point in reliability oriented settlement computations.

8 Concluding Remarks

Choosing the limit state function necessary for reliability calculations in its closed form is preferred. Unfortunately, the closed form often does not exist. The response surface method is chosen most often when the set of points of settlement being a function of variables considered as random variables is relatively large. Such a situation may arise if the number of field measurements is large. However, the approximated form of the limit state function is most often determined by means of a certain numerical procedure, for example the finite element method. Using a polynomial of the second degree to approximate the set of data points is considered as a standard procedure. In order to reduce labour consumption of reliability computations, it is advisable to modify this function. This allows for extending the area where it gives a satisfactory approximation of the shape of the function being analysed. An additional term to the approximation by a polynomial of the second degree can be added. The form of this term depends on the problem under consideration. In the case of shallow foundations (also raft foundations), it could be the oedometric term described in section 4.2. Inserting the oedometric term improves the quality of the fitting and enables the range of approximation to be extended.

The examples analysed enable one to reach the conclusion that the response surface method is good enough to examine the reliability of the serviceability limit of foundations. However, it should be stressed that in these cases we are dealing only with two random variables. If the number of random variables increases the problem will become seriously complicated, because the amount of computations has to be increased substantially to estimate the coefficients of the assumed hypersurface. In such a case, the appropriate regression algorithm, which affects significantly the fitting quality, is of vital importance. The algorithm based on Marquart's compromise may be recognised as highly effective. The approximation efficiency can be improved by means of successive approximations, but in the steps following the first one the surface fitting takes place in the neighbourhood of the design point.

The characteristic feature of the response surface in the form of a polynomial of the second degree manifested itself as a possibility for obtaining the surface in the form of two branches. Both branches can have local minima which, in turn, may result in false design points. In order to avoid errors, we take advantage of the MES computations to verify the minima and then we introduce the appropriate constraints on the variation interval of random parameters. A more efficient way seems application the response surface originated from neural networks. It allows avoiding difficulties in finding "true" design point when the FORM and SORM methods are in use.

Results of the reliability computations done for shallow foundation settlement allows one to draw the following conclusions:

In the case of soils with small values of Poisson's ratio (expected value $E(\nu)= 0.2$, or smaller), the effect of random variability of this ratio on settlement variation is very poor and usually can be disregarded.

Reliability index β corresponding to the probability of exceeding the settlement threshold increases with the augmentation of the negative correlation (its absolute value) between parameters E and ν. Because there is no available information about the value of the correlation coefficient between these parameters, it seems safe to assume their stochastic independence. Scant information about this correlation in geotechnical literature constrains us to presume that if there is a correlation between these parameters, it is negative. The possible existence of a positive correlation between the above parameters could complicate the problem, since the reliability index decreases with the increase of a positive correlation.

The probability distribution of parameters E and v should be fitted in such a way that the appropriate random variables take the values admissible from a physical viewpoint. In the first place, it refers to Poisson's ratio whose interval of possible values is usually considerably narrow, which also puts a constraint on the value of the coefficient of variation random variable ν.

Probabilistic modelling of parameters E and ν is difficult, because statistical data on these parameters for soils are sadly lacking. Such a modelling is of primary importance in reliability analyses of close accuracy.

References

Bauer J., Puła W. (2000). Reliability with respects to settlement limit-states of shallow foundation on linearly-deformable subsoil. *Computers & Geotechnics*, 25 No.3-4: 281-308.

Becker, E.D. (1996). An overview of the foundation design process (part I). Development for the National Building Code of Canada (part II). *Canadian Geotechnical Journal,* 33, 1996: 956-1007.

Box, G.P., Draper, N.R. (1996). *Empirical Model-Building and Response Surface*. New York : J. Wiley & Sons.

Brząkała W., Puła W.(1996). A probabilistic analysis of foundation settlements, *Computers and Geotechnics,* 18: 291-309.

Brząkała W., Puła W., El-Meligy M. (1995). Settlement analysis by non-linear finite element method, In *Applications of Statistics and Probability, Proc. of the ICASP 7 Conference,Paris 1995*,M. Lemaire, J-L. Favre and A. Mebarki eds.. Rotterdam, Balkema: 55-63.

Bucher, C.G., Bourgund, U. (1990). A fast and efficient response surface approach for structural reliability problems. *Structural Safety,*7: 57-66.

Ditlevsen, O., Madsen, H.O. (1996). *Structural Reliability Methods.* Chichester: J. Wiley & Sons.

El-Meligy M., Harraz A., Rackwitz R. (1997). Reliability of tunnel using a response surface methodology. In *Proc. of the Euromech 372 „Reliability in Nonlinear Structural Mechanics,* O. Ditlevsen and J.C. Mitteau eds., Clermond-Ferrand: 119-122.

Engelund S., Rackwitz R. (1992). Experiences with experimental design schemes for failure surface estimation and reliability. In *Proc. 6th Speciality Conf. Probabilistic Mechanics and Structural and Geotechnical Reliability*, Denver 1992, 252-255.

Faravelli, L.A. (1989). A response surface approach for reliability analysis, *Journal of the Engineering Mechanics Division*, ASCE, 115(12): 2763-2781.

Hecht-Nielson, R. (1991). *Neurocomputing*. Amsterdam: Addison Wesley.

Hoek E., Carvalho J. L., Corkum B. T. (1992). *PHASES, a 2D Hybrid FE/BE Program for* Calculating Stresses and Estimation Support around Underground Excavations, *Program* Manual, 1992.

Liu, P-L., Der Kiureghian A. (1986). Multivariate distribution models with prescribed marginals and covariances. *Probabilistic Engineering Mechanics,* **1**(2), 1986: 105-112.

Marquardt D. W. (1963), An algorithm for least-squares estimation of non-linear parameters, *J. Soc. Indust. Appl. Math*, 11, No. 2.

Marquardt D. W. (1966). *Least-squares estimation of non-linear parameters computer code*, NLIN2, Distribution No. 309401, IBM Share Library.

Meyerhof, G.G. (1995). Development of geotechnical limit state design. *Canadian Geotechnical Journal,* 32, 1995: 128-136.

Myers, R.F. (1971). *Response Surface Methodology.* Boston: Allyn and Bacon, Inc.

PN-81/B-03020. (1981). Polish Standard. Foundation bases. Static computations and design.

Rackwitz R. (1982). Response surfaces in structural reliability. *Berichte zur Zuverlässigkeitstheorie der Bauwerke,* Heft 67 LKI: Technische Universität München.

STRUREL (1995). *A Structural Reliability Analysis Program System,* COMREL & SYSREL. Users Manual, RCP Consultant, Munich

Tsytovitch N. A. (1973). *Mechanika gruntov*, Izd.2, Moskwa: Vysšaja Škola (in Russian).

Wong, F.S. (1985). Slope reliability and response surface method. *Journal of Geotechnical Engineering,* 111, No 1, ASCE: 32-53.

Reliability of Laterally Loaded Rigid Piles

Wojciech Puła[1]

Institute of Geotechnics and Hydrotechnics, Wrocław University of Technology, Wrocław, Poland

Abstract. Due to modernisation of main railway tracks in Poland there was a need to change the existing overhead electrical transmission lines together with their supports and foundations of the supports. One of possible way to construct a new foundation of such support is the direct connection of the support with a single pre-cast concrete pile embedded in soil. The piles used are usually short, then in a certain soil conditions have to be treated as rigid piles and the ultimate soil lateral resistance has to be considered. Computations of rigid piles by Brinch Hansen method demonstrates high sensitivity of ultimate lateral loading to precise determination of the rotation centre of the pile under consideration. The position of the centre is affected by some random factors, for example random variability of soil properties and loading applied. due to very complex nature of solution of equilibrium equations in the case of Brinch Hansen method being in use any evaluation of safety measure constitutes rather difficult problem. Two alternative approaches are suggested. one is an algorithm supported by some symbolic computations combined with some power series expansions. The second one bases on the response surface method and can be applied both for cohesive and non cohesive soils. Within the presentation some important numerical aspects will be discussed. Additionally computational examples allow the study a relationship between "classical" safety factor versus reliability index as well as an effect of spatial averaging.

1 Introduction

Due to modernisation of main railway tracks in Poland there is a need to change the existing overhead electrical transmission lines together with their supports and foundations of the supports. One of possible way to construct a new foundation of such support is the direct connection of the support with a single pre-cast concrete pile embedded in soil (as presented in Figure 1). This kind of foundations was successfully applied in some important railway lines in Europe. A special equipment allows to install these piles in a fast way, which is important advantage (idle time reduction) in comparison with traditional block (massive) foundations. In the design of such pile foundations lateral forces and moments are of the vital importance. The piles used are usually short, then in a certain soil conditions have to be treated as rigid piles and the ultimate soil resistance must be considered. For rigid piles one of the most precise method to evaluate ultimate lateral resistance (Poulos and Davies, 1980) is the procedure recommended by Brinch Hansen (1961). In this procedure, the centre of rotation of a rigid pile has to be found. In can be demonstrated that the value of the ultimate horizontal loading H_u , which can be applied in the head of the pile is extremely sensitive to a position of the rotation centre z_r (Figure 1). Then the accuracy

[1] Co-author of the sections 3.3 and 3.4 is Adrian Różański, Wrocław University of Technology.

of determining the rotation centre is very important for a precise computation of an allowable horizontal force H_a . On the other hand the position of the centre is affected by some random factors. Among them the friction

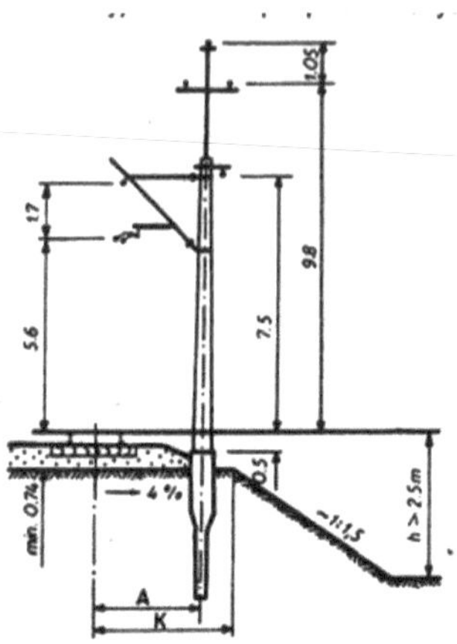

Figure 1. An example of overhead electrical transmission line support founded on a single pile (railway line near Wrocław, Poland).

angle of the subsoil as well as the uncertainty of geotechnical recognition (the number of bore-hols along the railway track can be not sufficient enough) are of the prime importance. Then an important problem is to investigate for such a value of the total safety factor which can guarantee a small probability that the applied load H_a exceed the ultimate loading H_u . This problem is a typical one within the framework of the structural reliability theory. However, due to the nature of the solution of the equilibrium equations, which could not be written in a closed analytical form, the existing structural reliability procedures could not be applied straightforward. The solution of the reliability problem will be demonstrated and discussed in the section 4 of this paper.

2 General assumptions and the Brinch Hansen method

The following general assumptions are imposed (see Figure 2):

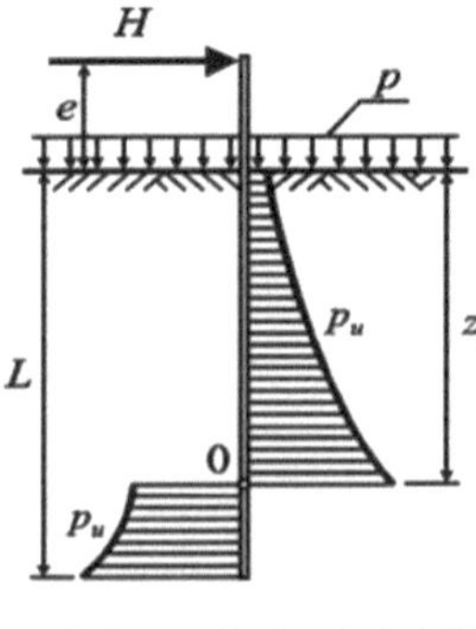

Figure 2. Laterally-loaded rigid pile.

A short rigid pile is considered with an unrestrained head. Any elastic deformations within the pile material are neglected.

- In the state of failure the pile is assumed to rotate as a rigid body about a rotation centre at the depth z_r.
- The pile is subjected to a horizontal force H_a and the ultimate soil pressure at any depth z below the soil surface is p_u .
- The limiting values H_u and M_u , to cause failure - that is, to mobilise the ultimate soil resistance along the pile, may be obtained by considering equilibrium of horizontal forces and moments, and solving the following resulting simultaneous equations for the unknown depth of rotation z_r , and the ultimate horizontal load H_u :

$$H_u = \int_0^{z_r} p_u(z)Ddz - \int_{z_r}^{L} p_u(z)Ddz \tag{1}$$

$$M_u = H_u e = -\int_0^{z_r} p_u(z)Dzdz + \int_{z_r}^{L} p_u(z)Dzdz \quad , \tag{2}$$

where L is the embedding of the pile in soil, D is its diameter or width and e is the eccentricity of loading.

According to the Brinch Hansen approach (Brinch Hansen, 1961) the resultant (passive minus active) ultimate lateral soil resistance (per unit area), based essentially on limit state and earth-pressure theory, on an arbitrary depth z can be expressed by the following equation:

$$p(z) = q(z)K_q(z) + cK_c(c) \quad , \tag{3}$$

where $q(z)$ is the effective vertical overburden pressure at the depth z

$$q(z) = p + \gamma z_d + \gamma' z_s ; z = z_d + z_s \tag{4}$$

and γ, γ' are unit weights above ground water table and below, respectively and c is the cohesion of the soil. The coefficients $K_q(z)$, $K_c(z)$ are the pressure coefficients which depend on the friction angle φ of the soil, diameter of the pile and the depth z. They can be obtained by means of non-linear interpolation formulae, where solutions for the ground level and solutions at great depth are utilised. Namely

$$K_q = \frac{K_q^0 + K_q^\infty a_q \dfrac{z}{D}}{1 + a_q \dfrac{z}{D}} \tag{5}$$

$$K_c = \frac{K_c^0 + K_c^\infty a_c \dfrac{z}{D}}{1 + a_c \dfrac{z}{D}} \tag{6}$$

where

$$a_q = \frac{K_q^0}{K_q^\infty - K_q^0} \frac{K_0 \sin\varphi}{\sin\left(\frac{\pi}{4} + \frac{\varphi}{2}\right)} \tag{7}$$

$$a_c = \frac{K_c^0}{K_c^\infty - K_c^0} 2\sin(\frac{\pi}{4} + \frac{\varphi}{2}) \tag{8}$$

$$K_0 = 1 - \sin\varphi \tag{9}$$

K_q^0 and K_c^0 are the solutions for the ground level and can be expressed by the following equations:

$$K_q^0 = e^{(\frac{1}{2}\pi+\varphi)\tan\varphi} \cos\varphi \tan(\frac{\pi}{4} + \frac{\varphi}{2}) - e^{-(\frac{1}{2}\pi-\varphi)\tan\varphi} \cos\varphi \tan(\frac{\pi}{4} - \frac{\varphi}{2}) \tag{10}$$

$$K_c^0 = \left[e^{(\frac{1}{2}\pi+\varphi)\tan\varphi} \cos\varphi \tan(\frac{\pi}{4} + \frac{\varphi}{2}) - 1 \right] \cot\varphi \tag{11}$$

The coefficients K_q^∞ and K_c^∞ are the coefficients for great depth. Namely:

$$K_c^\infty = N_c d_c^\infty \tag{12}$$

$$K_q^\infty = K_c^\infty K_0 \tan\varphi = N_c d_c^\infty K_0 \tan\varphi \tag{13}$$

$$N_c = \left[e^{\pi\tan\varphi} \tan^2(\frac{\pi}{4} + \frac{\varphi}{2}) - 1 \right] \cot\varphi \tag{14}$$

$$d_c^\infty = 1.58 + 4.09 \tan^4\varphi \tag{15}$$

Brinch Hansen has proposed the "total safety factor" as a measure of safety defined as the ratio of the ultimate lateral force H_u and the applied lateral force H_a

$$F = \frac{H_u}{H_a} \tag{16}$$

Now the equations (1) and (2) have to be solved with respect to unknown parameters z_r and H_u taking into account relationship defined by equations (3) – (15). It can be proved (Puła 1997, 2004) that the equation for the rotation centre z_r can be written as follows:

$$a_o + a_o'c + (a_1 + a_1'c)z_r + (a_2 + a_2')z_r^2 + (a_3 + a_3')z_r^3 + b_1 \ln(D + a_q z_r) + b_1'c \ln(D + a_c z_r) = 0 \tag{17}$$

where c denotes the cohesion of soil and a_o, a'_o, a_1, a'_1, a_2, a'_2, a_3, a'_3, b_1, b'_1, a_q, a_c are coefficients depending on soil properties, namely the friction angle φ and the unit weight γ, as well as load parameters like overburden pressure p and eccentricity e (see Figure 2). For example the coefficients a_o takes the form

$$a_0 = -\left\{\frac{1}{3}\gamma\left(K_q^\infty + cK_c^\infty\right)L^3 + \left[\frac{(e\gamma + p)}{2}\left(K_q^\infty + cK_c^\infty\right) + \frac{\gamma D}{2a_q}\left(K_q^0 - K_q^\infty\right) + c\frac{D\gamma}{2a_c}\left(K_c^0 - K_c^\infty\right)\right]L^2\right\} +$$
$$-\left\{\left[ep\left(K_q^\infty + cK_c^\infty\right) + \frac{D}{a_q}\left(K_q^0 - K_q^\infty\right)\left(e\gamma + p - \frac{\gamma D}{a_q}\right) + \frac{cD}{a_c}\left(K_c^0 - K_c^\infty\right)\left(e\gamma + p - \frac{\gamma D}{a_c}\right)\right]L\right\} +$$
$$-\left\{\frac{D}{a_q}\left(K_q^0 - K_q^\infty\right)\left(ep - \frac{\gamma De}{a_q} - \frac{pD}{a_q} + \frac{\gamma D^2}{a_q^2}\right)\ln\left(D^2 + a_q LD\right)\right\} +$$
$$-c\left\{\frac{D}{a_c}\left(K_c^0 - K_c^\infty\right)\left(ep - \frac{\gamma De}{a_c} - \frac{pD}{a_c} + \frac{\gamma D^2}{a_c^2}\right)\ln\left(D^2 + a_c LD\right)\right\} \tag{18}$$

The complicated nature of the above dependencies produces significant difficulties in further reliability computations. By substitution of the equation (17) into (1) one gets (Puła 2004)

$$H_u = a_o'' + a_o'''c + \left(a_1'' + a_1'''c\right)z_r + \left(a_2'' + a_2'''c\right)z_r^2 + b_1''\ln\left(D + a_q z_r\right) + b_1'''c\ln\left(D + a_c z_r\right) \tag{19}$$

As before the coefficients in eqn. (19) are also functions of soil properties and load parameters.

3 Random fluctuations and the reliability approach

Applying the Brinch Hansen method in ultimate lateral resistance estimation it can be easily observed a high sensitivity of the ultimate loading value H_u to a position of pile's rotation centre z_r. Even small fluctuations in the position of the centre z_r (of a range of 3.5 cm) can produce large variation in the value of the ultimate lateral force H_u (of a range of 5 kN to12 kN). This imposes that the coordinate z_r must be determined with high precision by solving the equation (2). Of course it can be done with the help of a computer. On the other hand it is well known that random fluctuations of soil properties in natural deposits are usually significant and very important in the context of engineering computations. It is also follows from equations (1)-(15) that any random fluctuation of the angle of internal friction can caused substantial changes in the position z_r of the rotation centre (strongly non-linear dependencies) and hence in the value of the ultimate loading H_u. An example of such a phenomenon is presented in the Table 1. If some soil properties and loads are a subject of random variability then a natural question arises how reliable the total safety factor (eqn. (16)) is. An appropriate reliability problem can be formulated as follows: find the probability p_F that the applied loading H_a exceeds the ultimate lateral loading H_u:

$$p_F = P\{H_a > H_u\} = P\left\{\frac{H_u}{H_a} < 1\right\} = P\{F < 1\} \tag{20}$$

Table 1. Changes of the ultimate lateral loading H_u due to friction angle variability.

Angle of internal friction φ [°]	Position of the rotation centre z_r [m]	Ultimate lateral loading H_u [kN]
26.4	2.106	13.30
27.5	2.110	14.61
28.6	2.113	16.07
29.8	2.117	17.67
30.9	2.121	19.45
32.1	2.125	21.43
33.2	2.129	23.62

The main difficulty in evaluating probability (17) is the lack of dependencies between the random soil parameters and the ultimate force H_u in an explicit form. As it can be seen from eqns (17) and (19) evaluation of H_u requires the expression for z_r. Additionally relationships between basic random variables, like friction angle, and ultimate lateral force are of the complicated nature involving many non-linear functions. In the framework of this chapter two ways of overcoming this difficulty will be suggested, namely: applying some symbolic computations and utilising the response surface method. In both cases computations are supported by the SORM method (see e.g., Hohenbichler et al, 1987 or Ditlevsen and Madsen, 1995).

In the following instead of probability of failure p_F the reliability index β is utilised, which is a measure more frequently used in structural reliability. Both measures are equivalent and correspond one to another by the relationship

$$p_F = \Phi_o(-\beta) \quad , \tag{21}$$

provided that $p_F < 0.5$, where Φ_0 is the standard normal one-dimensional distribution function.

In order to evaluate reliability index corresponding to a given value of total safety factor F (eqn. (21)) the following computational steps should be applied:

Select a value of the total safety factor F.

Assume that all random variables involved are constant and equal to their expected values. Applying these values evaluate the ultimate load H_u utilising equilibrium equations (1) and (2).

For the value of F selected in step 1 find, by means of eqn. (16) value of the load H_a.

Assuming that H_a is a random variable with the expected value computed in step 3 and the friction angle is a random variable with characteristics given in Table 1 evaluate reliability index β corresponding to probability of failure p_F given by (21).

Repeat above steps for several values of total safety factor F.

3.1 Solution supported by symbolic computations

The equation (17) can not be solved with respect to z_r in the closed form. To obtain a closed formula for z_r and consequently for H_u, the logarithmic terms in eqn. (17) can be expanded into a power series usually up to the term of degree three. This kind of approximation is easier in the case of cohesionless soils when only one logarithmic term appears in the equations (17), then

$$\ln(D+a_q z_r) \approx \ln(D+\theta \cdot a_q) + \frac{a_q(-\theta+z_r)}{D+\theta \cdot a_q} - \frac{a_q^{\,2}(-\theta+z_r)^2}{2(D+\theta \cdot a_q)^2} + \frac{a_q^{\,3}(-\theta+z_r)^3}{3(D+\theta \cdot a_q)^3} \quad (22)$$

The substitution of the right-hand side of eqn. (22) into (17) allows to solve (17) with respect to z_r. It is clear that the approximation point θ is a function of the soil friction angle φ as it is demonstrated in the Figure 3. The suggested way of approximation can be efficient for probabilistic

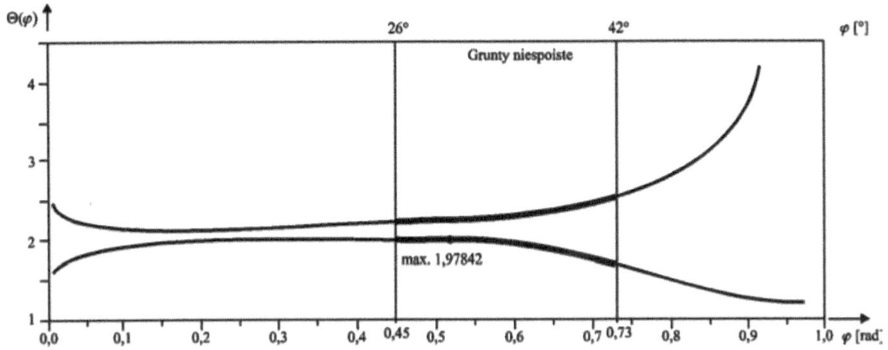

Figure 3. Approximation point θ versus friction angle φ. The selected value is $\theta = 1.97842$

computations if one approximation point θ can serve relatively wide interval of values of φ. In the Figure 3 there are two branches of the function $\theta(\varphi)$. The point selected for further examples has been marked in the central part of the lower branch. Having found the approximate equation for z_r one finds equation for the ultimate force H_u utilizing equation (19). The resulting formula is very complex and long, however all transformations can be carried out automatically by a computer systems doing symbolic computations, like *Mathematica* (2003) for example. Finally the resulting equation can be transferred to a code executing reliability computations according to FORM and SORM procedures. This way the probability (20) can be evaluated. It is necessary to mention, however, that in the case of cohesive soils the attempt to find an approximation point θ serving relatively wide range of friction angle fluctuations has failed.

3.2 Numerical example

As an example consider a rigid pile embedded into cohesionless soil as presented in Figure 2. The input data for this example are given in the Table 2. The computations have been carried out by utilizing the procedure described above. This approach will be in the sequel called the "symbolic algorithm". The reliability index β has been evaluated by means of the SORM method. Repeating the computations allows for investigations the reliability index β versus the "total safety factor" F. The corresponding graph is presented in Figure 4.

Table 2. Parameters of the problem.

Parameter	Probability distribution	Expected value	C.O.V. (%) *
Friction angle φ	Lognormal	33.6 °	10; 15
Applied load H_a	Normal	8 – 25 kN	10; 15; 20
Surface overburden p	Normal	8.8 kNm^{-2}	5
Eccentricity e	Normal	8.64 m	5
Unit weight γ	Constant (nonrandom)	20.15 kNm^{-3}	
Pile's diameter D	Constant (nonrandom)	0.36 m	
Pile's length L	Constant (nonrandom)	2.9 m	

* C.O.V. = coefficient of variation = (standard deviation / expected value) 100%

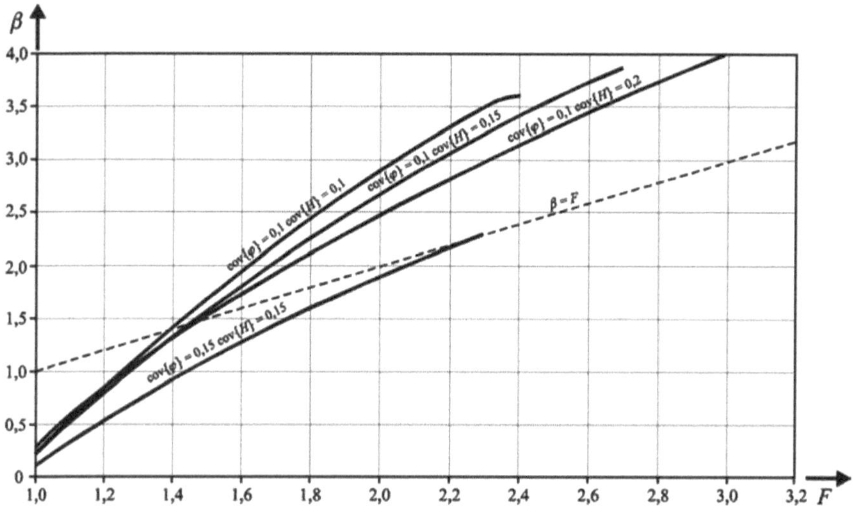

Figure 4. Reliability index β versus "total safety factor" F.

Sensitivity analysis has shown the friction angle random fluctuations are the most important among all parameters for the value of reliability index β ($\alpha_\varphi = 0.88$, provided that c.o.v.$\{\varphi\}$ = 10% and c.o.v.$\{H_a\}$ = 10% - the sensitivity coefficients α are defined in the Chapter 12). Another important random variable in the performed analysis appears the external lateral load applied H_a (α_H = = -0.37 provided that c.o.v.$\{\varphi\}$ = 10% and c.o.v.$\{H_a\}$ = 10%).

It is easy to observe (Figure 4) that in the case of the friction angle variation coefficient of c.o.v.$\{\varphi\}$ = 15% the total safety factor F must be assumed at least 3 to guarantee the reliability index β on the level 3. For the value of c.o.v.$\{\varphi\}$ = 10%, total safety index of a range 2.1-2.4 can be satisfactory to get level of β near 3. However it is worth mentioning that values less than 15% of c.o.v.$\{\varphi\}$ are rather not often observed in the case of inherent random variability of friction angle. This shows the necessity of the use of high values of total safety factors F when the Brinch Hansen method is applied in engineering computations.

3.3 Solution basing on response surface method

As a second solution of the problem considered the application of the response surface method is proposed. The response surface method has been described in the Chapter 13. In the case of numerical computations, relationships between the model parameters $x_1, x_2, \ldots x_n$, which are introduced as the input data (random parameters of the problem), and the values obtained as output data $y = f(x_1, x_2, \ldots x_n)$ is defined. In the case considered within this study $y = H_u$. The fitting of the function f is done by means of the non-linear regression analysis. Having the dependence $H_u = \mathrm{f}(x_1, x_2, \ldots x_n)$ the SORM method can be straightforward utilised. This way to each deterministically obtained (by substitution of mean values of random variables) value of the total safety factor F we are able to prescribed a certain probability p_F in accordance to (20) according to the algorithm specified in the beginning of the section3.

Let us now consider the example analysed in the section 3.2. In the beginning let us assume that the friction angle and the external lateral load H_a both of them having the same coefficients of variations $\text{c.o.v.}\{\varphi\} = \text{c.o.v.}\{H_a\} = 15\%$. In order to determine the response surface $y = H_u$, a series of evaluations of the force H_u have been carried out for different values of φ by means of equations (17) and (19). In this case for each value of φ equation (17) has been solved numerically. Next by non-linear regression algorithm based on the Marquardt compromise (Marquardt 1963) a dependence has been set of the following form

$$H_u = \left(a + b\sqrt{\varphi}\right)^c + d \tag{23}$$

where the coefficients have been found as equal to $a = 1.113$, $b = -0.1123$, $c = -4.208$ and $d = 0.5829$. The above curve fits very well φ - H_u dependence as it can be seen in Figure 5.

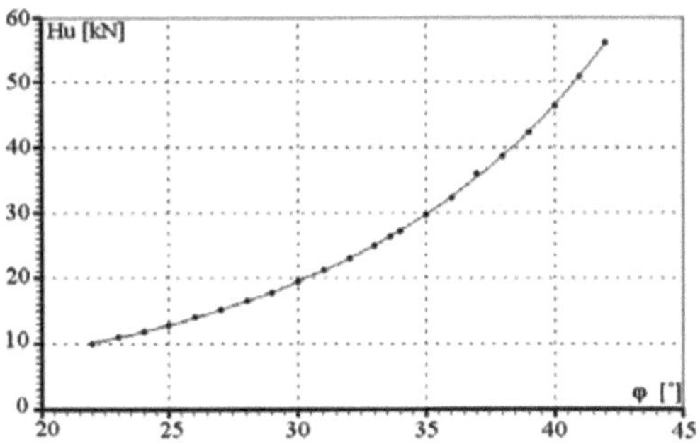

Figure 5. The response surface Pile length: $L = 2.9$ m

Having the eqn. (23), reliability computations were carried out utilising two different approaches, namely the "symbolic algorithm" in conjunction with the SORM method and the response surface in conjunction with the SORM method. The results in the form of reliability indices β are summarised in Table 3.

Table 3. Reliability indices β corresponding to specified values of total safety factor F.

Total safety factor F	Expected value of the lateral force H_a [kN]	Reliability index β by „symbolic algorithm"	Reliability index β by response surface
1.2	21.92	0.37	0.38
1.4	18.79	0.74	0.75
1.6	16.44	1.08	1.09
1.8	14.61	1.39	1.41
2.0	13.15	1.69	1.70
2.2	11.95	1.96	1.97
2.4	10.96	2.21	2.22
2.6	10.12	2.45	2.46
2.8	9.39	2.68	2.69
3.0	8.77	2.89	2.91
3.2	8.22	3.10	3.12

It is easy to see that β indices evaluated by support of the "symbolic algorithm" and the response surface method, corresponding to the same value of total safety factor F only slightly differ one to another. Rounding results to one digit after point (which is quite reasonable in geotechnics) both indices (corresponding to the same value of F) are the same. This demonstrates good quality of evaluation by applying response surface method. Note that in order to have the reliability index greater than 3 (the ISO 2394 standard requires 3.8 for moderate consequences of a failure) it is necessary to fix the total safety factor equal to 3.2.

Results demonstrated in Table 3 are associated with the length of the pile equal to $L = 2.9$ m. In order to involve errors in precision of the pile embedding (imprecision of technology) response surfaces have been fitted to ultimate loading H_u in the case of different pile's length, namely $L = 2.4$ m, $L = 2.8$ m, $L = 3.0$ m, $L = 3.2$ m and $L = 3.4$ m. This approach allows to involve three random variables into the problem, namely the friction angle φ, the external lateral load H_a and the pile length L. After some numerical simulations a new response surface has been found in the following form:

$$H_u = aL^b(c-\varphi)^d \tag{24}$$

As before, four unknown coefficients were determined by means of non-linear regression. It was found that:

$$H_u = 0.17\cdot 10^{20} L^{2.868}(141.6-\varphi)^{-9.411} \tag{25}$$

The surface (25) fits computed values of H_u very well, as it is demonstrated in Figure 6. It is easy to see that the greatest values of the ultimate force H_u are obtained for the cases of large values both friction angle φ and length of the pile L. In next steps the series of reliability computations have been carried out.

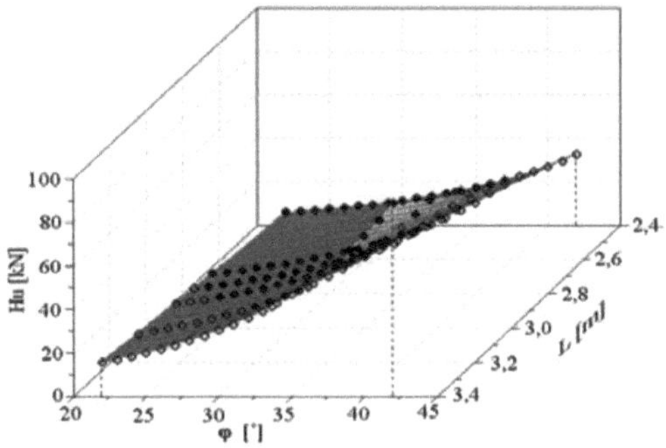

Figure 6. The response surface obtained in two-dimensional case.

Table 4 summarises the results. Evaluations have been made by the "symbolic algorithm" with the coefficient of variation of length, c.o.v.$\{L\}$ = 2% and by the response surface for two values of coefficient of variation: cov$\{L\}$ = 2% (5.8 cm) and c.o.v.$\{L\}$ = 5% (14.5 cm). As in the case of earlier computations the results of reliability indices almost coincide for both methods (regarding the same value of F). Comparing the results, presented in columns third and fourth of the Table 4, one can see rather small effect of length`s random variability, provided that the values of cov$\{L\}$ are reasonable.

Table 4. Reliability indices received from analysis with three random variables (φ, L, H_a)

Total safety factor F	Reliability index β by „symbolic algorithm" cov {L} = 0.02	Reliability index β by response surface cov {L} = 0.02	Reliability index β by response surface cov {L} = 0.05
1.2	0.36	0.37	0.35
1.4	0.73	0.73	0.70
1.6	1.07	1.07	1.01
1.8	1.38	1.38	1.30
2.0	1.66	1.67	1.56
2.2	1.93	1.94	1.81
2.4	2.18	2.19	2.04
2.6	2.42	2.44	2.26
2.8	2.64	2.67	2.46
3.0	2.87	2.90	2.66

Finally it is worth mentioning that the numerical effort in the response surface method significantly grows with increasing the number of random variables involved in the problem.

3.4 Spatial averaging application

By now reliability computations in the framework of this example have been performed without any spatial averaging. On the other hand it is well-known that spatial averaging of soil properties random fields, proposed by Vanmarcke (1977), leads to more realistic values of reliability indices. Then in the next step the spatial averaging of the internal friction angle random field has been applied. According to this, a new random variable $\varphi_{\overline{L}}$ is defined as follows:

$$\varphi_{\overline{L}} = \frac{1}{\overline{L}} \int_{\overline{L}} \varphi(z) dz \tag{26}$$

where $\varphi(z)$ is the random function, which describes random variability of the friction angle φ with the depth z and L is the pile length (in cases where L is treated as random variable $\overline{L}$ is understood as the expected value of L). The function $\varphi(z)$ is assumed to be stationary with constant mean value m_φ and constant point variance $\sigma_\varphi{}^2$. The variance of $\varphi_{\overline{L}}$ can be computed as:

$$\text{VAR}\left[\varphi_{\overline{L}}\right] = \sigma_{\overline{L}}^2 = \gamma\left(\overline{L}\right)\sigma_\varphi^2 \tag{27}$$

In the present example the Gaussian autocorrelation function has been selected, which is considered as one of the most suitable for describing soil properties (Rackwitz 2000). The Gaussian autocorrelation function implies the following variance function

$$\gamma\left(\overline{L}\right) = \frac{\frac{\pi}{\delta}\overline{L}\,\text{erf}\left(\frac{\sqrt{\pi}}{\delta}\overline{L}\right) - 1 + \exp\left(-\frac{\pi}{\delta^2}\overline{L}^2\right)}{\frac{\pi}{\delta^2}\overline{L}^2}, \tag{28}$$

where δ is a fluctuation scale. Here, in the case of non-cohesive soils in vertical direction it was assumed as about 1 m or less (Cherubini 1997). Next computations were carried out for three different values of δ. Resulting values of the variance function are given in Table 5. Table 6 shows resulting reliability indices. In the second column indices without spatial averaging are presented. This column is the same as the third column in Table 4.

Table 5. Values of variance function for three different values of fluctuation scale

Averaging length $\overline{L}$ [m]	Values of the fluctuation scale		
	δ = 0.6 [m]	δ = 0.8 [m]	δ = 1.0 [m]
	Values of the variance function		
2.9	0.1933	0.2516	0.3070

Columns: third, fourth and fifth (in Table 6) demonstrate indices obtained when spatial averaging has been applied. The results are also presented in graphical form in Figure 7.

Table 6. The effect of spatial averaging on reliability indices (cov{L}=0.02)

Total safety factor F	Reliability index β without spatial averaging	Reliability index β with spatial averaging (δ = 0.6 m)	Reliability index β with spatial averaging (δ = 0.8 m)	Reliability index β with spatial averaging (δ = 1.0 m)
1.2	0.37	0.76	0.70	0.65
1.4	0.73	1.42	1.30	1.22
1.6	1.07	2.01	1.85	1.73
1.8	1.38	2.56	2.35	2.20
2.0	1.67	3.06	2.82	2.64
2.2	1.94	3.53	3.25	3.05
2.4	2.19	3.96	3.66	3.43
2.6	2.44	4.37	4.04	3.79
2.8	2.67	4.77	4.40	4.13
3.0	2.90	5.13	4.75	4.46

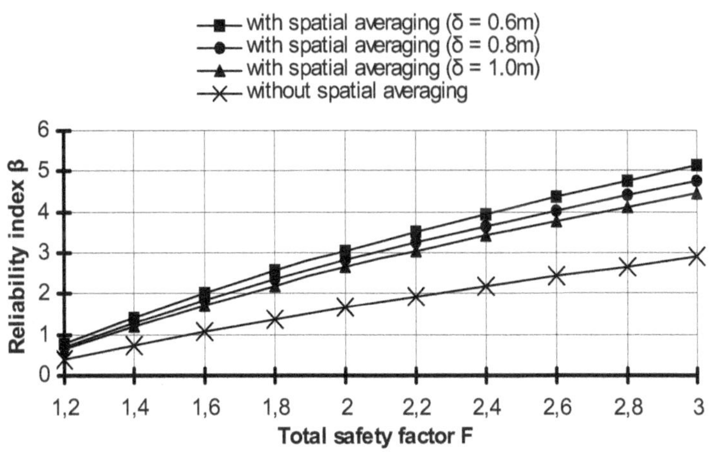

Figure 7. The influence of spatial averaging on the values of reliability indices β

Obtained results evidently demonstrate high influence of spatial averaging on reliability measures when safety of rigid piles is considered. Note that the total safety factor F of level 2.4 gives relatively high values of reliability indices (greater than 3.4), if the averaging is applied.

On the other hand it is worth mentioning that the effect of the value of fluctuation scale is remarkable. This means that the value of fluctuation scale has to be carefully selected and supported by laboratory testing.

4 Concluding Comments

Random fluctuations of soil properties can caused significant changes in the value of ultimate lateral loading determined according to the Brinch Hansen method. Then reliability analysis can be an important tool indicating a save way in selecting parameters for design. The difficulties in applying commonly used structural reliability methods, due to complex nature of solution of equilibrium equations, can be overcame by applying the symbolic computations in conjunction with structural reliability FORM/SORM methods or by utilizing the response surface method associated with FORM/SORM. Both approaches can lead to practically identical results. It is necessary to remind that the "symbolic algorithm" produces significant numerical difficulties in the case of cohesive soil. This is caused that by the fact that the optimal approximation point for logarithm function expansion has not been obtained in this case. The response surface approach, however avoids some important numerical difficulties in "symbolic algorithm". In the present study only non-cohesive soils are considered. But due to high "flexibility" of the response surface method generalisation for the case of cohesive soils can be obtained in the same manner as it is presented in section 3.3. On the other hand an important drawback of the response surface approach is the fact that numerical effort significantly grows with increasing the number of random variables involved in the problem.

The effect of inherent variability of soil friction angle as well as uncertainty in the precise pile embedding have been considered within numerical examples. Comparison of the total safety factors versus reliability indices shows a vital role of spatial averaging as well as the importance of precise evaluation of the fluctuation scale value. Reliability computations in conjunction with spatial averaging give evident growth of reliability indices. In author's opinion the spatial averaging along the length of the pile should be incorporated into reliability computations of a pile foundation, provided a reasonable evaluation of the fluctuation scale is available.

References

Brinch Hansen J. (1961). The ultimate resistance of rigid piles against transversal force. *The Danish Geotechnical Institute, Bulletin.* 1961, No.12.

Cherubini C. (1997) Data and Consideration on the variability of geotechnical properties of soils. *Proc. of the ESREL Conf, Lisboa 1997:* 1538-1591.

Ditlevsen, O., Madsen, H.O. (1996). *Structural Reliability Methods.* Chichester: J. Wiley & Sons.

Hohenbichler M., Gollwitzer S., Kruse W., Rackwitz R. (1987). New light on first and second-order reliability methods. *Structural Safety*, 4: 267-284.

Marquardt, D.W. (1963). An algorithm for least-squares estimation of non-linear parameters, *J. Soc. Indust. Appl. Math,* 11, No. 2., 1963.

Mathematica (2003). A system for doing mathematics by computer. Ver. 5.0, Wolfram Research Inc.

Poulos H.G., Davis E.H. (1980). *Pile foundation analysis and design.* New York: J.Wiley.

Puła W. (1997). Reliability analysis of rigid piles subjected to lateral loads. In: *Numerical Models in Geomechanics*, S. Pietruszczak and G.N. Pande eds. Balkema Rotterdam, s. 521-526, 1997

Puła W.: *Applications of Structural Reliability Theory in Safety Evaluation of Foundations* (in Polish). Oficyna Wydawnicza Politechniki Wrocławskiej, Wrocław, 2004.

Rackwitz, R. (2000). Reviewing probabilistic soils modeling. *Computers & Geotechnics* 25 (3-4): 199-223.

Vanmarcke E.H. (1977). Probabilistic Modeling of Soil Profiles. *Journ. of the Geoechnical Eng. Div.*, ASCE, Vol. 103, No. GT11: 1227-1246.

Case history: Pile Foundations of a Large Gantry Crane

Wojciech Puła and Jarosław Rybak

Institute of Geotechnics and Hydrotechnics, Wrocław University of Technology, Wrocław, Poland

Abstract. Direct motivation for the present study was the reconstruction of the existing gantry crane foundation situated in one of the major factories in the Northern part of Poland. The problem has been analysed in connection with the excessive settlement of the part of the track founded on the approximately 9-meter thick layer of non-compact embankment. One of suggestions of repair works assumed the carrying out of the jet-grouting columns under the beam. It seemed that random fluctuations in flexibility or bearing capacity of piles had resulted in large random fluctuations of bending moments of the beam. A static scheme as well as random model has been proposed. Then a procedure of evaluating random fluctuations of bending moments, based on Monte Carlo method, has been elaborated. The procedure can serve the cases when piles work within the elastic range, as well as in the cases when some piles under the beam are over-loaded and the load has to be carried by other piles. The computations have provided evidence for a considerable influence of the fluctuation of the pile bearing capacities on the bending moments. The extent of that effect, of course, depends, first of all, on the bearing capacity fluctuation range, measured by the coefficient of variation. It can be demonstrated that in the situation when the limiting bearing capacities of the piles have been exceeded, the positive correlation causes a considerable increase of the absolute value of the bending moments. When accompanied by the high values of the reliability index, the values of bending moments obtained from the probabilistic analysis may considerably exceed the moments from the deterministic computations.

1 Introduction

The estimation of the random variability of bending moments inside the beam of a pile foundation constitutes a very important problem from the point of view of civil engineering. The problem was firstly considered by Evangelista et al (1977). The case study of the reconstruction of the existing gantry crane foundation was the direct motivation for the present analysis.

Due to large subsoil deformations it was necessary to reinforce the existing foundation. One of suggestions of repair works assumed the carrying out of the jet-grouting columns under the beam. It seemed that random fluctuations in flexibility or bearing capacity of piles might result in large random fluctuations of bending moments of the beam. This effect can be especially important in the case of long foundation beams supported by large numbers of piles. The procedure of evaluating random fluctuations of bending moments, based on Monte Carlo method, has been elaborated. As basic random variables the flexibilities and bearing capacities of individual piles were considered, while their random changes can be roughly evaluated by means of data collected from load testing. The procedure can serve the cases where piles work within the elastic range as well as in the cases where some piles under the beam are over-loaded and the load has to be carried by other piles. Thereafter, the computations concerning already mentioned case of gantry crane foundation have been carried out. These are commented largely within the present paper. It turned out that all piles have worked inside an elastic range. Therefore an additional example is considered to show some interesting features of random bending moments when the limiting bearing capacity of selected piles under the beam has been exceeded.

2 The Foundation of the Gantry Crane

As it has been already mentioned, the underpinning of the track foundation of a large gantry crane in one of major factories in the Northern part of Poland was the starting point for the further analysis. In this work the method of structural reliability analysis has been applied to test the correctness of the suggested improvement design. The reconstruction of the foundation has been analysed in connection with the excessive settlement of the part of the track founded on the approximately 9-meter thick layer of non-compact embankment. The fill was deposited in the place of an excavation left after the exploitation of sands. On the track section founded on the made ground occurred a distinct subsiding trough, corresponding in its extent, approximately, to the backfilled excavation, that is circa 80 m. The total length of the gantry crane track amounts to 120 m and consists of six 20-meter shallow foundations. The beam had the T-section with the footing width of $B = 1.4$ m, the web width $b = 0.5$ m and the total height of $h = 1.30$ m.

Geodesic measurements made revealed considerable differences in the settlement on particular track sections, maximally reaching about 100 mm. Such large settlements, increasing in time and impossible to be immediately rectified, precluded safe operating of the gantry crane and demanded undertaking of the repair work. Geotechnical conditions of the crane track foundation are basically not differentiated. The southern crane beam has been founded on the subsoil, whereas the northern beam has only partially been founded on the unconsolidated embankment. The thickness of the embankment according to the geological documentation, however, amounts to 9,3 m, starting from the level of the track beam foundation. Below the made ground there are non-cohesive subsoil: sands, gravels and run-of-mines with density factor I_D=0.55. The suggested repair work design assumed the carrying out of the jet-grouting columns under the beam, on the section where considerable settlement occurred. The type of piles involved in this technology permits to transfer the load from the crane track to the subsoil deposited below the embankment (sands, gravels and run-of-mines), without the additional elements transferring the load from the crane track to the pile.

The designed piles were supposed to be made under the track beam footing, through the holes bored in the beam flange. Under each of the three strengthened 20-meter track sections 14 piles have been designed, with the spacing of 1.45 m (see Fig. 1). The technology of the piling demands the carrying out of the rebores with circa 150 mm diameter. That results in cutting of some of the reinforcement bars and thus the weakening of the section in the fulcrum. Therefore, it has been assumed that the key points of the beam, demanding a detailed analysis, are not the sections in which the maximal bending moments occur, but the sections weakened by the rebores.

The loads are transferred from the gantry crane to the beam by eight wheels (from the two travelling crabs of the crane – see Figure 2) with the following wheel spaces: d_1 = 0.85m, d_2 = 1.15m, d_3 =0. 85m, d_4 = 3.45m, d_5 = 0.85m, d_6 = 1.15m, d_7 =0.85m. Particular wheels transfer the following loads: $R_1 = 140.58$ kN, $R_2 = 94.95$ kN. The gantry crane may occur at any place of the beam. The diagram of the load has been shown in Figure 2.

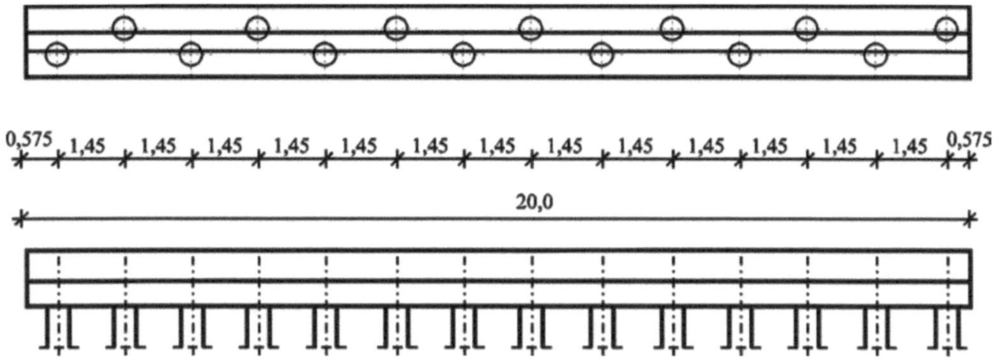

Figure 1. Positions of piles under the foundation beam.

The computations have been carried out for all the sections above the piles and selected sections in between, in the process of shifting the crane wheels by 0.50 m. That permitted to mark the influence line for the average values of the bending moment at selected points of the beam and the influence line of the standard deviation of the bending moment.

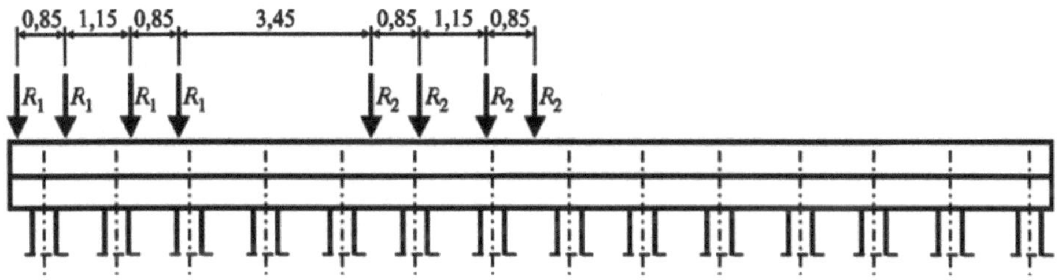

Figure 2. An example of foundation loaded by the travelling crab.

The preliminary analyses have proved that the load quantity and the load-bearing capacity of the assumed piles provide evidence that the piles work in the elastic range, so the case has to be studied as a random stiffness problem. The probabilistic model for solving the problem of that type has been suggested below.

3 Deterministic Model

3.1 Elastic Range

The static diagram of the continuous beam has been assumed, where the beam is founded on a pile series, which has been investigated according to the method of forces. Each pile is represented by the support loaded along the longer centre line of the beam, in accordance with the linear dependence of the load and the settlement, until the limiting bearing capacity has been achieved. For the amount of piles larger than two, as it happens hereby, the structure becomes statically indeterminable. In such cases the relevant determinable structure was used, replacing the actual one (see Figure 3). In the equivalent structure the left end of the beam has been restrained, and the piles – removed and replaced by the unknown reactions, thus creating a cantilever beam.

The reactions in r_1, r_2, ..., r_N piles, as well as the displacement y_0 and the angle of rotation φ_0 of the left end of the beam make for N+2 unknowns. According to the condition of the method of forces, when the relative displacement of the point connecting the beam to each pile must be equal to zero, one gets N canonical equations. The additional two equations, necessary for the solving of the structure, are obtained from the conditions of the static equilibrium conditions. In that case, the sum of vertical forces and moments has to be equal to zero. The canonical equation can be written in the following way:

$$r_1\delta_{k,1}+r_2\delta_{k,2}+...+r_i\delta_{k,i}+...+r_N\delta_{k,N}-y_0-l_k\varphi_0-\Delta_{k,F}=0 \qquad (1)$$

where : $k = 1,2,...,N$, r_i is the reaction in the i-th pile, y_0 denotes vertical displacement of the left end of the beam of the pile structure, φ_0 is the angle of rotation of the left end of the beam of the pile structure, l_k denotes the distance between the pile k and the left end of the beam, $\delta_{k,i}$ is relative displacement of the point of tangency of the k-pile with the beam, caused by the unitary reaction of the pile i, and $\Delta_{k,F}$ denotes vertical displacement of the point of tangency of the pile k with the beam caused by all the actual vertical loads affecting the beam. Static equilibrium equations have the following form (see Figure 3):

$$\sum_{i=1}^{N} r_i - \sum_{j=1}^{n} F_j = 0 \qquad (2)$$

$$\sum_{i=1}^{N} r_i - \sum_{j=1}^{n} F_j = 0 \qquad (3)$$

where: F_j is the concentrated load j affecting the foundation, n is the number of the forces, λ_j denotes the distance between the force F_j and the left end of the beam. For $i = k$ the coefficients of the system of canonical equations are equal to $\delta_{k,k} = y_{k,k} + s_{k,k}$, where $y_{k,k}$ is the displacement of the point of tangency of the pile k and the beam, caused by the unitary reaction in the k-pile, and $s_{k,k}$ is the k-pile settlement under the influence of a unitary force. For the pile k (regardless of the influence of neighbouring piles) the following dependences hold:

$$y_{k,k} = \frac{l_k^3}{3EI} \qquad (4)$$

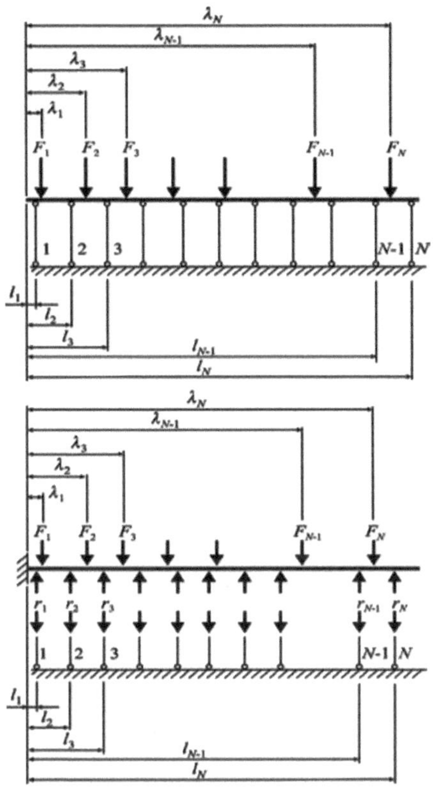

Figure 3. The scheme of the pile system a) pile structure b) equivalent structure.

$$s_{k,k} = \frac{w}{P_{all-k}} \tag{5}$$

where w is the pile settlement, when its bearing reaction equals the allowable bearing capacity P_{all-k} (e.g. suggested in the design). When we take into account the influence of the neighbouring pile settlement for $i \neq k$, $\delta_{k,i} = y_{k,i}$, then:

$$y_{k,i} = \frac{1}{6EI} l_i^2 (3l_k - l_i) \qquad \text{for} \quad (l_k > l_i) \tag{6}$$

$$y_{k,i} = \frac{1}{6EI} l_k^2 (3l_i - l_k) \qquad \text{for} \quad (l_k < l_i) \tag{7}$$

where EI is the stiffness of the pile capping beam. $\Delta_{k,F}$ is calculated in a similar way:

$$\Delta_{k,F} = \frac{1}{6EI} \sum_{j=1}^{n} V_{kj} F_j \tag{8}$$

where:

$$V_{kj} = l_j^2 (3l_k - l_j) \qquad \text{for} \quad l_k > l_j \tag{9}$$

$$V_{kj} = l_k^2 (3l_j - l_k) \qquad \text{for} \quad l_k < l_j \tag{10}$$

In matrix formulation the system of equations can be written as follows:

$$\boldsymbol{\delta}\mathbf{R} = \boldsymbol{\Delta} \tag{11}$$

where $\boldsymbol{\delta}$ is the matrix of canonical equations coefficients

$$\boldsymbol{\delta} = \begin{bmatrix} \delta_{1,1} & \delta_{1,2} & \ldots.. & \delta_{1,N} & -1 & -l_1 \\ \delta_{2,1} & \delta_{2,2} & \ldots.. & \delta_{2,N} & -1 & -l_2 \\ \ldots.. & \ldots.. & \ldots.. & \ldots.. & \ldots.. & \ldots.. \\ \delta_{N,1} & \delta_{N,2} & \ldots.. & \delta_{N,N} & -1 & -l_N \\ 1 & 1 & \ldots.. & 1 & 0 & 0 \\ l_1 & l_2 & \ldots.. & l_N & 0 & 0 \end{bmatrix} \tag{12}$$

$\mathbf{R}$ and $\boldsymbol{\Delta}$ are the following vectors:

$$\mathbf{R} = [\, r_1, r_2, \ldots, r_N, y_0, \varphi_0]^{\mathrm{T}} \tag{13}$$

$$\boldsymbol{\Delta} = \left[\Delta_{1,F}, \Delta_{2,F}, \ldots, \Delta_{N,F}, \sum_{j=1}^{n} F_j, \sum_{j=1}^{n} F_j \lambda_j \right]^{\mathrm{T}} \tag{14}$$

3.2 Plastic range

The following procedure should be applied, if we take into consideration the exceeding of the limiting bearing capacity. If in formula (5) $P_{all\text{-}k}$ stands for a certain arbitrary value (allowable capacity), which will not be exceeded by the reaction r_k, therefore pile k may be treated as an elastic support, and the parameter defined in formula (5) may be treated as its flexibility. Whereas, when the reaction r_t in the pile t exceeds limiting bearing capacity of the pile $P_{gr\text{-}t}$, the dependence of the pile load and its settlement is non-linear. If from the above-presented system of equations results that the reaction in pile t exceeds the value of $P_{gr\text{-}t}$, the equation (11) must be transformed. In the matrix $\boldsymbol{\delta}$ of the system of equations coefficients, the row t and the column t are erased, as well as the row t in the vectors $\mathbf{R}$ and $\boldsymbol{\Delta}$, whereas the indices of the matrix elements are reduced by 1 if they are larger than t.

Additionally, the right-hand side vector $\boldsymbol{\Delta}$ of eq. (11) is replaced by the vector $\boldsymbol{\Delta}$':

$$\Delta' = \begin{bmatrix} \Delta_{1,F} - P_{Pgr\text{-}t}\,\delta_{1,t} \\ \Delta_{2,F} - P_{Pgr\text{-}t}\,\delta_{2,t} \\ \cdots\cdots \\ \cdots\cdots \\ \Delta_{N-1,F} - P_{Pgr\text{-}t}\,\delta_{N-1,t} \\ \sum_{j=1}^{n} F_j - P_{Pgr\text{-}t} \\ \sum_{j=1}^{n} F_j l_j - P_{Pgr\text{-}t}\,\lambda_t \end{bmatrix} \tag{15}$$

Finally, a new system of equations is formulated:

$$\boldsymbol{\delta}'\mathbf{R}' = \boldsymbol{\Delta}' \tag{16}$$

with N+1 unknowns. If the above-presented system of equations does not indicate that any of the consequent pile load has exceeded its limiting bearing capacity $P_{gr\text{-}t}$, then the indices of the $\mathbf{R}$ -vector assume their original values, and the reaction r_t equals $P_{gr\text{-}t.}$ Otherwise, the equation (16) must still be modified.

4 Stochastic Model

In general case a vector of basic random variables is considered to be a vector of capacities of piles $(P_1, P_2, ..., P_n) = (P_{all-1}, P_{all-2}, ..., P_{all-n})$. If piles run solely in the elastic range, stiffness of individual piles are considered as random variables. If, however, for any pile the elastic range is exceeded and a pile works in the plastic range, then the capacity of such pile is assumed as random variable. When the plastic behaviour is considered, it is always necessary to define a criterion, which separates elastic and plastic behaviour. The criterion can be defined as a function of random capacity of pile. Within present study random bending moments $M(P_1, P_2, ..., P_n)$ in the foundation beam as function of random capacities (or flexibilities) are the subject of interest. Then the limit state function can be written as:

$$g(P_1, P_2, ..., P_n) = M_{max} - M(P_1, P_2, ..., P_n) \tag{17}$$

where $M_{\max}$ denotes a threshold moment value, which is defined among assumptions of the problem. It can be, for example, the design value. The eq. (17) can be associated with several cross-section of the beam, namely these, which are most important form the design point of view. In the case of negative values of bending moments the limit state function takes the following form,

$$g(P_1, P_2, ..., P_n) = M(P_1, P_2, ..., P_n) - M_{min} \tag{18}$$

Hence probability of failure can be evaluated as

$$p_F = \mathrm{P}\{M_{max} - M(P_1, P_2, ..., P_n) < 0\} \text{ or } p_F = \mathrm{P}\{M(P_1, P_2, ..., P_n) - M_{min} < 0\} \tag{19}$$

In this paper, however, the task will be rather formulated in the inverse form. Namely, for given value of probability of failure p_F (or, equivalently, the reliability index $\beta = -\Phi_0^{-1}(p_F)$, where Φ_0 is the standard normal distribution function) the threshold value $M_{\max}$ (or $M_{\min}$) will be investigated. Such value should be assumed for the calculation of the reinforcement, which means the dimensioning for the pre-defined reliability level. The application of the efficient methods of structural reliability analysis, like FORM/SORM (Ditlevsen and Madsen 1996), is limited by the number of piles (the number of basic random variables) or by the possibility that the piles exceed the elastic range. In that case, the application of the Monte Carlo simulation method appears to be the optimal solution.

Let us now consider a foundation of the gantry crane described in the previous sections. Let $y \in [0, L]$ denote the coordinate of an arbitrary selected cross-section on the foundation beam with respect of the beginning of the beam (the left end of the beam). The random bending moment in y-section will be described by the random function $M(y, F(x), \mathbf{Q})$, where $F(x)$ is a random variable of load exerted by the travelling crab wheel, depending on the location of the crab on the beam x, and Q = $(Q_1, ..., Q_n)$ is the vector of random flexibilities of piles. Beneath, some preliminary assumptions are specified:

1. The location of the eight crab wheels is a random variable with uniform distribution along the length of the beam, which means that any location on the beam is equally probable. That assumption corresponds with the conditions of the gantry crane operating.
2. The coordinate x is defined as the location of the first travelling crab wheel (the wheels may be located beyond the beam, and then they do not affect the its loading)
3. For any value of the coordinate x the random variable $F(x)$ is stochastically independent of the random vector $\mathbf{Q} = (Q_1, ..., Q_n)$.

Within this and the next section we confine ourselves to the assumption that for given value of coordinate y the value of loading $F(x)$ is constant and non-random. The loading value $F(x)$ is assumed as design load calculated in accordance with the Polish standards. The location of $F(x)$ remains random according to the assumption 1. Let us consider two-dimensional random vector (M_y, X) at the point y, whose coordinates are random variables of the bending moment at the point y and the location of the crab. The variables M_y and X are not stochastically independent. Let $f(m,x)$ denote joint probability density function of vector (M_y, X). Then, the probability that the bending moment M_y exceeds certain threshold value M_0, defined *a priori*, equals:

$$p_F = P(M_y > M_0) = \int_a^b \int_{M_0}^{\infty} f(m,x)\,dm\,dx \tag{20}$$

The joint density $f(m,x)$ may be written as:

$$f(m,x) = f_x(m)\,r(x) \tag{21}$$

where $f_x(m)$ is the conditional density of the bending moment M, on condition that the first wheel of the crab is located at the point x on the beam. Substituting (20) to (21) and bearing in mind assumption 1, one gets:

$$p_F = P(M_y > M_0) = \frac{1}{b-a}\int_a^b \int_{M_0}^{\infty} f_x(m)\,dm\,dx \tag{22}$$

where a and b are, correspondingly, the coordinates of the two ends of the beam track. The conditional density $f_x(m)$ is computed by Monte Carlo simulation. However, in the situation when the time-invariant reliability analysis is carried out with all the crab locations on the beam are equally probable, it is indispensable to concentrate on finding such crab location which brings about the most unfavourable bending moment in the considered section y. In course of gantry crane operation the above-described crab location is a certain event, which generates the highest failure probability, then by applying eq. (22) one gets

$$p_F = P(M_y > M_0) = \int_{M_0}^{\infty} f_{x_0}(m)\,dm \tag{23}$$

where x_0 is the most unfavourable travelling crab location on the beam. That point can be determined by the analysis of the influence lines.

5 Case Study

In order to apply the previously presented procedure for selected beam sections, the influence lines for the expected value and the standard deviation of the bending moment were determined. Those lines allowed to identify the points x_0 for the analysed sections (x_0 is, of course, different for various sections). Figure 4 provides an example of influence lines at selected points of the beam, namely above the 5th and the 4th pile. Further computations have been carried out under the following assumptions:

a. Only elastic range of pile work has been considered.
b. Flexibilities of individual piles are assumed to be independent, lognormally distributed random variables. It could be demonstrated by examples (Puła 2004) that in the case of solely linear pile work, the assumption of independence leads to more conservative evaluation of safety indices.
c. By analysis of reports from piles load testing (for example Gwizdała 1997, Rybak 1999) the coefficient of variation in each case of flexibility has been assumed as equal to 20%. This seems to be reasonable for the piles in jet-grouting technology.
d. The sample size in the simulation process was $N_{ss} = 10000$ for each cross-section y.

The distributions of bending moments (with sample size of N_{ss}=10000) for several cross-sections y have been obtained as results of such computation. These distributions were estimated by normal distributions by means of least squares method. Next, goodness of fit was confirmed by statistical testing (Kolmogorov-Smirnoff and Anderson-Darling).

In case of the section above the pile no. 5 Gaussian distribution was obtained, with the parameters: $E\{M\}$ = 510.3 kNm and $\sigma\ \{M\}$ = 32.0 kNm, when the wheels of the travelling crabs had the following positions: the first wheel was located at the distance of 5.5 m from the beginning of the beam, whereas for the section above the pile no. 4 - $E\{M\}$ = 508.7 kNm and $\sigma\ \{M\}$ = 36.6 kNm when the first wheel of the crab was located at the distance of 3.5 m from the beginning of the beam.

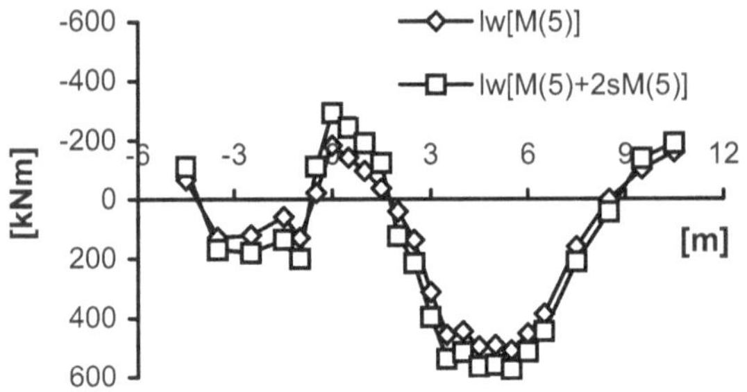

a.)

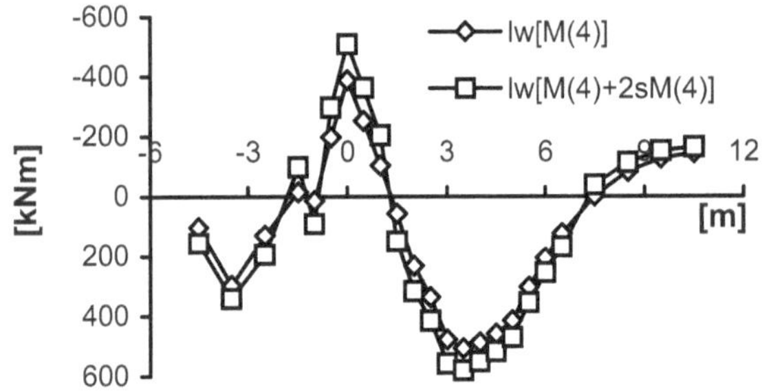

b.)

Figure 4. The influence lines of the bending moment mean value and the mean value increased by the doubled standard deviation above the 5th pile (Figure a) and 4th pile (Figure b).

Additionally, above the pile no. 4 for negative values of bending moments when the first crab wheel was placed at the beginning of the beam, the estimated Gaussian distribution was characterised by the parameters: E$\{M\}$ = - 386.6 kNm and σ $\{M\}$ = 61.2 kNm. The threshold values of the bending moments for the existing foundation beam have been estimated on the basis of the accessible documentation. They equalled: M_{max}= 540 kNm and for the negative values of bending moments M_{min}= - 420 kNm. The corresponding reliability indices failure probabilities are for the pile no. 5:

$$p_F = P\{M > M_{max}\} = 0.177 \Rightarrow \beta = -\Phi_0^{-1}(p_F) = 0.93$$

and for the pile no. 4:

$$p_F = P\{M > M_{max}\} = 0.197 \Rightarrow \beta = -\Phi_0^{-1}(p_F) = 0.86$$

$$p_F = P\{M < M_{min}\} = 0.292 \Rightarrow \beta = -\Phi_0^{-1}(p_F) = 0.55$$

If we want to obtain the reliability indices at the satisfactory level of β = 3.0, the following threshold values must be presupposed: M_{max}= 606.3 kNm above the pile no. 5, M_{max}= 618.5 kNm above the pile no. 4 and M_{min}= - 570.2 kNm in the case of pile no. 4. It becomes evident that the above-mentioned values are considerably larger than the ones that have occurred in the design. The relation between the reliability index β and the section bending strength above the pile has also been analysed (Figure 5.). In the case of designing a "new" object such procedure allows for the dimensioning of the structure that leads to the identical reliability index for each bent section.

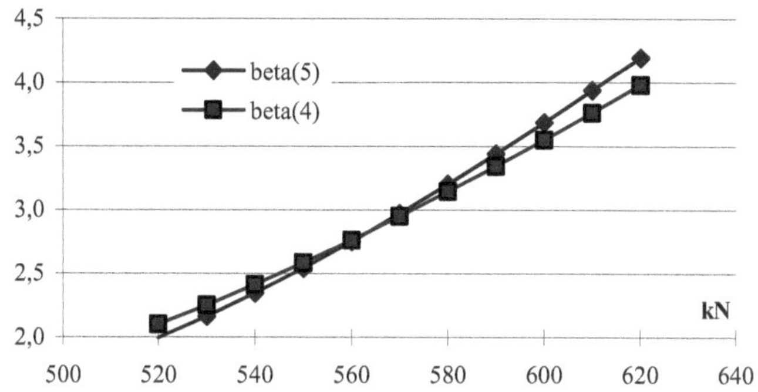

Figure 5. The relation between the reliability index β and the section bending strength above the pile.

The computations carried out with respect to flexibility of piles lead to the following conclusions:
- the previously presented computation of the failure probability permits to verify the traditional computation method
- random pile flexibility may result in a considerable bending moments increase
- the beam sections above the piles, which are additionally weakened by the rebores, may be especially subject to failure caused by bending
- although random pile bearing capacities were characterised by the coefficient of variation equal to 20%, the resulting coefficient of variation for the bending moments was circa 8%.

6 Remarks on plastic behaviour

The case study presented in section 6 deals solely with the elastic behaviour of piles. It was a natural consequence of the bearing capacities and loads of the piles, assumed for the needs of the design. Therefore an additional example is considered to demonstrate features of random bending moments when the limiting bearing capacity of selected piles under the beam has been exceeded. As it has been already mentioned in section 4 it was always necessary to define a criterion which separates elastic and plastic behaviour, when the plastic behaviour is considered. Let us assume that

$$P_{gr-t} = 2P_t \tag{24}$$

in which P_t is the random bearing capacity of the pile t, treated here as the limiting bearing capacity accepted for the purpose of the design. In the beneath considered example the following assumptions have been accepted:

- The row of 11 identical piles covered by a capping beam transfers the load from 6 columns to the subsoil (Figure 6).

– The value of axial force in each column equals $F_i = 1800$ kN, $i = 1,2,\ldots,6$.
– The bearing capacity of each pile is a random variable of the expected value $E\{P_t\} = 683.4$ kN and the coefficient of variation cov{P_t} = 0,15 (standard deviation $\sigma_{Pt} = 102.5$ kN, t = 1,2,…,11).
– Constant stiffness of the capping beam along the whole length EI=653386 kNm^2.

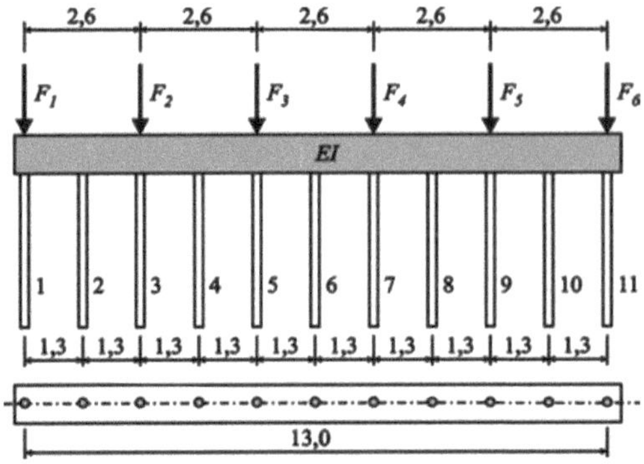

Figure 6. The considered design scheme (The dimensions have been given in meters).

The computations have been carried out for 4 different series of probabilistic data:
Series 1. The bearing capacity of a single pile is a random variable of Gaussian distribution. The bearing capacities of each pile are independent random variables.
Series 2. The bearing capacity of a single pile is a random variable of Gaussian distribution. The bearing capacities are, however, correlated in accordance with the following mutual relations: for the piles located at the distance of d = 1.3 m the correlation coefficient equals ρ = 0.4; if d = 2.6 m then ρ = 0.3; if d =3.9 m then ρ = 0.2; if d = 5.2 m then ρ = 0.1; for the larger distances, ρ = 0.
Series 3 The bearing capacity of a single pile is a random variable of lognormal distribution. The bearing capacities of each pile are independent random variables.
Series 4. The bearing capacity of a single pile is a random variable of lognormal distribution. Assumptions concerning correlation structure are identical as in Series 2.
The sample size in the simulation procedure was N_{ss} = 15000. Such number is unlikely to be sufficient for the estimation of the failure probability of the order of p_F = 0,001 (the required number is circa M = 1000000). Thus, the obtained histograms were further processed, after which relevant probability density functions were matched. For such hypothetical probability density functions, the threshold values of the bending moment were assigned. In course of computations, the results for particular series of probabilistic data were monitored in order to determine the frequency of exceeding the limiting bearing capacity of piles. Table 1 presents the number of tests in which the limiting bearing capacity of one pile, at the least, was exceeded.

It is evident that, in the considered case, correlation between the bearing capacity of single piles has a larger impact than the type of distribution. The expected values and standard deviations of the bending moment in 11 capping beam sections, calculated on the basis of the simulated series of data, have been presented in Table 2.

It must be noticed that standard deviations along the whole length of the capping beam vary insignificantly. The variability of the mean value effects in considerable fluctuation of the coefficient of variability. Moreover, the expected values of the bending moments depend on the correlation of the bearing capacity of the piles, to a larger extent than on the distribution type (Gaussian or lognormal).

Table 1. The percentage of samples in which the limiting bearing capacity of one pile, at the least, was exceeded (15 000 series of considered data).

	Gaussian distribution	Lognormal distribution
independent	5873 (39%)	6053 (40%)
correlated	6798 (45%)	7043 (47%)

The introduction of the correlation between the bearing capacity of piles increases the absolute value of the expected values of the negative bending moments. In the case of the positive bending moments – contrarily – the absolute value decreases when the correlation occurs. Moreover, correlation, on the one hand, triggers the increase of the standard deviation in the central part of the beam, and, on the other, it brings about the decrease of the same property at the both beam ends. The threshold values of the bending moments have been calculated according to the formula 19 for the reliability index $\beta = 3.0$, which corresponds to failure probability $p_F = \Phi(-\beta) = 1{,}35 \cdot 10^{3}$. As it was previously mentioned, due to small probability values p_F, the accuracy of threshold values calculation depends, to a large extent, on finding of the appropriate probability distribution functions. Those probability distribution functions are estimated according to the moment method. It has been verified that Gaussian distributions have been well fitted to these previously obtained by simulation.

The threshold moments evaluated in this way, for earlier mentioned four series of calculations are presented in Figure 7. When we compare the threshold values for the four computation series demonstrated above, we mustn't overlook the considerable increase of their absolute value in the central part of the beam after the correlation of the pile bearing capacities has been taken into account. Once the positive correlation has occurred, there's a larger chance for the same phenomenon to take place in the neighbouring pile or even piles. It is interesting that this effect is much more visible when the Gaussian, and not lognormal, distribution is taken into account.

When we compare the threshold values for the four computation series demonstrated above, we mustn't overlook the considerable increase of their absolute value in the central part of the beam after the correlation of the pile bearing capacities has been taken into account. Once the positive correlation has occurred, there's a larger chance for the same phenomenon to take place in the neighbouring pile or even piles. It is interesting that this effect is much more visible when the Gaussian, and not lognormal, distribution is taken into account.

Table 2. The expected values, standard deviations and the coefficient of variations (cov) of the bending moments in capping beam sections for 4 series.

Series	M [kNm]	Section 1	2	3	4	5	6	7	8	9	10	11
1	E{*M*}	0	-733.7	-76.5	-519.9	172.7	-368.3	173.9	-518.1	-75.1	-733.7	0
	σ_M	0	150.2	194.3	177.8	144.4	125.3	142.0	174.3	191.5	149.4	0
	cov{*M*}	-	20%	254%	34%	84%	34%	82%	34%	255%	20%	-
2	E{*M*}	0	-742.4	-97.8	-551.0	135.9	-407.3	135.9	-551.0	-97.8	-743.5	0
	σ_M	0	136.3	215.9	251.6	257.9	255.4	255.2	247.0	211.8	134.7	0
	cov{*M*}	-	18%	221%	46%	190%	63%	188%	45%	217%	18%	-
3	E{*M*}	0	-730.1	-70.4	-513.4	177.1	-365.1	177.0	-515.6	-70.7	-731.3	0
	σ_M	0	137.4	172.7	155.4	124.5	109.1	124.6	154.5	173.3	138.6	0
	cov{*M*}	-	19%	245%	30%	70%	30%	70%	30%	245%	19%	-
4	E{*M*}	0	-740.0	-90.0	-537.4	153.9	-388.1	153.9	-537.0	-90.2	-740.8	0
	σ_M	0	123.1	183.7	197.5	187.0	177.5	181.8	192.4	179.0	120.9	0
	cov{*M*}	-	17%	204%	37%	122%	46%	118%	36%	198%	16%	-

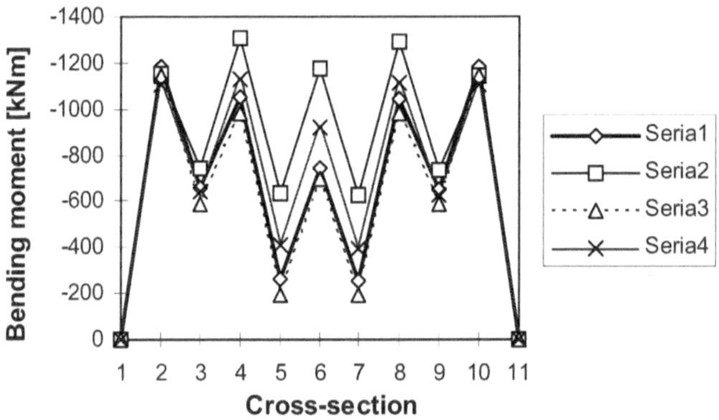

Figure 7. Threshold bending moments corresponding to β= 3.0 reliability index

7 Concluding remarks

The computations have provided evidence for a considerable influence of the fluctuation of the pile bearing capacities on the bending moments. The extent of that effect, of course, depends, first of all, on the bearing capacity fluctuation scale/range, measured by the coefficient of variation. The coefficient of variation, in turn, may be determined on the basis of trial loads, carried out each time in case of large pile foundation execution. The selection of the bearing capacity distribution type does not affect the estimation of threshold values in any significant manner.

It is noteworthy, however, what follows if the potential positive correlation between the bearing capacities of single piles is taken into account. The calculations have proved that in the situation when the limiting bearing capacities of the piles have been exceeded the positive correlation causes a considerable increase of the absolute value of the bending moments (in the sense of threshold values). It may be explained by the increase of chance for the situation in which the bearing capacities of two or three neighbouring piles are exceeded. If, nonetheless, we deal only with the piles working in the elastic range, the positive correlation results in the decrease of the absolute value of the bending moments. In that case, the positive correlation effects in the reduction of variation.

It is also worth mentioning that the obtained standard deviations of the moments vary in a rather narrow range along the section of the beam. That permits to extrapolate the solution on the other sections, in which no *a priori* calculations have been carried out. When accompanied by the high values of the reliability index, the values of bending moments obtained from the probabilistic analysis may considerably exceed the moments from deterministic computations.

References

Evangelista A., Pellegrino A., Viggiani C. (1977): Variability among piles of the same foundation, *Proceedings of International Conference on Soil Mechanics and Foundation Engineering*, Tokyo, session 2, 493-500.

Ditlevsen O., Madsen H.O. (1996). *Structural Reliability Methods*. John Wiley & Sons, Chichester.

Puła W. (2004): *Applications of Structural Reliability Theory to Foundations Safety Evaluation* (in Polish). Oficyna Wydawnicza Politechniki Wrocławskiej, Wrocław.

Gwizdała K. (1997). Design by static load tests. In: *ISSMFE European Regional Technical Committee 3 (ERTC3)*, Brussels.

Rybak Cz. (1999). Load tests of piles. Private communication.

Random Field Generation and the Local Average Subdivision Method

Gordon A. Fenton* and D.V. Griffiths†

* Department of Engineering Mathematics, Dalhousie University, Canada
† Division of Engineering, Colorado School of Mines, U.S.A.

Abstract The use of multi-dimensional random fields to model real soils is becoming ever more important, simply because soils are spatially random. Such random field models allow the rational quantification of the behaviour of spatially variable soils, which are inherently uncertain, and lead to reliability estimates, decision analysis, and, ultimately, optimal designs. Random models are commonly used either in analytical studies employing theoretical results or in Monte Carlo simulations. Since theoretical results do not exist for many problems of interest to geotechnical engineers, the Monte Carlo approach is often the practical choice.

In that the accuracy of such models depends directly on the accuracy of the algorithm used to generate realizations of the representative random fields, there is a need to evaluate and compare various random field generators. To address this issue, three common random field generators are considered in this chapter; 1) the FFT method, 2) the Turning Bands Method (TBM), and 3) the Local Average Subdivision (LAS) method. For each, an ensemble of realizations of a two-dimensional homogeneous Gauss-Markov process is generated and the field mean, variance, and covariance structures are checked for statistical accuracy. Concerns such as ease of use and efficiency are also considered. It is shown that all three methods have distinct advantages and disadvantages, and the choice of algorithm will depend on the particular application. A number of guidelines and suggestions are made to help avoid or minimize problems associated with each method.

1 Introduction

Random field models of complex engineering systems having spatially variable properties are becoming increasingly common. This trend is motivated by the widespread acceptance of reliability methods in engineering design and is made possible by the increasing power of personal computers. It is no longer sufficient to base designs on best estimate or mean values alone. Information quantifying uncertainty and variability in the system must also be incorporated to allow the calculation of failure probabilities associated with various limit state criteria. To accomplish this, a probabilistic model is required. In that most engineering systems involve loads and materials spread over some spatial extent, their properties are appropriately represented by random fields. For example, to estimate the failure probability of a highway bridge, a designer may represent both concrete strength and input earthquake ground motion using independent random fields, the latter time varying. Subsequent analysis using a Monte Carlo approach and a dynamic finite element package would lead to the desired statistics.

In this chapter, a number of different algorithms which can be used to produce scalar multi-dimensional random fields are evaluated in light of their accuracy, efficiency, ease of implementation, and ease of use. Many different random field generator algorithms are available of which the following are perhaps the most common:
1) Moving Average (MA) methods,
2) Discrete Fourier Transform (DFT) method,
3) Covariance Matrix Decomposition,
4) Fast Fourier Transform (FFT) method,
5) Turning Bands Method (TBM),
6) Local Average Subdivision (LAS) method,
In all of these methods, only the first two moments of the target field may be specified, namely the mean and covariance structure. Since this completely characterizes a Gaussian field, attention will be restricted in the following to such fields. (Non-Gaussian fields may be created through non-linear transformations of Gaussian fields, however some care must be taken since the mean and covariance structure will also be transformed.) In addition, only weakly homogeneous fields, whose first two moments are independent of spatial position, will be considered in this chapter.

Although potentially very accurate, both the Moving Average and the DFT methods tend to be computationally slow. At each point $\underset{\sim}{x}$ in space, the MA technique constructs the random field $Z(\underset{\sim}{x})$ as a weighted average of a white noise process

$$Z(\underset{\sim}{x}) = \int_{-\infty}^{\infty} f(\underset{\sim}{\xi} - \underset{\sim}{x})dW(\underset{\sim}{\xi}) \tag{1}$$

where $dW(\underset{\sim}{\xi})$ is a mean zero incremental white noise process with variance $d\underset{\sim}{\xi}$ and f is a weighting function. In practice (1) is computed as a sum, its extent and resolution directly affecting the accuracy of the resulting field. Because the sum is computed separately at each point $\underset{\sim}{x}$, the moving average technique can be very computationally expensive. For reasonable sized fields in two and higher dimensions it can be orders of magnitude slower than some of the other methods to be discussed. This, along with the fact that the weighting function f can be difficult to find for an arbitrary covariance structure, renders the method unwieldy in practice and it will not be considered further. In a sequence of two papers, Mignolet and Spanos (1992) and Spanos and Mignolet (1992) discuss in considerable detail the moving average (MA), autoregressive (AR) and ARMA approaches to simulating two-dimensional random fields. In their examples, they obtain accurate results at the expense of running about 10 or more times slower than the fastest of the methods to be considered here.

The Fourier Transform methods are based on the spectral representation of homogeneous mean square continuous random fields, $Z(\underset{\sim}{x})$, which can be expressed as follows (Yaglom, 1962)

$$Z(\underset{\sim}{x}) = \int_{-\infty}^{\infty} e^{i\underset{\sim}{x}\cdot\underset{\sim}{\omega}}\, W(d\underset{\sim}{\omega}) \tag{2}$$

where $W(d\underset{\sim}{\omega})$ is an interval white noise process with mean zero and variance $S(\underset{\sim}{\omega})\,d\underset{\sim}{\omega}$. This representation is in terms of the physically meaningful spectral density function, $S(\underset{\sim}{\omega})$, and so is intuitively attractive. In practice, the n-dimensional integral becomes an

n-dimensional sum. In the case of the Discrete Fourier Transform, the sum is evaluated separately at each point $\underset{\sim}{x}$ which, although potentially accurate, is computationally slow for reasonable field sizes and typical spectral density functions – the DFT is generally about as efficient as the MA discussed above. Its major advantage over the MA approach is that the spectral density function is estimated in practice using standard techniques. However, because of its inefficiency, the DFT approach will not be considered further in this chapter.

Covariance matrix decomposition is a direct method of producing a homogeneous random field with prescribed covariance structure $B(\underset{\sim}{x}_i - \underset{\sim}{x}_j) = B(\underset{\sim}{\tau}_{ij})$, where $\underset{\sim}{x}_i$, $i = 1, 2, \ldots, n$ are discrete points in the field and $\underset{\sim}{\tau}_{ij}$ is the lag vector between the points $\underset{\sim}{x}_i$ and $\underset{\sim}{x}_j$. If $\underset{\approx}{B}$ is a positive definite covariance matrix with elements $B_{ij} = B(\underset{\sim}{\tau}_{ij})$, then a mean zero discrete process $Z_i = Z(\underset{\sim}{x}_i)$ can be produced (using vector notation) according to

$$\underset{\sim}{Z} = \underset{\approx}{L}\underset{\sim}{U} \tag{3}$$

where $\underset{\approx}{L}$ is a lower triangular matrix satisfying $\underset{\approx}{L}\underset{\approx}{L}^T = \underset{\approx}{B}$ (typically obtained using Cholesky decomposition) and $\underset{\sim}{U}$ is a vector of n independent mean zero, unit variance Gaussian random variables. Although appealing in its simplicity and accuracy, this method is only useful for small fields. In two dimensions, the covariance matrix of a 128×128 field would be of size $16,384 \times 16,384$ and the Cholesky decomposition of such a matrix would be both time consuming and prone to considerable round-off error.

In the remainder of the chapter, attention will be focused on the last three methods mentioned above, namely the Fast Fourier Transform (FFT), the Turning Bands Method (TBM), and the Local Average Subdivision (LAS) algorithms. Some of the discussion will be based on observations of simulations, which in all cases will be of a homogeneous isotropic Gaussian random field with Gauss-Markov covariance structure

$$B(\underset{\sim}{\tau}) = \sigma^2 e^{-2|\underset{\sim}{\tau}|/\theta} \tag{4}$$

where σ^2 is the variance of the process (in this case unity), and θ is the scale of fluctuation. Reference will be made to the estimated mean and variance fields which are simply the mean and variance estimated at each field point over an ensemble of realizations. At each point in the field, the mean and variance are expected to follow that predicted by theory for random samples of a Gaussian process and the fields are inspected for the presence of patterns indicating errors and/or bias.

The FFT, TBM and LAS methods are typically much more efficient than the first three methods discussed above. However, the gains in efficiency do not come without some loss in accuracy, as is typical in numerical methods. In the next few sections, the chapter proposes an implementation strategy for the FFT method and reviews the TBM and LAS methods. The types of errors associated with each method and ways to avoid them will be discussed in some detail. Finally the methods will be compared and guidelines as to their use suggested.

2 The Fast Fourier Transform Method

In order to apply the Fast Fourier Transform method (see, for example, Cooley and Tukey, 1965) to compute Eq. (2), a number of additional assumptions are needed. First, the process $Z(\underset{\sim}{x})$ is assumed to be mean zero, real and discrete. For the purposes of this development, only the one-dimensional case will be considered and multi-dimensional results will be stated subsequently. For discrete $Z(x_j)$, $j = 1, 2, \dots, N$, Eq. (2) becomes

$$\begin{aligned} Z(x_j) &= \int_{-\pi}^{\pi} e^{ix_j\omega}\, W(d\omega) \\ &= \lim_{K\to\infty} \sum_{k=-K}^{K} e^{ix_j\omega_k}\, W(\Delta\omega_k) \\ &= \lim_{K\to\infty} \sum_{k=-K}^{K} \left\{ \mathcal{A}(\Delta\omega_k)\cos(x_j\omega_k) + \mathcal{B}(\Delta\omega_k)\sin(x_j\omega_k) \right\} \end{aligned} \tag{5}$$

where $\omega_k = k\pi/K$, $\Delta\omega_k$ is an interval of length π/K centered at ω_k, and the last step in (5) follows from the fact that Z is real. The functions $\mathcal{A}(\Delta\omega_k)$ and $\mathcal{B}(\Delta\omega_k)$ are independent identically distributed random interval functions with mean zero and $\mathrm{E}\left[\mathcal{A}(\Delta\omega_k)\mathcal{A}(\Delta\omega_m)\right] = \mathrm{E}\left[\mathcal{B}(\Delta\omega_k)\mathcal{B}(\Delta\omega_m)\right] = 0$ for all $k \neq m$ in the limit as $\Delta\omega \to 0$. At this point, the simulation involves generating realizations of $\mathcal{A}_k = \mathcal{A}(\Delta\omega_k)$ and $\mathcal{B}_k = \mathcal{B}(\Delta\omega_k)$ and evaluating (5). Since the process is real, $S(\omega) = S(-\omega)$, and the variances of $\mathcal{A}_k$ and $\mathcal{B}_k$ can be expressed in terms of the one-sided spectral density function $G(\omega) = 2S(\omega)$, $\omega \geq 0$. This means that the sum in (5) can have lower bound $k = 0$. Note that an equivalent way of writing (5) is

$$Z(x_j) = \sum_{k=0}^{K} \mathcal{C}_k \cos(x_j\omega_k + \Phi_k), \tag{6}$$

where Φ_k is a random phase angle uniformly distributed on $[0, 2\pi]$ and $\mathcal{C}_k$ follows a Rayleigh distribution. Shinozuka and Jan (1972) take $\mathcal{C}_k = \sqrt{2G(\omega_k)\Delta\omega}$ to be deterministic, an approach not followed here since it gives an upper bound on Z over the space of outcomes of $Z \leq \sum_{k=0}^{K} \sqrt{2G(\omega_k)\Delta\omega}$ which may be an unrealistic restriction, particularly in reliability calculations.

Next, the process $Z_j = Z(x_j)$ is assumed to be periodic, $Z_j = Z_{K+j}$, with the same number of spatial and frequency discretization points ($N = K$). As will be shown later, the periodicity assumption leads to a symmetric covariance structure which is perhaps the major disadvantage to the FFT approach. If the physical length of the one-dimensional process under consideration is D and the space and frequency domains are discretized according to

$$x_j = j\Delta x = \frac{jD}{K-1} \tag{7}$$

$$\omega_j = j\Delta\omega = \frac{2\pi j(K-1)}{KD} \tag{8}$$

for $j = 0, 1, \ldots, K-1$, then the Fourier transform

$$Z_j = \sum_{k=0}^{K-1} \mathcal{X}_k \, e^{i(2\pi jk/K)} \tag{9}$$

can be evaluated using the FFT algorithm. The Fourier coefficients, $\mathcal{X}_k = \mathcal{A}_k - i\mathcal{B}_k$, have the following symmetries due to the fact that Z is real,

$$\mathcal{A}_k = \frac{1}{K}\sum_{j=0}^{K-1} Z_j \cos 2\pi\tfrac{jk}{K} = \mathcal{A}_{K-k} \tag{10}$$

$$\mathcal{B}_k = \frac{1}{K}\sum_{j=0}^{K-1} Z_j \sin 2\pi\tfrac{jk}{K} = -\mathcal{B}_{K-k} \tag{11}$$

which means that $\mathcal{A}_k$ and $\mathcal{B}_k$ need only be generated randomly for $k = 0, 1, \ldots, K/2$ and that $\mathcal{B}_0 = \mathcal{B}_{K/2} = 0$. Note that if the coefficients at $K-k$ are produced independently of the coefficients at k, the resulting field will display aliasing. Thus there is no advantage to taking Z to be complex, generating all the Fourier coefficients randomly, and attempting to produce two independent fields simultaneously (the real and imaginary parts), or in just ignoring the imaginary part.

As far as the simulation is concerned, all that remains is to specify the statistics of $\mathcal{A}_k$ and $\mathcal{B}_k$ so that they can be generated randomly. If Z is a Gaussian mean zero process, then so are $\mathcal{A}_k$ and $\mathcal{B}_k$. The variance of $\mathcal{A}_k$ can be computed in a consistent fashion (Fenton, 1990) by evaluating $\mathrm{E}\left[\mathcal{A}_k^2\right]$ using (10)

$$\mathrm{E}\left[\mathcal{A}_k^2\right] = \frac{1}{K^2}\sum_{j=0}^{K-1}\sum_{\ell=0}^{K-1} \mathrm{E}\left[Z_j Z_\ell\right] \cos 2\pi\tfrac{jk}{K} \cos 2\pi\tfrac{\ell k}{K} \tag{12}$$

This result suggests using the covariance function directly to evaluate the variance of $\mathcal{A}_k$, an approach that was investigated by Fenton (1990), however the implementation is complex and no particular advantage in accuracy was noticed. A simpler approach involves the discrete approximation to the Wiener-Khinchine relationship

$$\mathrm{E}\left[Z_j Z_\ell\right] \simeq \Delta\omega \sum_{m=0}^{K-1} G(\omega_m) \cos 2\pi\tfrac{m(j-\ell)}{K} \tag{13}$$

which when substituted into (12) leads to

$$\begin{aligned}
\mathrm{E}\left[\mathcal{A}_k^2\right] &= \frac{\Delta\omega}{K^2}\sum_{j=0}^{K-1}\sum_{\ell=0}^{K-1}\sum_{m=0}^{K-1} G(\omega_m) \cos 2\pi\tfrac{m(j-\ell)}{K} C_{kj} C_{k\ell} \\
&= \frac{\Delta\omega}{K^2}\sum_{m=0}^{K-1} G(\omega_m) \sum_{j=0}^{K-1} C_{mj} C_{kj} \sum_{\ell=0}^{K-1} C_{m\ell} C_{k\ell}
\end{aligned}$$

$$+\frac{\Delta\omega}{K^2}\sum_{m=0}^{K-1} G(\omega_m)\sum_{j=0}^{K-1} S_{mj}C_{kj}\sum_{\ell=0}^{K-1} S_{m\ell}C_{k\ell}, \tag{14}$$

where $C_{kj} = \cos 2\pi\frac{kj}{K}$ and $S_{kj} = \sin 2\pi\frac{kj}{K}$.
To reduce (14) further, use is made of the following two identities

1) $$\sum_{k=0}^{K-1} \sin 2\pi\tfrac{mk}{K} \cos 2\pi\tfrac{jk}{K} = 0$$

2) $$\sum_{k=0}^{K-1} \cos 2\pi\tfrac{mk}{K} \cos 2\pi\tfrac{jk}{K} = \begin{cases} 0, & \text{if } m \neq j \\ \frac{K}{2}, & \text{if } m = j \text{ or } K - j \\ K, & \text{if } m = j = 0 \text{ or } \frac{K}{2} \end{cases}$$

By identity (1), the second term of (14) is zero. The first term is also zero, except when $m = k$ or $m = K - k$, leading to the results

$$\mathrm{E}\left[\mathcal{A}_k^2\right] = \begin{cases} \frac{1}{2}G(\omega_k)\Delta\omega, & \text{if } k = 0 \\ \frac{1}{4}\{G(\omega_k) + G(\omega_{K-k})\}\Delta\omega, & \text{if } k = 1, \ldots, \frac{K}{2} - 1 \\ G(\omega_k)\Delta\omega, & \text{if } k = \frac{K}{2} \end{cases} \tag{15}$$

remembering that for $k = 0$ the frequency interval is $\frac{1}{2}\Delta\omega$. An entirely similar calculation leads to

$$\mathrm{E}\left[\mathcal{B}_k\right]^2 = \begin{cases} 0, & \text{if } k = 0 \text{ or } \frac{K}{2} \\ \frac{1}{4}\{G(\omega_k) + G(\omega_{K-k})\}\Delta\omega, & \text{if } k = 1, \ldots, \frac{K}{2} - 1 \end{cases} \tag{16}$$

Thus the simulation process is as follows;

1) generate independent normally distributed realizations of $\mathcal{A}_k$ and $\mathcal{B}_k$ having mean zero and variance given by (15) and (16) for $k = 0, 1, \ldots, K/2$ and set $\mathcal{B}_0 = \mathcal{B}_{K/2} = 0$,
2) use the symmetry relationships, (10) and (11), to produce the remaining Fourier coefficients for $k = 1 + K/2, \ldots, K - 1$
3) produce the field realization by Fast Fourier Transform using Eq. (9).

In higher dimensions a similar approach can be taken. To compute the Fourier sum over non-negative frequencies only, the spectral density function $S(\underset{\sim}{\omega})$ is assumed to be even in all components of $\underset{\sim}{\omega}$ (quadrant symmetric) so that the 'one-sided' spectral density function, $G(\underset{\sim}{\omega}) = 4S(\underset{\sim}{\omega})\ \forall\ \omega_i \geq 0$, can be employed. Using $L = K_1 - \ell$, $M = K_2 - m$, and $N = K_3 - n$ to denote the symmetric points in fields of size $K_1 \times K_2$ in 2-D or $K_1 \times K_2 \times K_3$ in 3-D, the Fourier coefficients yielding a real two-dimensional process must satisfy

$$\begin{aligned} \mathcal{A}_{LM} &= \mathcal{A}_{\ell m}, & \mathcal{B}_{LM} &= -\mathcal{B}_{\ell m} \\ \mathcal{A}_{\ell M} &= \mathcal{A}_{Lm}, & \mathcal{B}_{\ell M} &= -\mathcal{B}_{Lm} \end{aligned} \tag{17}$$

for $\ell, m = 0, 1, \ldots, \frac{K_\alpha}{2}$ where K_α is either K_1 or K_2 appropriately. Note that these relationships are applied modulo K_α, so that $\mathcal{A}_{K_1-0,m} \equiv \mathcal{A}_{0,m}$ for example. In two dimensions, the Fourier coefficients must be generated over two adjacent quadrants of the field, the rest of the coefficients obtained using the symmetry relations. In three dimensions, the symmetry relationships are

$$
\begin{aligned}
\mathcal{A}_{LMN} &= \mathcal{A}_{\ell mn}, & \mathcal{B}_{LMN} &= -\mathcal{B}_{\ell mn} \\
\mathcal{A}_{\ell MN} &= \mathcal{A}_{Lmn}, & \mathcal{B}_{\ell MN} &= -\mathcal{B}_{Lmn} \\
\mathcal{A}_{LmN} &= \mathcal{A}_{\ell Mn}, & \mathcal{B}_{LmN} &= -\mathcal{B}_{\ell Mn} \\
\mathcal{A}_{\ell mN} &= \mathcal{A}_{LMn}, & \mathcal{B}_{\ell mN} &= -\mathcal{B}_{LMn}
\end{aligned}
\tag{18}
$$

for $\ell, m, n = 0, 1, \ldots, \frac{K_\alpha}{2}$. Again, only half the Fourier coefficients are to be generated randomly.

The variances of the Fourier coefficients are found in a manner analogous to the one-dimensional case, resulting in

$$
\mathrm{E}\left[\mathcal{A}_{\ell m}^2\right] = \tfrac{1}{8}\delta_{\ell m}^{A}\Delta\underset{\sim}{\omega}\Big(G_{\ell m}^d + G_{\ell N}^d + G_{Ln}^d + G_{LN}^d\Big) \tag{19}
$$

$$
\mathrm{E}\left[\mathcal{B}_{\ell m}^2\right] = \tfrac{1}{8}\delta_{\ell m}^{B}\Delta\underset{\sim}{\omega}\Big(G_{\ell m}^d + G_{\ell N}^d + G_{Ln}^d + G_{LN}^d\Big) \tag{20}
$$

for two-dimensions and

$$
\mathrm{E}\left[\mathcal{A}_{\ell mn}\right]^2 = \tfrac{1}{16}\delta_{\ell mn}^{A}\Delta\underset{\sim}{\omega}\Big(G_{\ell mn}^d + G_{\ell mN}^d + G_{\ell Mn}^d + G_{Lmn}^d + G_{\ell MN}^d + G_{LmN}^d + G_{LMn}^d + G_{LMN}^d\Big) \tag{21}
$$

$$
\mathrm{E}\left[\mathcal{B}_{\ell mn}\right]^2 = \tfrac{1}{16}\delta_{\ell mn}^{B}\Delta\underset{\sim}{\omega}\Big(G_{\ell mn}^d + G_{\ell mN}^d + G_{\ell Mn}^d + G_{Lmn}^d + G_{\ell MN}^d + G_{LmN}^d + G_{LMn}^d + G_{LMN}^d\Big) \tag{22}
$$

in three-dimensions, where for p dimensions,

$$
\Delta\underset{\sim}{\omega} = \prod_{i=1}^{p} \Delta\omega_i, \tag{23}
$$

$$
G^d(\underset{\sim}{\omega}) = \frac{G(\omega_1, \ldots, \omega_p)}{2^d}, \tag{24}
$$

and d is the number of components of $\underset{\sim}{\omega} = (\omega_1, \ldots, \omega_p)$ which are equal to zero. The factors $\delta_{\ell mn}^{A}$ and $\delta_{\ell mn}^{B}$ are given by

$$
\delta_{\ell mn}^{A} = \begin{cases} 2 & \text{if } \ell = 0 \text{ or } \frac{K_1}{2} \text{ \underline{and} } m = 0 \text{ or } \frac{K_2}{2} \text{ \underline{and} } n = 0 \text{ or } \frac{K_3}{2} \\ 1 & \text{otherwise} \end{cases} \tag{25}
$$

$$
\delta_{\ell mn}^{B} = \begin{cases} 0 & \text{if } \ell = 0 \text{ or } \frac{K_1}{2} \text{ \underline{and} } m = 0 \text{ or } \frac{K_2}{2} \text{ \underline{and} } n = 0 \text{ or } \frac{K_3}{2} \\ 1 & \text{otherwise} \end{cases} \tag{26}
$$

(ignoring the index n in the case of two dimensions). Thus, in higher dimensions, the simulation procedure is almost identical to that followed in the 1-D case – the only difference being that the coefficients are generated randomly over the half plane (2-D) or the half volume (3-D) rather than the half line of the 1-D formulation.

It is appropriate at this time to investigate some of the shortcomings of the method. First of all it is easy to show that regardless of the desired target covariance function, the covariance function $\hat{B}_k = \hat{B}(k\Delta x)$ of the real FFT process is always symmetric about the midpoint of the field. In one-dimension, the covariance function is given by (using complex notation for the time being),

$$\begin{aligned}\hat{B}_k &= \mathrm{E}\left[Z_{\ell+k}\overline{Z_\ell}\right] \\ &= \mathrm{E}\left[\sum_{j=0}^{K-1} \mathcal{X}_j \exp\left\{i\left(\tfrac{2\pi(\ell+k)j}{K}\right)\right\} \sum_{m=0}^{K-1} \overline{\mathcal{X}_m} \exp\left\{-i\left(\tfrac{2\pi\ell m}{K}\right)\right\}\right] \\ &= \sum_{j=0}^{K-1} \mathrm{E}\left[\mathcal{X}_j\overline{\mathcal{X}_j}\right] \exp\left\{i\left(\tfrac{2\pi jk}{K}\right)\right\},\end{aligned} \tag{27}$$

where use was made of the fact that $\mathrm{E}\left[\mathcal{X}_j\overline{\mathcal{X}_m}\right] = 0$ for $j \neq m$ (overbar denotes the complex conjugate). Similarly one can derive

$$\begin{aligned}\hat{B}_{K-k} &= \sum_{j=0}^{K-1} \mathrm{E}\left[\mathcal{X}_j\overline{\mathcal{X}_j}\right] \exp\left\{-i\left(\tfrac{2\pi jk}{K}\right)\right\} \\ &= \overline{\hat{B}_k}\end{aligned} \tag{28}$$

since $\mathrm{E}\left[\mathcal{X}_j\overline{\mathcal{X}_j}\right]$ is real. The covariance function of a real process is also real in which case (28) becomes simply

$$\hat{B}_{K-k} = \hat{B}_k. \tag{29}$$

In one dimension, this symmetry is illustrated by Figure 1. Similar results are observed in higher dimensions. In general, this deficiency can be overcome by generating a field twice as long as required in each coordinate direction and keeping only the first quadrant of the field. Figure 1 also compares the covariance, mean, and variance fields of the LAS method to that of the FFT method (the TBM method is not defined in one dimension). The two methods give satisfactory performance with respect to the variance and mean fields, while the LAS method shows superior performance with respect to the covariance structure.

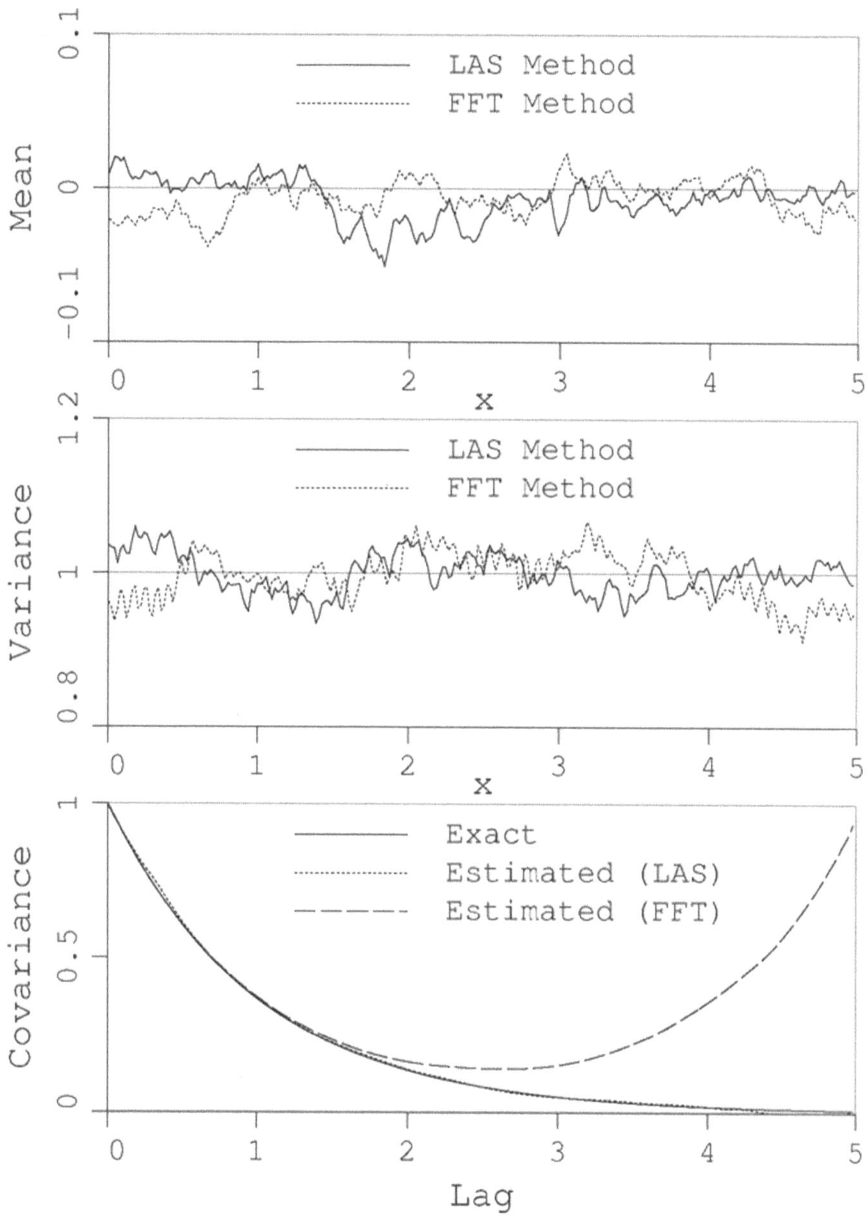

Figure 1. Mean, variance, and covariance of a 1-D 128 point Gauss-Markov process estimated over an ensemble of 2000 realizations.

The second problem with the FFT method relates primarily to its ease of use. Because of the close relationship between the spatial and frequency discretization, considerable care must be exercised when initially defining the spatial field and its discretization. First of all the physical length of the field D must be large enough that the frequency increment $\Delta\omega = 2\pi(K-1)/KD \simeq 2\pi/D$ is sufficiently small. This is necessary if the sequence $\frac{1}{2}G(\omega_0)\Delta\omega, G(\omega_1)\Delta\omega, \ldots$ is to adequately approximate the target spectral density function. Figure 2 shows an example where the frequency discretization is overly coarse. Secondly, the physical resolution Δx must be selected so that the spectral density above the frequency $2\pi/\Delta x$ is negligible. Failure to do so will result in an underestimation of the total variance of the process. In fact the FFT formulation given above folds the power corresponding to frequencies between $\pi/\Delta x$ and $2\pi/\Delta x$ into the power at frequencies below the Nyquist limit $\pi/\Delta x$. This results in the point variance of the simulation being more accurate than if the power above the Nyquist limit were ignored, however it leads to a non-uniqueness in that a family of spectral density functions, all having the same value of $G(\omega_k) + G(\omega_{K-k})$, yield the same process. In general it is best to choose Δx so that the power above the Nyquist limit is negligible. The second term involving the symmetric frequency $G(\omega_{K-k})$ is included here because the point variance is the most important second-order characteristic.

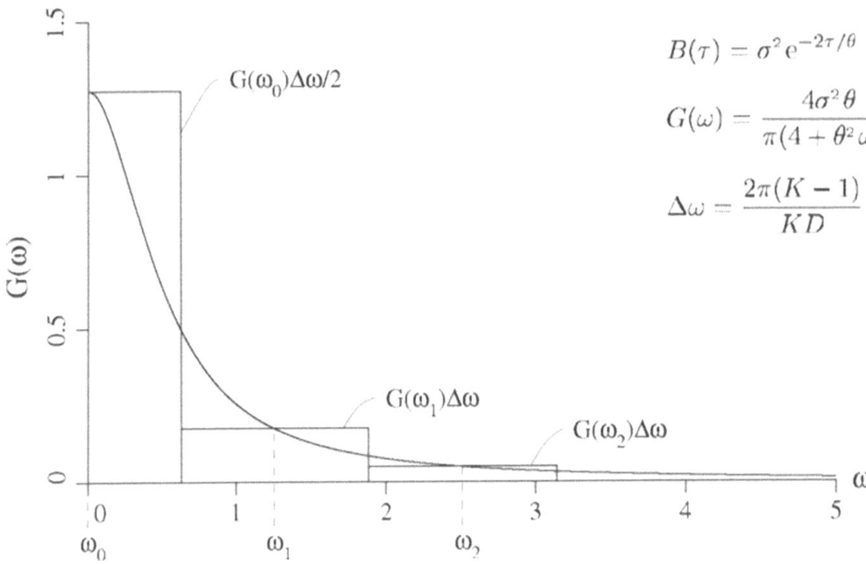

Figure 2 Example of overly coarse frequency discretization resulting in a poor estimation of point variance ($D = 5$ and $\theta = 4$).

Unfortunately, many applications dictate the size and discretization of the field *a-priori* or the user may want to have the freedom to easily consider other geometries or spectral density functions. Without a good deal of careful thought and analysis, the FFT approach can easily yield highly erroneous results.

A major advantage of the FFT method is that it can easily handle anisotropic fields with no sacrifice in efficiency. The field need not be square, although many implementations of the FFT require the number of points in the field in any coordinate direction to be a power of two. Regarding efficiency, it should be pointed out that the time to generate the first realization of the field is generally much longer than that required to generate subsequent realizations. This is because the statistics of the Fourier coefficients must be calculated only once (see Eq.'s 15 and 16).

3 The Turning Bands Method

The Turning Bands Method (TBM), as originally suggested by Matheron (1973), involves the simulation of random fields in two- or higher-dimensional space by using a sequence of one-dimensional processes along lines crossing the domain. With reference to Figure 3, the algorithm can be described as follows,

1) choose an arbitrary origin within or near the domain of the field to be generated,
2) select a line i crossing the domain having a direction given by the unit vector $\underset{\sim}{u}_i$ which may be chosen either randomly or from some fixed set,
3) generate a realization of a one-dimensional process, $Z_i(\xi_i)$, along the line i having zero mean and covariance function $B_1(\tau_i)$ where ξ_i and τ_i are measured along line i,
4) orthogonally project each field point $\underset{\sim}{x}_k$ onto the line i to define the coordinate ξ_{ki} ($\xi_{ki} = \underset{\sim}{x}_k \cdot \underset{\sim}{u}_i$ in the case of a common origin) of the one-dimensional process value $Z_i(\xi_{ki})$,
5) add the component $Z_i(\xi_{ki})$ to the field value $Z(\underset{\sim}{x}_k)$ for each $\underset{\sim}{x}_k$,
6) return to step (2) and generate a new one-dimensional process along a subsequent line until L lines have been produced,
7) normalize the field $Z(\underset{\sim}{x}_k)$ by dividing through by the factor $\sqrt{L}$.

Essentially, the generating equation for the zero-mean discrete process $Z(\underset{\sim}{x})$ is given by

$$Z(\underset{\sim}{x}_k) = \frac{1}{\sqrt{L}} \sum_{i=1}^{L} Z_i(\underset{\sim}{x}_k \cdot \underset{\sim}{u}_i), \tag{30}$$

where if the origins of the lines and space are not common, the dot product must be replaced by some suitable transform. This formulation depends on knowledge of the one-dimensional covariance function, $B_1(\tau)$. Once this is known, the line processes can be produced using some efficient 1-D algorithm.

The covariance function $B_1(\tau)$ is chosen such that the multi-dimensional covariance structure $B_n(\underset{\sim}{\tau})$ in R^n is reflected over the ensemble. For two-dimensional isotropic processes, Mantoglou and Wilson (1981) give the following relationship between $B_2(\underset{\sim}{\tau})$ and $B_1(\eta)$ for $r = |\underset{\sim}{\tau}|$,

$$B_2(r) = \frac{2}{\pi} \int_0^r \frac{B_1(\eta)}{\sqrt{r^2 - \eta^2}}\, d\eta, \tag{31}$$

which is an integral equation to be solved for $B_1(\eta)$. In three dimensions, the relationship between the isotropic $B_3(r)$ and $B_1(\eta)$ is particularly simple,

$$B_1(\eta) = \frac{d}{d\eta}\Big(\eta B_3(\eta)\Big). \tag{32}$$

Mantoglou and Wilson supply explicit solutions for either the equivalent one-dimensional covariance function or the equivalent one-dimensional spectral density function for a variety of common multi-dimensional covariance structures.

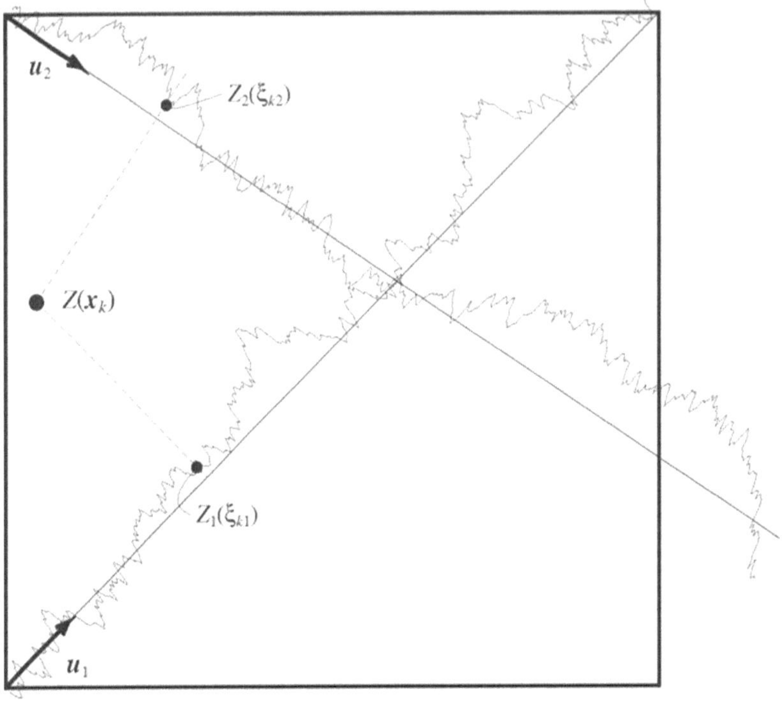

Figure 3. The Turning Bands Method: contributions from the line process $Z_i(\xi_i)$ at the closest points are summed into the field process $Z(\underset{\sim}{x})$ at $\underset{\sim}{x}_k$.

In this implementation of the TBM, the line processes were constructed using a 1-D FFT algorithm as discussed in the previous section. The LAS method was not used for this purpose because the local averaging introduced by the method would complicate the resulting covariance function of (**31**). Line lengths were chosen to be twice that of the field diagonal to avoid the symmetric covariance problem inherent with the FFT method. To reduce errors arising due to overly coarse discretization of the lines, the ratio between the incremental distance along the lines, $\Delta\xi$, and the minimum incremental distance in the field along any coordinate, Δx, was selected to be $\Delta\xi/\Delta x = \frac{1}{2}$.

Figure 4 represents a realization of a 2-D process. The finite number of lines used, in this case **16**, results in a streaked appearance of the realization. A number of origin locations were experimented with to mitigate the streaking, the best appearing to be the use of all four corners as illustrated in Figure **3** and as used in Figure 4. The corner selected as an origin depends on which quadrant the unit vector $\underset{\sim}{u}_i$ points into. If one considers the spectral representation of the one-dimensional random processes along each line (see 2) it is apparent that the streaks are a result of constructive/destructive interference between randomly oriented traveling plane waves. The effect will be more pronounced for narrow band processes and for a small number of lines. For this particular covariance function (see 4), the streaks are still visible when **32** lines are used, but, as shown in Figure 5, are negligible when using 64 lines (the use of number of lines which are powers of 2 is arbitrary). While the 16 line case runs at about the same speed as the 2-D LAS approach, the elimination of the streaks in the realization comes at a price of running about 4 times slower. The streaks are only evident in an average over the ensemble if non-random line orientations are used, although they still appear in individual realizations in either case. Thus, with respect to each realization, there is no particular advantage to using random versus non-random line orientations.

Figure 4. Sample function of a 2-D field via TBM using 16 lines.

Since the streaks are present in the field itself, this type of error is generally more serious than errors in the variance or covariance field. For example, if the field is being used to represent soil conductivity, then the streaks could represent paths of reduced resistance to flow, a feature which may not be desirable in a particular study. Crack propagation studies may also be very sensitive to such linear correlations in the field. For applications such as these, the Turning Bands method should only be used with a sufficiently large number of lines. This may require some preliminary investigation for arbitrary covariance functions. In addition, the minimum number of lines in 3 and higher dimensions is difficult to determine due to visualization problems.

Note that the Turning Bands Method does not suffer from the symmetric covariance structure that is inherent in the FFT approach. The variance field and covariance structure are also well preserved. However, the necessity of finding an equivalent 1-D covariance or spectral density function through an integral equation along with the streaked appearance of the realization when an insufficient number of lines are used makes the method less attractive. Using a larger number of lines, TBM is probably the most accurate of the three methods considered, at the expense of decreased efficiency. TBM can be extended to anisotropic fields, although there is an additional efficiency penalty associated with such an extension since the 1-D process statistics must be recalculated for each new line orientation (see Mantaglou and Wilson, 1981, for details).

Figure 5. Sample function of a 2-D field via TBM using 64 lines.

4 The Local Average Subdivision Method

Of the three methods considered, the LAS method is probably the most difficult to implement. The details are given by Fenton and Vanmarcke (1990). The one- and two-dimensional implementation of LAS used for this study differs slightly from that reported by Fenton and Vanmarcke in that an initial set of cells are generated directly from the associated covariance matrix according to Eq. 3. Specifically, in the 1-D case, a positive integer k_1 is found so that the total number of cells, N_1, desired in the final field can be expressed as

$$N_1 = k_1(2^m) \tag{33}$$

where m is the number of subdivisions to perform and k_1 is as large as possible with $k_1 \leq 16$. In one-dimension, this modification was implemented to reduce the variance field errors discussed later. In two dimensions, two positive integers k_1 and k_2 are found such that $k_1 k_2 \leq 256$ and the field dimensions can be expressed as

$$N_1 = k_1(2^m) \tag{34a}$$

$$N_2 = k_2(2^m) \tag{34b}$$

from which the first $k_1 \times k_2$ lattice of cell values are simulated directly using covariance matrix decomposition (3). Since the number of subdivisions, m, is common to the two parameters, one is not entirely free to choose N_1 and N_2 arbitrarily. It does, however, give a reasonable amount of discretion in generating non-square fields, as is also possible with both the FFT and TBM methods.

Perhaps the major advantage of the LAS method is that it produces a field of local average cells whose statistics are consistent with the field resolution. As such, it is well suited to problems where the system is represented by a set of elements and average properties over each element are desired. Changing the element size automatically results in changes in the statistics of the element average, as dictated by random field theory. This is appealing since almost all measured engineering properties are based on local averages (concrete strength, for example, is based on a finite volume cylinder).

To enable a discussion of some of the features and problems with the LAS method, a brief overview of the 2-D implementation will be repeated here. The 2-D LAS method involves a subdivision process in which a 'parent' cell is divided into 4 equal sized cells. In Figure 6, the parent cells are denoted P_i, $i = 1, 2, \ldots$ and the subdivided, or child cells are denoted Q_j, $j = 1, 2, 3, 4$. Although each parent cell is eventually subdivided in the LAS process, only P_5 is subdivided in Figure 6 for simplicity. Using vector notation, the values of $\underset{\sim}{Q}^T = \{Q_1, Q_2, Q_3, Q_4\}$ are obtained by adding a mean term to a random component. The mean term derives from a best linear unbiased estimate using a 3×3 neighborhood of the parent values, in this case $\underset{\sim}{P}^T = \{P_1, \ldots, P_9\}$.

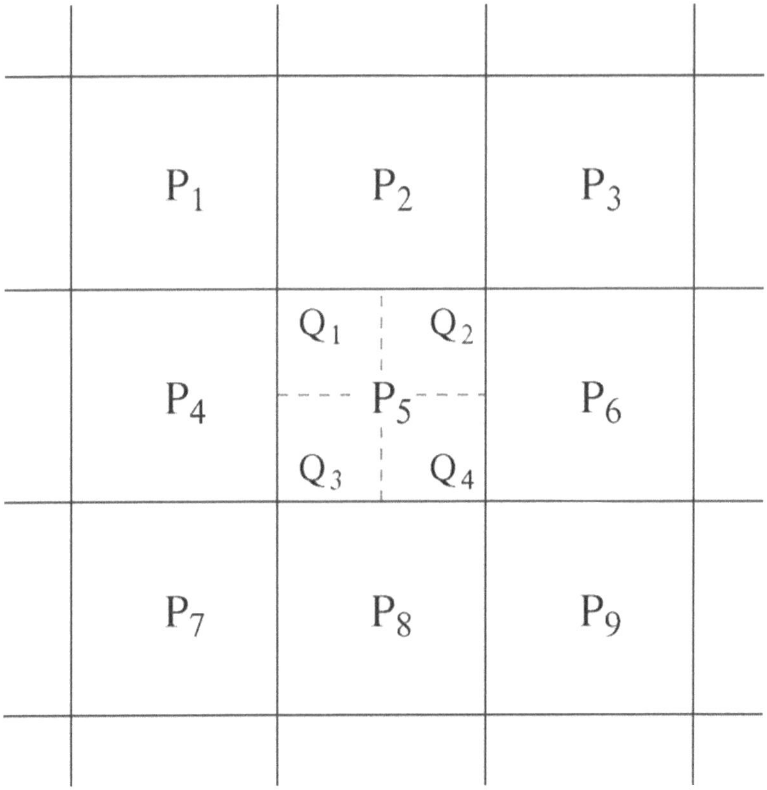

Figure 6. Local Average Subdivision in two-dimensions.

Specifically

$$\underset{\sim}{Q} = \underset{\approx}{A}^T \underset{\sim}{P} + \underset{\approx}{C}\underset{\sim}{U} \tag{35}$$

where $\underset{\sim}{U}$ is a random vector with independent $N(0, 1)$ elements. This is essentially an ARMA model in which the 'past' is represented by the previous coarser resolution stages. Defining the covariance matrices

$$\underset{\approx}{R} = \mathrm{E}\left[\underset{\sim}{P}\underset{\sim}{P}^T\right], \tag{36a}$$

$$\underset{\approx}{S} = \mathrm{E}\left[\underset{\sim}{P}\underset{\sim}{Q}^T\right], \quad \text{and} \tag{36b}$$

$$\underset{\approx}{B} = \mathrm{E}\left[\underset{\sim}{Q}\underset{\sim}{Q}^T\right], \tag{36c}$$

then the matrix $\underset{\approx}{A}$ is determined by

$$\underset{\approx}{A} = \underset{\approx}{R}^{-1}\underset{\approx}{S} \tag{37}$$

while the lower triangular matrix $\underset{\approx}{C}$ satisfies

$$\underset{\approx}{C}\underset{\approx}{C}^T = \underset{\approx}{B} - \underset{\approx}{S}^T\underset{\approx}{A} \tag{38}$$

Note that the matrix on the right hand side of (38) is only rank 3, so that the 4×4 matrix $\underset{\approx}{C}$ has a special form with columns summing to zero (thus $C_{44} = 0$). While this results from the fact that all the expectations used in Eq.'s (36) are derived using local average theory over the cell domains, the physical interpretation is that upwards averaging is preserved, ie. that $P_5 = \frac{1}{4}(Q_1 + Q_2 + Q_3 + Q_4)$. This means that one of the elements of $\underset{\sim}{Q}$ is explicitly determined once the other three are known.

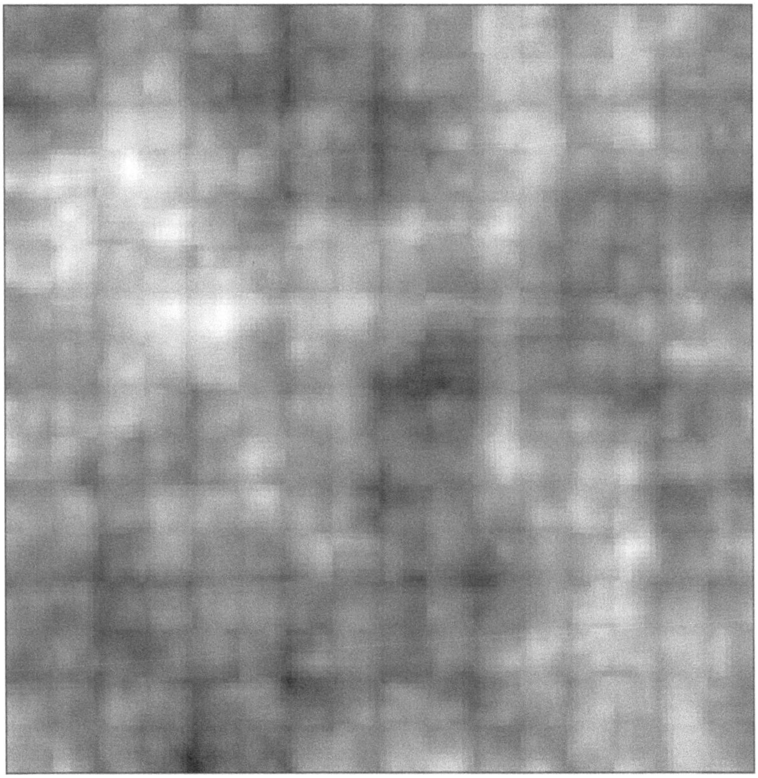

Figure 7. Two-dimensional LAS variance field estimated over 200 realizations.

Although Figure 1 illustrates the superior performance of the LAS method over the FFT method in one dimension, a systematic bias in the variance field is observed in two dimensions. Figure 7 shows a grey scale image of the estimated cell variance in a two-dimensional field obtained by averaging over the ensemble. There is a definite pattern in the variance field – the variance tends to be lower near the major cell divisions, that is at the 1/2, 1/4, 1/8, etc. points of the field. This is because the actual diagonal, or variance, terms of the 4×4 covariance matrix corresponding to a subdivided cell are affected by the truncation of the parent cell influence to a 3×3 neighborhood. The error in the variance is compounded at each subdivision stage and cells close to 'older' cell divisions show more error than do 'interior' cells. The magnitude of this error varies with the number of subdivisions, the scale of fluctuation, and type of covariance function governing the process and can be obtained by evaluating the transfer function of Eq. (35). Such a quantitative analysis is yet to be performed.

Figure 8 depicts the estimated variances along a line through the plane for both the LAS and TBM methods. Along this line, the pattern in the LAS estimated variance is not particularly noticeable and the values are about what would be expected for an estimate over the ensemble. Figure 9 compares the estimated covariance structure in the vertical and horizontal directions, again for the TBM (64 lines) and LAS methods. In this respect, both the LAS and the TBM methods are reasonably accurate. In addition, the LAS method yields good quality realizations and an accurate mean field.

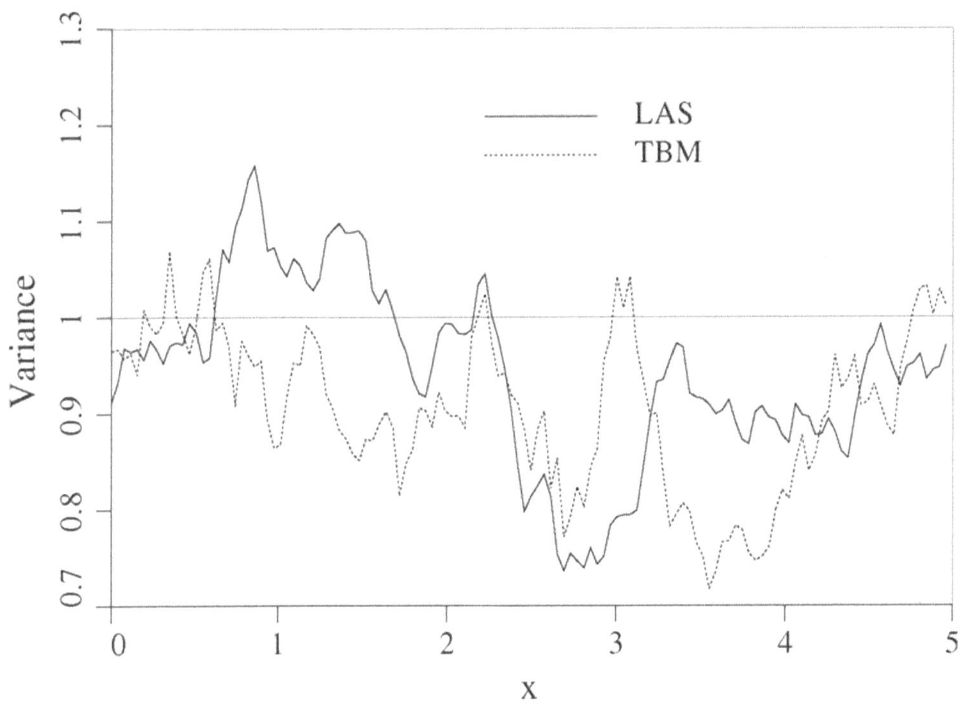

Figure 8. Variance along a horizontal line through the two-dimensional LAS and TBM fields estimated over 200 realizations.

The LAS method can be used to produce anisotropic random fields with no code change and no loss in efficiency, however the anisotropic nature of such a field is due entirely to the initial production of a $k_1 \times k_2$ field via covariance matrix decomposition. The subdivision algorithm itself is incapable of preserving anisotropy, the directional scales of fluctuation tending toward the minimum for the field. Thus, although the field may be globally anisotropic, small neighborhoods of cells at the final resolution tend to follow an isotropic correlation structure. One notes that ellipsoidally anisotropic random fields can always be produced from isotropic random fields by suitably stretching the coordinate axes and so this error is not always a problem.

It may be possible to improve the LAS covariance approximations by extending the size of the parent cell neighborhood. A 3×3 neighborhood is used in the current implementation of the 2-D LAS algorithm, as shown in Figure 6, but any odd sized neighborhood could be used to condition the statistics of the subdivided cells. Larger neighborhoods have not been tested in two and higher dimensions, although in one dimension increasing the neighborhood size to 5 cells resulted in a more accurate covariance function representation, as would be expected.

The LAS method also depends on knowledge of the variance function which governs the variance reduction when a process is locally averaged. Although this function can be obtained through a direct, possibly numerical, integration of the covariance function, analytically exact variance functions can be difficult to find for some covariance functions in higher dimensions. Vanmarcke (1984) has derived the variance functions for a number of common random processes.

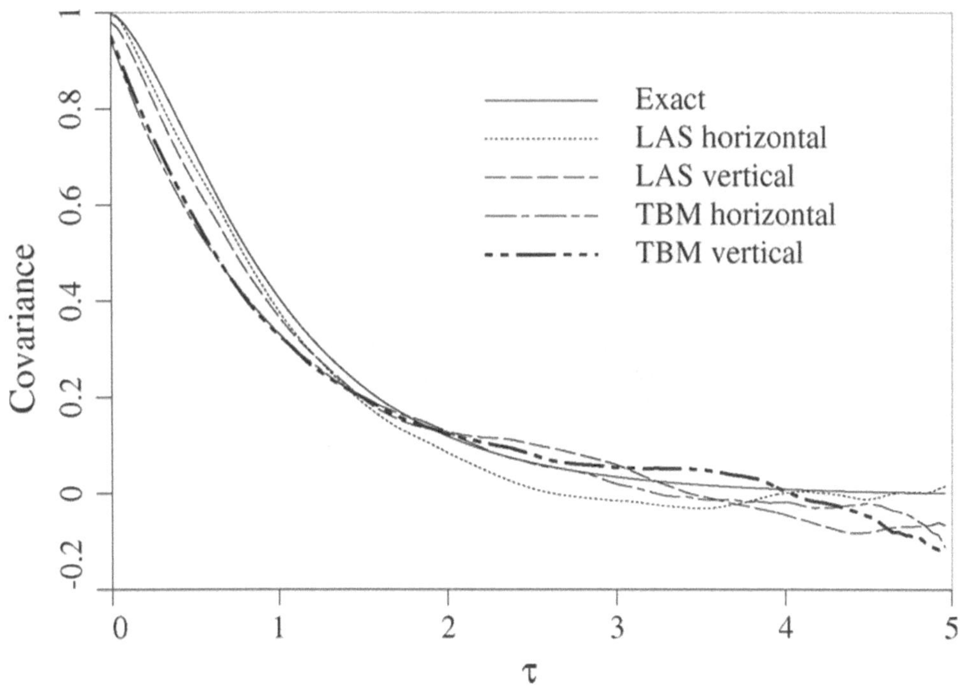

Figure 9. Covariance structure of the LAS and TBM two-dimensional random fields estimated over 200 realizations.

5 Comparison and Conclusions

The choice of a random field generator to be used for a particular problem or in general depends on many issues. Table 1 shows the relative run times of the three algorithms to produce identically sized fields. The times have been normalized with respect to the FFT method so that a value of 2 indicates that the method took twice as long. If efficiency alone were the selection criteria, then either the TBM with a small number of lines or the LAS methods would be selected, with probably the LAS a better choice if streaking is not desired. However, efficiency of the random field generator is often not an overriding concern – in many applications, the time taken to generate the field is dwarfed by the time taken to subsequently process or analyze the field. Substantial changes in generator efficiency may be hardly noticed by the user.

Table 1. Comparison of run-times of the FFT, TBM and LAS algorithms in one and two-dimensions.

Dimension	FFT	LAS	TBM	
			16 lines	64 lines
1-D	1.0	0.70	–	–
2-D	1.0	0.55	0.64	2.6

As a further comparison of the accuracy of the FFT, TBM, and LAS methods, a set of 200 realizations of a 128×128 random field were generated using the Gauss-Markov covariance function with a scale of fluctuation $\theta = 2$ and a physical field size of 5×5. The mean and variance fields were calculated by estimating these quantities at each point in the field (averaging over the ensemble) for each algorithm. The upper and lower 90% quantiles are listed in Table 2 along with those predicted by theory assuming a normal distribution. To obtain these numbers, the mean and variance fields were first estimated, then upper and lower bounds were found such that 5% of the field exceeded the bounds above and below, respectively. Thus 90% of the field is observed to lie between the bounds. It can be seen that all three methods yield very good results with respect to the expected mean and variance quantiles. The TBM results were obtained using 64 lines. Although these results are strictly only valid for the particular covariance function used, they are believed to be generally true over a wider variety of covariance functions and scales of fluctuation.

Table 2. Upper and lower 90% quantiles of the estimated mean and variance fields for the FFT, TBM, and LAS methods (200 realizations).

Algorithm	Mean	Variance
FFT	$(-0.06, 0.12)$	$(0.87, 1.19)$
TBM	$(-0.11, 0.06)$	$(0.83, 1.14)$
LAS	$(-0.12, 0.09)$	$(0.82, 1.13)$
Theory	$(-0.12, 0.12)$	$(0.84, 1.17)$

Purely on the basis of accuracy in the mean, variance and covariance structures, the best algorithm of those consider here is probably the TBM method using a large number of lines. The TBM method is also one of the easiest to implement once an accurate 1-D generator has been implemented. Unfortunately, there is no clear rule regarding the minimum number of lines to be used to avoid streaking. In two dimensions using the Gauss-Markov covariance function, it appears that at least 50 lines should be employed. However, as mentioned, narrow band processes may require more. In three dimensions, no such statements can be made due to the difficulty in studying the streaking phenomena off a plane. Presumably one could use a 'density' of lines similar to that used in the two-dimensional case, perhaps subtending similar angles, as a guide. The TBM method is reasonably easy to use in practice as long as the equivalent 1-D covariance or spectral density function can be found.

The FFT method suffers from symmetry in the covariance structure of the realizations. This can be overcome by generating fields twice as large as required in each coordinate direction and ignoring the surplus. This correction results in slower run times (a factor of 2 in 1-D, 4 in 2-D, etc.). The FFT method is also relatively easy to implement and the algorithm is similar in any dimension. Its ability to easily handle anisotropic fields makes it the best choice for such problems. Care must be taken when selecting the physical field dimension and discretization interval to ensure that the spectral density function is adequately approximated. This latter issue makes the method more difficult to use in practice. However, the fact that the FFT approach employs the spectral density function directly makes it an intuitively attractive method, particularly in time dependent applications.

The LAS method has a systematic bias in the variance field, in two and higher dimensions, which is not solvable without changing the algorithm. However, the error generally does not result in values of variance that lie outside what would be expected from theory – it is primarily the pattern of the variance field which is of concern. Of the three methods considered, the LAS method is the most difficult to implement and it depends on the variance function representation of the field. It is, however, one of the easiest to use once coded since it requires no decisions regarding its parameters, and it is generally the most efficient. If the problem at hand requires or would benefit from a local average representation, then the LAS method is the logical choice.

6 Acknowledgements

The authors would like to thank the Natural Sciences and Engineering Research Council of Canada for their financial support under Grant OPG0105445. Any opinions, findings, and conclusions and recommendations are those of the author and do not necessarily reflect the views of the aforementioned organization.

Notation

The following symbols are used in this chapter:

$\mathcal{A}_k$ = Fourier coefficient (real part)

$\mathcal{B}_k$ = Fourier coefficient (imaginary part)

$B(\cdot)$ = covariance function

$\hat{B}$ = algorithmic covariance function

$\underset{\approx}{B}$ = covariance matrix

C_k = random amplitude

D = physical dimension of the field

$\mathrm{E}[\cdot]$ = expectation operator

f = weighting function

$G(\cdot)$ = one-sided spectral density function

i = square root of negative one

K = number of frequency and spatial points

$\underset{\approx}{L}$ = lower triangular matrix

N = number of spatial points

$\underset{\sim}{P}$ = vector of parent cell values (LAS)

$\underset{\sim}{Q}$ = vector of child cell values (LAS)

$S(\cdot)$ = two-sided spectral density function

$\underset{\sim}{U}$ = vector of independent random numbers

W = white noise process

$\mathcal{X}_k$ = complex Fourier coefficient

$\underset{\sim}{x}$ = spatial coordinate

Z = random field

$\underset{\sim}{Z}$ = vector of random field values at discrete points or cells

Φ_k = random phase angle

θ = scale of fluctuation

σ = standard deviation

$\underset{\sim}{\tau}$ = lag vector

$\underset{\sim}{\omega}$ = frequency

Bibliography

J.W. Cooley and J.W. Tukey . An algorithm for the machine calculation of complex Fourier Series," *Mathematics of Computation*, **19**(90), 297-301, 1965.

G.A. Fenton and E.H. Vanmarcke . Simulation of random fields via Local Average Subdivision, *ASCE J. Engrg. Mech.*, **116**(8), 1733–1749, 1990.

G.A. Fenton . *Simulation and Analysis of Random Fields*, Ph.D. Thesis, Princeton University, Princeton, New Jersey, 1990.

A. Mantoglou and J.L. Wilson . Simulation of random fields with the Turning Bands Method, MIT, Dept. Civil Engrg., Report #264, Cambridge, MA, 1981.

G. Matheron . The intrinsic random functions and their applications, *Adv. in Appl. Probab.*, **5**, 439-468, 1973.

M.P. Mignolet and P.D. Spanos . Simulation of homogeneous two-dimensional random fields: Part I – AR and ARMA Models, *ASME J. Appl. Mech.*, **59**, S260–S269, 1992.

M. Shinozuka and C.M. Jan . Digital simulation of random processes and its applications, *J. Sound Vibration*, **25**(1), 111-128, 1972.

P.D. Spanos and M.P. Mignolet . Simulation of homogeneous two-dimensional random fields: Part II – MA and ARMA Models, *ASME J. Appl. Mech.*, **59**, S270–S277, 1992.

E.H. Vanmarcke . *Random Fields: Analysis and Synthesis*, The MIT Press, Cambridge, Massachusetts, 1984.

A.M. Yaglom . *An Introduction to the Theory of Stationary Random Functions*, Dover, Mineola, NY, 1962.

The Random Finite Element Method (RFEM) in Settlement Analyses

Gordon A. Fenton*, D.V. Griffiths†, and W. Cavers‡

* Department of Engineering Mathematics, Dalhousie University, Canada
† Division of Engineering, Colorado School of Mines, U.S.A.
‡ Golder Associates Ltd., Ottawa, Canada

Abstract In order to control serviceability problems arising from excessive settlement of shallow footings, geotechnical design codes generally include specifications regarding maximum settlement which often govern the footing design. Once the footing has been designed and constructed, the actual settlement it experiences on a real three-dimensional soil mass can be quite different than expected, due to the soil's spatial variability. Because of this generally large variability (compared to other engineering materials such as concrete and steel) and because this particular serviceability limit state often governs the design, it makes sense to consider a reliability-based approach to settlement design. This chapter looks in some detail at a Load and Resistance Factor Design (LRFD) approach to limiting footing settlement. In particular, the resistance factors required to achieve a certain level of settlement reliability as a function of soil variability and site investigation intensity are determined analytically using random field theory. Simplified approximate relationships are proposed and tested using simulation via the Random Finite Element Method. It is found that the simplified relationships are validated both by theory and simulation and so can be used to augment the calibration of geotechnical LRFD code provisions with respect to shallow foundation settlement.

1 Reliability-Based Geotechnical Design Issues

In an effort to harmonize with structural codes, geotechnical design codes around the world are beginning to migrate towards some form of reliability-based design (RBD). Significant steps in this direction can be found in, for example, Eurocode 7, 2004, Australian Standard AS 4678, 2002, NCHRP Report 507, 2004, and the Canadian Foundation Engineering Manual, 1992. These RBD provisions are most often presented in the form of a Limit States Design (LSD), to define critical failure states, combined with load and resistance factors calibrated to achieve the target reliabilities associated with the various limit states. The use of load and resistance factors is generally referred to as Load and Resistance Factor Design (LRFD).

By and large, the random characteristics of loads, or "actions", in civil engineering projects, are fairly well known and so load factors are reasonably well established. On the resistance side, for most common structural materials representative tests can easily be performed, and have been, to establish material property distributions that apply with reasonable accuracy anywhere that the material is used. Thus, resistance factors for materials such as concrete, steel, and wood have been known for decades.

The Random Finite Element Method (RFEM) in Steady Seepage Analysis

D.V. Griffiths[*] and Gordon A. Fenton[†]

Division of Engineering, Colorado School of Mines, U.S.A.
Department of Engineering Mathematics, Dalhousie University, Canada

bstract The e ect of stochastic soil permeability on confined seepage occurring beneath water retaining structures has been studied. Random field concepts for the generation of soil permeability properties with a fixed mean, standard deviation and spatial correlation structure, have been combined with finite element methods to perform *Monte arlo* simulations of the seepage problem. Analyses have been performed for the case of a dam with two cut-o walls. The results of parametric studies to gauge the e ect of the standard deviation and correlation structure of the permeability on the output statistics relating to seepage quantities, exit gradients and uplift pressures are presented. In all cases, comparison is made with results that would be achieved on a deterministic basis. Flow rates and other quantities of interest are shown to be significantly a ected by both the standard deviation and the correlation structure of soil permeability.

1 Introduction

The great majority of geotechnical analysis is *deterministic* in that the soil properties used are assumed to be 'average' values. Variations in the soil properties are then accounted for by the use of safety factors which are often rather arbitrarily applied to the computed result.

This 'average' approach to the definition of soil properties has tended to be applied, not only to classical soil mechanics calculations, but also to numerical computations using sophisticated numerical techniques such as the finite element method. Properties are usually assigned on the basis of a limited number of laboratory tests. In reality, these properties vary from point to point and can only be determined deterministically through numerous field tests. Since this is both expensive and impractical, random field models can be used to represent the geomaterial. The parameters of these models can be estimated from a limited number of test results.

Mean soil properties are fairly well established, and recently there has been an improvement in the availability of data on second moment statistics (standard deviation and spatial correlation). In large part, the data gathering has been motivated by the availability of random field simulation algorithms and their potential for producing useful results. The increased performance of computers has also enabled more detailed discretisation of boundary value problems, and better modelling of the stastistical properties of the input parameters.

The finite element method is an ideal vehicle for modelling materials with a spatial variation in properties. Stochastic finite element analysis has been implemented in a number of areas of geotechnical interest. For example Beacher and Ingra (1981) and Righetti and Harrop-Williams (1988) for stress analysis and settlements of foundations, Ishii and Suzuki (1987) for slope stability and Smith and Freeze (1979a and 1979b) for confined seepage.

Stochastic finite elements can be interpreted in di erent ways. On the one hand statistical properties can be built into the finite element equations themselves (see e.g. Vanmarcke and Grigoriu 1983), or multiple analyses (*Monte Carlo*) can be performed, with each analysis stemming from a realisation of the soil properties treated as a multi-dimensional random field. In the present work the latter approach has been used to examine confined seepage, with particular reference to flow under a water retaining structure founded on stochastic soil. While the Monte Carlo approach tends to be computationally intensive, it has the distinct advantage of being able to model highly variable input properties.

2 Con ned seepage

In the study of seepage through soils beneath water retaining structures, three important quantities need to be assessed by the designers as shown in Figure 1): seepage quantity, exit gradients and uplift forces.

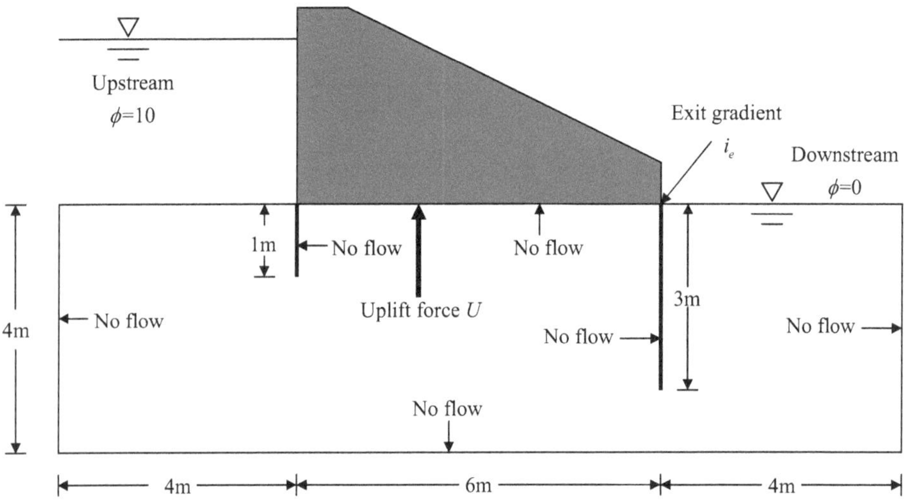

Figure 1. The boundary value problem

The classical approach used by Civil Engineers for estimating these quantities involves the use of carefully drawn flow nets (Casagrande 1940, Cedergren 1967, Verruijt 1970). Various alternatives to flow nets are available for solving the seepage problem, however in order to perform quick parametric studies, for example relating to the e ect of cut-o wall length, powerful approximate techniques such as the Method of Fragments (Pavlovsky 1933, Harr 1962, Griffiths 1984) are increasingly employed. The conventional methods are *deterministic*, in that the soil permeability is assumed to be constant and homogeneous, although anisotropic properties and stratification can be taken into account.

In this paper, a more rational approach to the modelling of soil properties is adopted, whereby the permeability of the soil underlying a structure such as that shown in Figure 1 is assumed to be *stochastic*, i.e. the soil property in question is assumed to be a 'random' field (e.g. Vanmarcke 1984) defined statistically . The best known statistics are the mean and standard deviation, however it is well known that spatial dependencies also exist – the soil properties at two points separated by 1 cm are likely to be more similar than those at two points separated by a metre or a kilometre. This spatial dependence is often characterized by a measure called the 'scale of fluctuation', which, loosely speaking, is the distance over which properties show appreciable correlation. In general for a site of fixed size, as the scale of fluctuation increases, the soil properties become more uniform over the site. These statistics will be discussed further in a later section.

The analyses in this paper use a technique called Local Average Subdivision (LAS) to generate realisations of the random permeability fields with given mean, standard deviation and correlation structure. This technique is fully described by Fenton (1990) and Fenton and Vanmarcke (1990). The resulting field of permeabilities is mapped onto a finite element mesh, and potential and stream function boundary conditions are specified. The governing elliptic equation for steady flow (Laplace) leads to a system of linear 'equilibrium' equations which are solved for the nodal potential values throughout the mesh using conventional Gaussian elimination.

Only deterministic boundary conditions are considered in this paper, the primary goal being to investigate the e ects of randomly varying soil properties on the engineering quantities noted above. The method is nevertheless easily extended to random boundary conditions corresponding to uncertainties in upstream and downstream water levels.

The next two sections give a brief description of the finite element technique and the method by which permeability values are assigned to the mesh. This is followed by a results section in which the statistics of the output quantities relating to flow rate, exit gradients and uplift are discussed.

3 Finite Element Analyses

The steady flow problem is governed in 2-d by Laplace's equation, in which the dependent variable ϕ is the piezometric head or potential at any point in the Cartesian field x-y.

$$k_x \frac{\partial^2 \phi}{\partial x^2} + k_y \frac{\partial^2 \phi}{\partial y^2} = 0 \tag{1}$$

where k_x and k_y are the permeabilities in the x and y directions. In the present work and at the element level, the permeability field is assumed to be isotropic ($k_x = k_y = k$). While the method discussed herein is simply extended to the anisotropic case (through the generation of a pair of correlated random fields) it was felt that such an extension is best restricted to a particular site of interest, the complexity introduced to a general discussion being unwarranted.

Note that equation (3.1) is strictly only valid for constant k. In this analysis the permeability is taken to be constant within each element, its value being given by the local geometric average of the permeability field within the domain of the element. From element to element, the value of k will vary, however, reflecting the random nature of the permeability. This approximation of the permeability field is consistent with the approximations made in the finite element method and is superior to most traditional approaches in which the permeability of an element is taken to be simply the permeability at some point within the element.

A typical finite element mesh used in this study is shown in Figure 2. It contains 1400 elements, and represents a model of 2-d flow beneath a dam which includes two cut-o walls. The upstream and downstream potential values are fixed at 10 and zero metres respectively. The cut-o walls are assumed to have zero thickness, and the nodes along those walls have two potential values corresponding to the right and left sides of the wall.

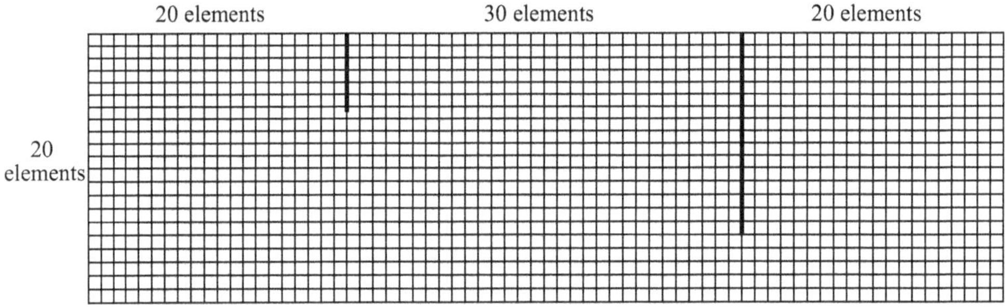

Figure 2. The FE mesh: all elements are 0.2m x 0.2m squares

The finite element code for the solution of Laplace's equation is broadly similar to that published by Smith and Griffiths (2004) . The element conductivity matrices are assembled into a global matrix in the usual way, resulting in a system of linear equations in the 'unknown' nodal potential values. The global conductivity relationship after assembly becomes

$$= \boldsymbol{Q} \qquad (2)$$

Once the global conductivity equations are solved leading to nodal potential values held in , the output quantities relating to flow rates, uplift pressures and exit gradients are easily deduced. More detail on how these values are obtained will be described in later sections.

For each boundary value problem considered, multiple solutions were obtained using successive realisations of the permeability field. The random permeability field is characterized by three parameters defining its first two moments, namely the mean μ_k, the standard deviation σ_k and the scale of fluctuation θ_k.

In order to obtain reasonably stable out statistics, it was decided that each parametric combination would be analysed using 1000 realisations.

4 Generation of permeability values

Field measurements of permeability have indicated an approximately log-normal distribution (see e.g. Hoeksma and Kitanidis 1985, Sudicky 1986) . The same distribution has therefore been adopted for the simulations presented in this Paper. Essentially, the permeability field is obtained through the transformation

$$k_i = \exp\{\mu_{\ln k} + \sigma_{\ln k}\, g_i\} \tag{3}$$

in which k_i is the permeability assigned to the ith element, g_i is the local average of a standard Gaussian random field, g, over the domain of the ith element, and $\mu_{\ln k}$ and $\sigma_{\ln k}$ are the mean and standard deviation of the logarithm of k_i (obtained from the 'target' mean and standard deviation μ_k and σ_k).

The LAS technique renders realisations of the local averages g_i which are derived from the random field g having zero mean, unit variance, and a spatial correlation controlled by the scale of fluctuation. As the scale of fluctuation goes to infinity, g_i becomes equal to g_j for all elements i and j – that is the field of permeabilities tends to become uniform on each realisation. At the other extreme, as the scale of fluctuation goes to zero, g_i and g_j become independent for all $i \neq j$ – the soil permeability changes rapidly from point to point.

In the two dimensional analyses presented in this paper, the scales of fluctuation in the vertical and horizontal directions are taken to be equal (isotropic) for simplicity. Although beyond the scope of this paper, it should be noted that for a layered soil mass the horizontal scale of fluctuation is generally larger than the vertical scale due to the natural stratification of many soil deposits. The 2-d model used herein implies that the out-of-plane scale of fluctuation is infinite – soil properties are constant in this direction – which is equivalent to specifying that the streamlines remain in the plane of the analysis. This is clearly a deficiency of the present model, however it is believed that useful information regarding the variability of flow quantities is still to be gained from the 2-d model.

5 Deterministic Solution

With regard to the seepage problem shown in Figure 2, a deterministic analysis was performed in which the permeability of all elements was assumed to be constant and equal to 10 ^{5} m/s. This value was chosen as it was to be the mean value of subsequent stochastic analyses. Both the potential and the inverse streamline problems were solved, leading to the flow net shown in Figure 3.

All output quantities were computed in non-dimensional form. In the case of the flow rate, the global flow vector $\boldsymbol{Q}$ was computed by forming the product of the potentials and the global conductivity matrix from equation (3.2). Assuming no sources or sinks in the flow regime, the only non-zero values in $\boldsymbol{Q}$ correspond to those freedoms on the

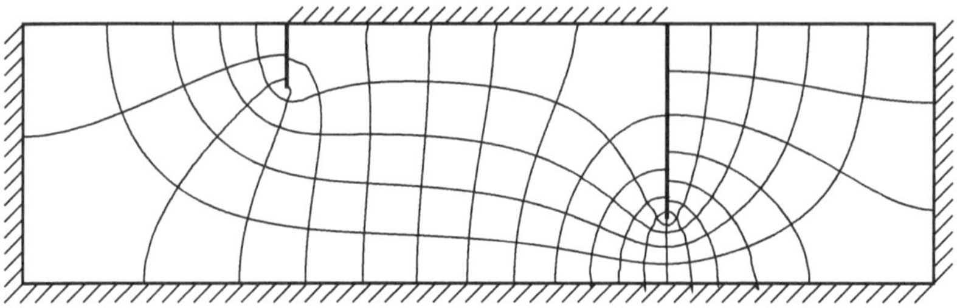

Figure 3. Deterministic flow net: $n_f = 5$, $n_d = 20$

upstream and downstream boundaries. These values were summed to give the flow rate Q in m^3/s/m, leading to a non-dimensional flow rate $\bar{Q}$ defined by

$$\bar{Q} = Q/(\mu_k H) \tag{4}$$

where μ_k is the (isotropic) mean permeability and H is the total head di erence between the up- and downstream sides.

The uplift force on the base of the dam U was computed by integrating the pressure distribution along the base of the dam between the cut-o walls. This quantity was easily deduced from the potential values at the nodes along this line together with a simple numerical integration scheme (Repeated Trapezium Rule). A non-dimensional uplift force $\bar{U}$ was defined:

$$\bar{U} = U/(H \quad _w L) \tag{5}$$

where $\quad _w$ is the unit weight of water and L is the distance between the cut-o walls. $\bar{U}$ is the uplift force expressed as a proportion of buoyancy force that would occur if the dam was submerged in water alone.

The exit gradient i_e is the rate of change of head at the exit point closest to the dam at the downstream end. This was calculated using a 4-point backward di erence numerical di erentiation formula of the form:

$$i_e \approx \frac{1}{6b}(11\phi_0 \quad 18\phi_{\ 1} + 9\phi_{\ 2} \quad 2\phi_{\ 3}) \tag{6}$$

where the ϕ_i values correspond to the piezometric head at the four nodes vertically below the exit point as shown in Figure 4, and b is the constant vertical distance between nodes. It may be noted that the downstream potential head is fixed equal to zero, thus $\phi_0 = 0.0$ m. The use of this four-point formula was arbitrary, and was considered a compromise between the use of very low order formulae which would be too sensitive to 'random' fluctuations in the potential, and high order formulae which would involve the use of correspondingly high order interpolation polynomials which would be hard to justify physically.

Referring to Figures 1 and 2, the constants described above were given the following values: $H = 10$m, $\mu_k = 10 \quad ^5$m/s, $\quad _w = 9.81$kN/m^3, $L = 6$m, $b = 0.2$m.

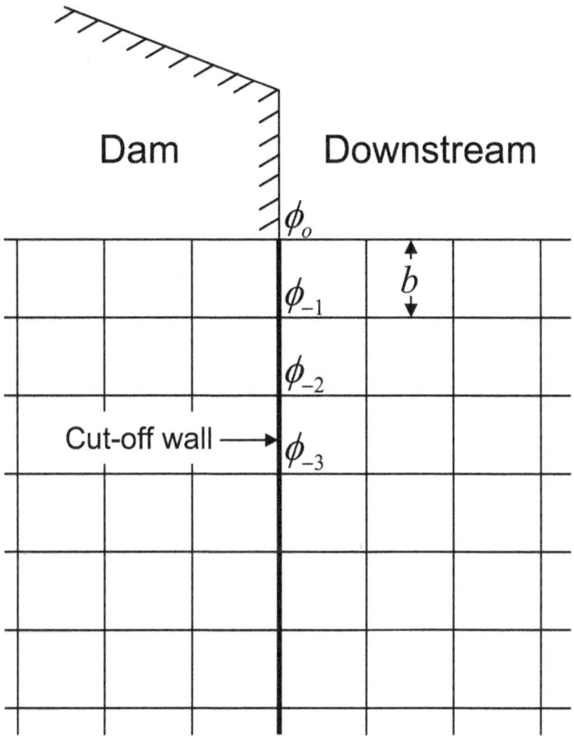

Figure 4. Detail of downstream cut-o wall

and a deterministic analysis using the mesh of Figure 2 led to the following output quantities: $\bar{Q} = 0.226$, $\bar{U} = 0.671$, $i_e = 0.688$.

This value of i_e would be considered unacceptable in a real design situation, bearing in mind that the critical hydraulic gradient for most soils approximately equals unity. The value of i_e is proportional to the head di erence H however, which in this case for simplicity and convenience of normalisation has been set to 10 m as shown above.

These results will be compared with output from the stochastic analyses described in the next section.

6 Stochastic Analyses

In all the 2-d stochastic analyses that follow, the soil was assumed to be isotropic with a mean permeability of $\mu_k = 10\ ^{5}$ m/s. More specifically, the random fields were generated such that the 'target' (geometric) mean permeability of each finite element was held constant at $10\ ^{5}$ m/s. Parametric studies were performed relating to the e ect of varying the standard deviation (σ_k) and the scale of fluctuation (θ_k) of the permeability field. Following 1000 realisations, statistics relating to output quantities $\bar{Q}, \bar{U}$ and i_e were calculated.

6.1 Single realisation

Before discussing the results from multiple realisations, an example of what a flow net might look like for a single realisation is given in Figure 5 for permeability statistics $\sigma_k/\mu_k = 1$, and $\theta_k = 1.0$m.

In Figure 5, the flow net is superimposed on a 'grey-scale' which indicates the spatial distribution of the permeability values. Dark areas correspond to low permeability and light areas to high permeability. The streamlines clearly try to 'avoid' the low permeability zones, but this is not always possible as some realisations may generate a complete 'blockage' of low permeability material in certain parts of the flow regime. This type of 'blockage' is most likely to occur where the flow route is compressed, such as under a cut-o wall. Flow in these (dark) low permeability zones is characterised by the streamlines moving further apart and the equipotentials moving closer together. Conversely, flow in the (light) high permeability zones is characterised by the equipotentials moving further apart and the streamlines moving closer together.

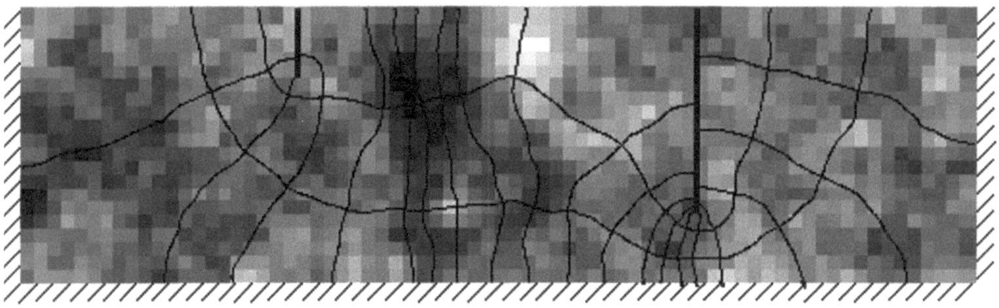

Figure 5. Stochastic flow net for a typical realisation

Although *local* variations in the permeability have an obvious e ect on the *local* paths taken by the water as it flows downstream, *globally* the stochastic and deterministic flow nets exhibit many similarities. The flow is predominantly in a downstream direction, with the fluid flowing down, under and around the cut-o walls. For this reason the statistics of the output quantities might be expected to be rather insensitive to the geometry of the problem (e.g. length of walls etc.), and qualitatively similar to the properties of a 1-d flow problem.

6.2 Statistics of the potential field

Figures 6 gives contours of the mean and standard deviation of the potential field following 1000 realisations for the case where $\theta_k = 1.0$m. Here we have follow an approach used by Smith and Freeze (1979a, 1979b) who presented the results of a series of numerical experiments on both 1-d and 2-d confined flow problems.

The mean potential values given in Figures 6a are very similar to those obtained in the deterministic analysis summarised in the flow net of Figure 3. The standard deviation of the potentials given in Figures 6b indicate the zones in which the greatest uncertainty

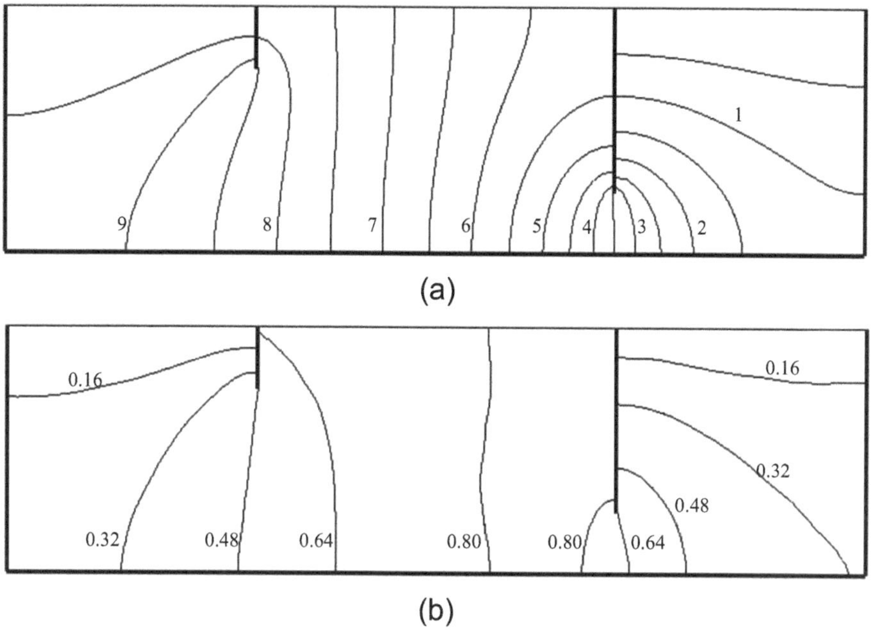

Figure 6. (a) Contours of mean potential values (contour interval=0.5m), (b) Contours of standard deviation of potential values (contour interval=0.16m), $\sigma_k/\mu_k = 1, \theta_k = 1$m

exists regarding the potential values. It should be recalled that the up- and down- stream (boundary) potentials are deterministic, so the standard deviation of the potentials on these boundaries equals zero. The greatest values of standard deviation occur in the middle of the flow regime, which in the case considered here represents the zone beneath the dam and between the cut-o walls. The standard deviation is virtually constant in this zone. The statistics of the potential field shown in Figure 6 are closely related to the statistics of the uplift force as will be considered in the next section.

6.3 Parameteric studies

The parametric studies based on the mesh of Figure 2 were designed to show the e ect of the permeability's standard deviation, σ_k, and scale of fluctuation, θ_k, on the output quantities $\bar{Q}$, $\bar{U}$ and i_e. In all cases the mean permeability, μ_k, was maintained constant at $10\ ^{5}$ m/s.

Instead of plotting σ_k directly, the dimensionless coefficient of variation of permeability was used, and the following values were considered: $\sigma_k/\mu_k = 0.125, 0.25, 0.50, 1, 2, 4, 8, 16.0$ together with scales of fluctuation values given by $\theta_k = 0, 1, 2, 4, 8,$ and ∞ m

All permutations of these values were analysed, and the results summarised in Figures 7,8 and 9 in the form of the logarithm (base 10) of σ_k/μ_k vs. the means and standard deviations of $\bar{Q}$, $\bar{U}$,and i_e, denoted $(\mu\ ,\sigma\ )$, (μ_U, σ_U), and $(\mu_i\ ,\sigma_i\ )$ respectively.

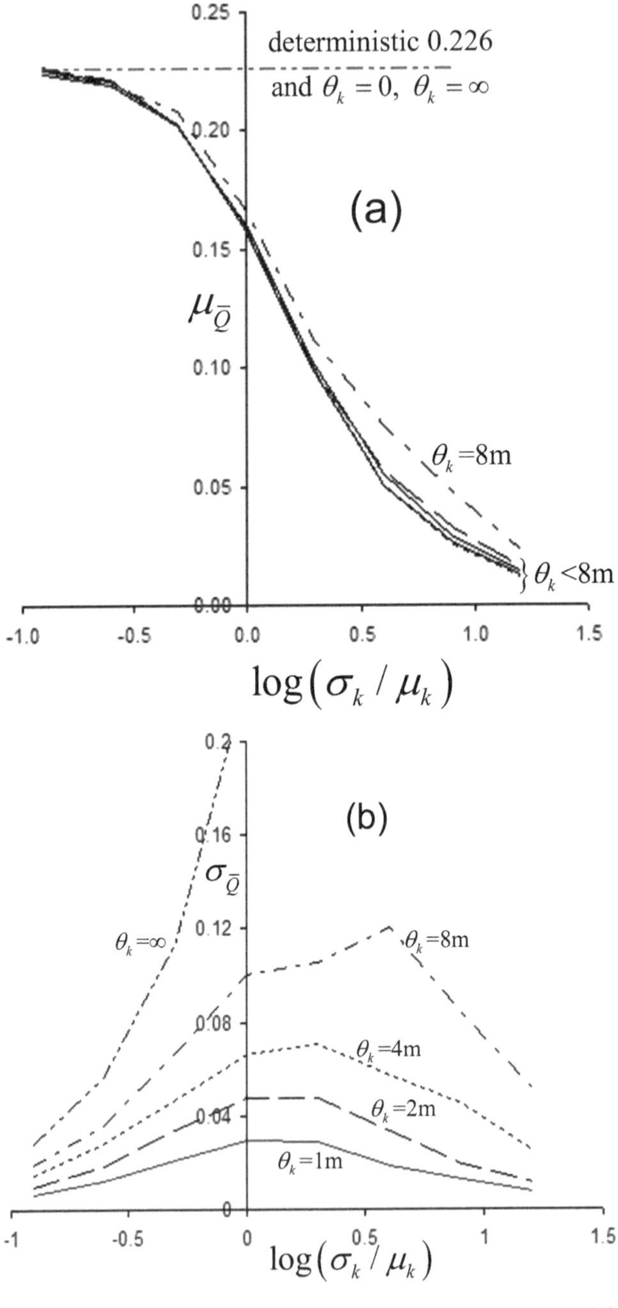

Figure 7. Coefficient of variation of permeability plotted against, (a) mean flow rate; (b) standard deviation of flow rate

Flow rate Figure 7a shows a significant fall in μ as σ_k/μ_k increases for $\theta_k < 8$ m. As the scale of fluctuation approaches infinity, the expected value of $\bar{Q}$ approaches the constant 0.226. This curve is also shown in Figure 7a, although it should be noted it has been obtained through theory rather than simulation. In agreement with this result, the curve $\theta_k = 8$ m shows a less marked reduction in μ with increasing coefficient of variation σ_k/μ_k. However, over typical scales of fluctuation, the e ect on average flow rate is slight. The decrease in flow rate as a function of the variability of the soil mass is an important observation from the point of view of design. Traditional design practice may very well be relying on this variability to reduce flow rates on average. It also implies that ensuring higher uniformity in the substrate may be unwarranted unless the mean permeability is known to be substantially reduced and/or the permeability throughout the site is carefully measured.

Figure 7b shows the behaviour of σ as a function of σ_k/μ_k. Of particular note is that σ reaches a maximum corresponding to σ_k/μ_k in the range 1.0 - 2.0 for $\theta_k \leq 8$ m. Again the theoretical result corresponding to $\theta_k = \infty$ has been plotted on the figure, showing a continuous increase with σ_k/μ_k.

In general, it appears that the greatest variability in $\bar{Q}$ occurs under rather typical conditions: scales of fluctuation between 1 and 4 m and coefficient of variation of permeability of around 1 or 2.

Uplift forces Figures 8a and 8b show the relationship between uplift force parameters μ_U and σ_U and input permeability parameters σ_k/μ_k and θ_k. From Figure 8a, μ_U is relatively insensitive to the parametric changes. There is a gradual fall in μ_U as both σ_k/μ_k and θ_k increase. The greatest reduction being about 10% of the deterministic value of 0.671 when $\sigma_k/\mu_k = 16.0$ and $\theta_k = 8.0$ m. The insensitivity of the uplift force to the permeability input statistics might have been predicted from Figures 3, and 6a in which the contours of (mean) piezometric head are virtually the same in both the deterministic and stochastic analyses.

Figure 8b shows that σ_U consistently rises as both σ_k/μ_k and θ_k increase. It is known that in the limit as $\theta_k \to \infty$, $\sigma_U \to 0$ since under those conditions, the permeability field becomes completely uniform. Some hint of this increase followed by a decrease is seen from Figure 8b in that the largest increases are for $\theta_k = 0$ to $\theta_k = 1$ while the increase from $\theta_k = 4$ to $\theta_k = 8$ is much smaller.

The actual value of σ_U for a given set of σ_k/μ_k and θ_k could easily be deduced from the standard deviation of the potential values. Figure 6b gave contours of the standard deviation of the potential values throughout the flow domain for the particular value $\sigma_k/\mu_k = 1.0$ and $\theta_k = 1.0$m. In Figure 6b, the potential standard deviation beneath the dam was approximately constant and equal to 0.8 m. After non- dimensionalisation by dividing by $H = 10$ m, these values closely agree with the corresponding values in the graphs of Figure 8b.

The magnitude of the standard deviation of the uplift force given in Figure 8b across the range of parameters considered was not very great. The implication is that this quantity can be estimated with a reasonable degree of confidence. The explanation lies in the fact that the uplift force is calculated using potential values over quite a large number of nodes beneath the dam. This 'averaging' process would tend to damp out

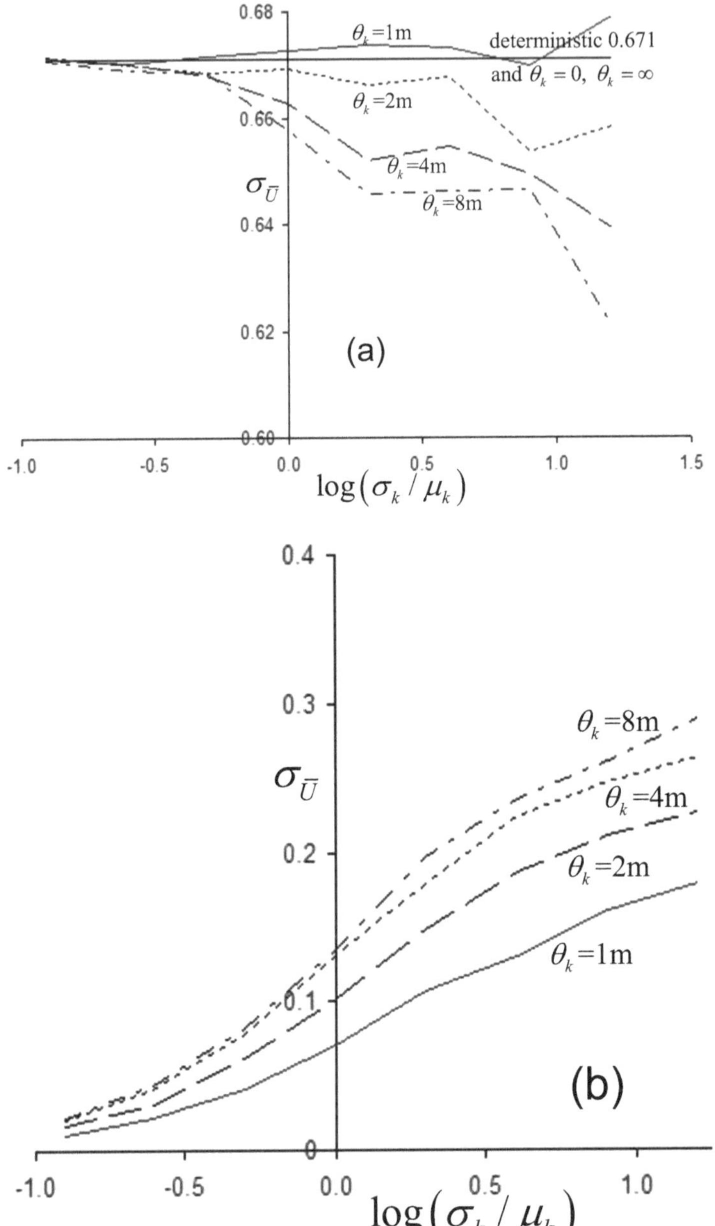

Figure 8. Coefficient of variation of permeability plotted against, (a) mean uplift force; (b) standard deviation of uplift force

fluctuations in the potential values that would be observed on a local scale, resulting in a variance reduction.

Exit gradients This quantity is based on the first derivative of piezometric head or potential with respect to distance at the exit point closest to the downstream end of the dam. It is well known that in a deterministic approach, the largest value of i_e, and hence the most critical, lies at the exit point of the uppermost (and shortest) streamline. While for a single realisation of a stochastic analysis this may not be the case, on average the location of the critical exit gradient is expected to occur at the 'deterministic' location.

As i_e is based on a first derivative at a *particular* location within the mesh (see Figure 4), it can be expected to be the most susceptible to local variations generated by the stochastic approach. In order to average the calculation of i_e over a few nodes, it was decided to use a 4-point (backward) finite di erence scheme as given previously in equation (5.3). This is equivalent to fitting a cubic polynomial over the potential values calculated at the four nodes closest to the exit point adjacent to the downstream cut-o wall. The cubic is then di erentiated at the required point to estimate i_e. Note then that the gradient is estimated by studying the fluctuations over a length of 0.6 m vertically (the elements are 0.2 m by 0.2 m in size). This length will be referred to as the 'di erentiation length' in the following.

The variation of μ_i and σ_i , over the range of parameters considered are given in Figures 9a and 9b. The sensitivity of i_e to σ_k/μ_k is clearly demonstrated. In Figure 9a, μ_i agrees quite closely with the deterministic value of 0.688 for values of σ_k/μ_k in the range 0.0 to 1.0, but larger values start to show significant instability and divergence. It is interesting to note that for $\theta_k \leq 1$, the tendency is for μ_i to fall below the deterministic value of i_e as σ_k/μ_k is increased, whereas for larger values of θ_k it tends to increase above the deterministic value. The scales 0 and 1 are less than and of the same magnitude as the di erentiation length of 0.6 m used to estimate the exit gradient, respectively, while the scales 2, 4, and 8 are substantially greater. If this has some bearing on the divergence phenomena seen in Figure 9a it calls into some question the use of a di erentiation length to estimate the derivative at a point. Suffice to say that there may be some conflict between the numerical estimation method and random field theory regarding the exit gradient that needs further investigation.

Figure 9b indicates the relatively large values of σ_i , which grow rapidly as σ_k/μ_k is increased. The influence of θ_k in this case is not so great, with the results corresponding to $\theta_k = 1.0, 2.0, 4.0$ and 8.0 m being quite closely grouped. It is noted that theoretically as $\theta_k \to \infty$ and $\theta_k \to 0$, $\mu_i \to 0.688$ and $\sigma_i \to 0$. Their appears to be some evidence of a reduction in σ_i as θ_k increases, which is in agreement with the theoretical result. For scales of fluctuation negligible relative to the di erentiation length, that is $\theta_k = 0$, the variability in i_e is much higher than that for other scales at all but the highest permeability variance. This is perhaps to be expected, since $\theta_k = 0$ yields large fluctuations in permeability within the di erentiation length.

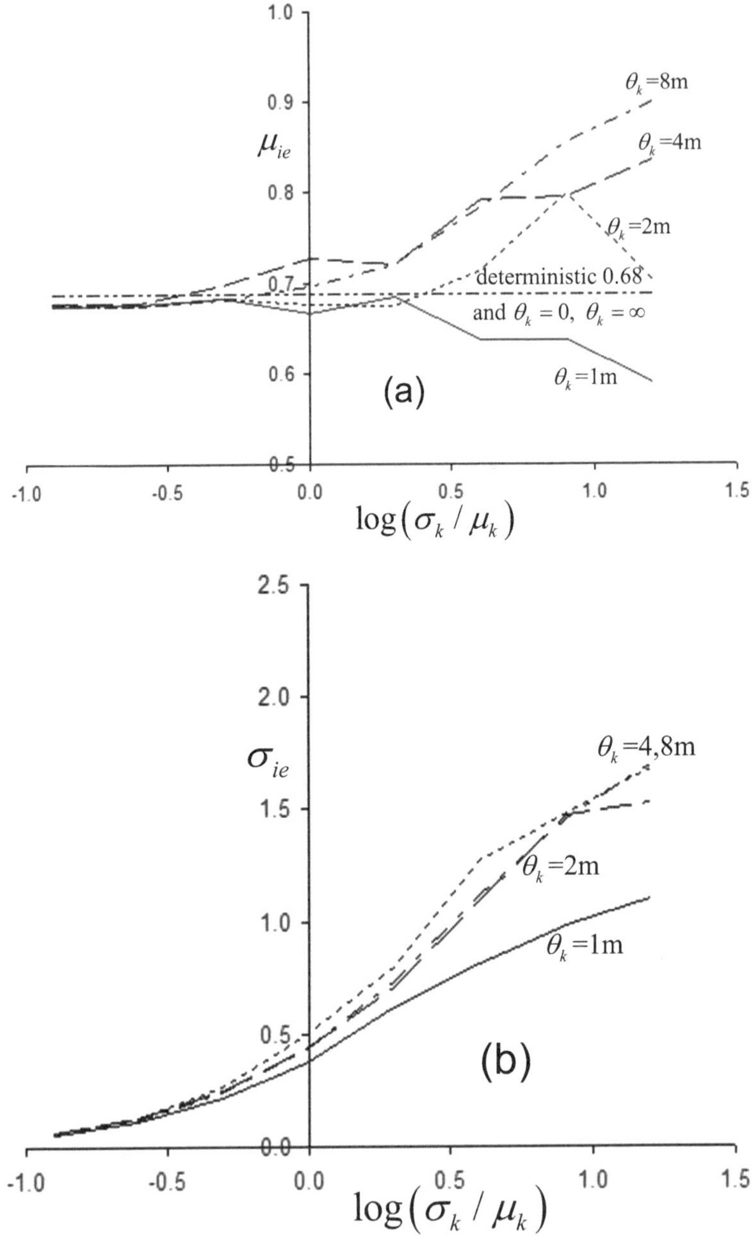

Figure 9. Coefficient of variation of permeability plotted against, (a) mean exit gradient; (b) standard deviation of exit gradient

7 Concluding Remarks

A range of parametric studies have been performed relating to flow beneath a water retaining structure with two cut-o walls founded on a stochastic soil. Random field concepts were used to generate permeability fields having predefined mean, standard deviation and correlation structure. These values were mapped onto a finite element mesh consisting of 1400 elements, and, for each set of parameters, 1000 realisations of the boundary value problem were analysed. In all cases, the 'target' mean permeability of each finite element was held constant and parametric studies were performed over a range of values of coefficient of variation and scale of fluctuation.

The three output quantities under scrutiny were the flow rate, the uplift force and the exit gradient; the first two of these being non-dimensionalised for convenience of presentation.

The mean flow rate was found to be relatively insensitive to typical scales of fluctuation, but fell consistently as the variance of of the permeability was increased. This observation may be of some importance in the design of such water retaining structures. The standard deviation of the flow rate consistently increased with the scale of fluctuation, but rose and then fell again as the coefficient of variation was increased. These maxima are currently the subject of further investigations by the Authors.

The mean uplift force was rather insensitive to the parametric variations, falling by only about 10% in the worst case. The relatively small variability of uplift force was due to a 'damping out' of local variations inherent in the random field by the averaging of potential values over the nodes along the full length of the base of the dam. Nevertheless, the standard deviation of the uplift force rose consistently with increasing scale of fluctuation and coefficient of variation, as was to be expected from the contour plots of the standard deviation of the potential values across the flow domain.

The mean exit gradient was much more sensitive to the statistics of the input field. Being based on a first derivative of piezometric head with respect to length at the exit point, this quantity is highly sensitive to local variations inherent in the potential values generated by the random field. Some local 'averaging' was introduced by the use of four-point numerical di erentiation formula, however the fluctuation in mean values was still considerable and the standard deviation values high.

8 Acknowledgement

The work described in this paper was supported in part by a travel grant (ref. 07-04-03-2) from the Association of Universities and Colleges of Canada.

This is a modified version of a paper that first appeared as, D.V. Griffiths and Gordon A. Fenton (1993). "Seepage beneath water retaining structures founded on spatially random soil" *Géotechnique* **43** , No. 4, 577-787

9 Notation

b	vertical spacing between nodes
g	standard Gaussian random field
g_i	local average of g over the ith element
H	head di erence between up- and downstream levels
i_e	exit hydraulic gradient
k	permeability
k_i	permeability assigned to the ith element
k_x, k_y	permeability in x- and y-directions
L	width of dam
n_f	number of flow channels
n_d	number of equipotential drops
$Q, \bar{Q}$	flow rate and non-dimensional flow rate
$U, \bar{U}$	uplift force and non-dimensional uplift force
x, y	Cartesian coordinates
$_w$	unit weight of water
θ_k	scale of fluctuation of permeability
μ_k, σ_k	mean and standard deviation of permeability
$\mu_{\ln k}, \sigma_{\ln k}$	mean and standard deviation of $\ln k_i$
$\mu_i\ , \sigma_i$	mean and standard deviation of exit gradient
$\mu\ , \sigma$	mean and standard deviation of dimensionless flow rate
μ_U, σ_U	mean and standard deviation of dimensionless uplift force
σ_k/μ_k	coefficient of variation of permeability
ϕ	potential or piezometric head
$\phi_0, \phi\ _1, \phi\ _2, \phi\ _3$	potential values near exit point
	global conductivity matrix
$\boldsymbol{Q}$	global flow vector
	global potential vector

Bibliography

G.B. Baecher and T.S. Ingra. Stochastic FEM in settlement predictions. *J Geotech Eng, ASCE*, 107(GT4):449–463, 1981.

A. Casagrande. Seepage through dams. *Journal of the New England Water Works Association*, L1(2), 1937.

H.R. Cedergren. *Seepage, drainage and flow nets.* John Wiley and Sons, Chichester, New York, 1967.

G.A. Fenton. *Simulation and analysis of random elds.* PhD thesis, Department of Civil Engineering and Operations Research, Princeton University, 1990.

G.A. Fenton and E.H. Vanmarcke. Simulation of random fields via local average subdivision. *J Eng Mech, ASCE*, 116(8):1733–1749, 1990.

D. V. Griffiths. Rationalised charts for the method of fragments applied to confined seepage. *Géotechnique*, 34(2):229–238, 1984.

M.E. Harr. *Groundwater and seepage.* McGraw Hill, London, New York, 1962.

R.J. Hoeksema and P.K. Kitanidis. Analysis of the spatial structure of properties of selected aquifers. *Water Resour Res*, 21(4):563–572, 1985.

K. Ishii and M. Suzuki. Stochastic finite element method for slope stability analysis. *Structural Safety*, 4:111–129, 1987.

N.N. Pavlovsky. Motion of water under dams. In *Proceedings of the 1st Congress on Large Dams, Stockholm*, pages 179–192. 1933.

G. Righetti and K. Harrop-Williams. Finite element analysis of random soil media. *J Geotech Eng, ASCE*, 114(GT1):59–75, 1981.

I. M. Smith and D. V. Griffiths. *Programming the Finite Element Method.* John Wiley and Sons, Chichester, New York, 4th edition, 2004.

L. Smith and R.A. Freeze. Stochastic analysis of steady state groundwater flow in a bounded domain, 1. One-dimensional simulations. *Water Resour Res*, 15(3):521–528, 1979a.

L. Smith and R.A. Freeze. Stochastic analysis of steady state groundwater flow in a bounded domain, 2. Two-dimensional simulations. *Water Resour Res*, 15(6):1543–1559, 1979b.

E.A. Sudicky. A natural gradient experiment on solute transport in a sand aquifer: Spatial variability of hydraulic conductivity and its role in the dispersion process. *Water Resour Res*, 22(13):2069–2083, 1986.

E.H. Vanmarcke. *Random elds: Analysis and synthesis.* The MIT Press, Cambridge, Mass., 1984.

E.H. Vanmarcke and M. Grigoriu. Stochastic finite element analysis of simple beams. *J Eng Mech, ASCE*, 109(5):1203–1214, 1983.

A. Verruijt. *Theory of groundwater flow.* Macmillan and Co. Ltd, London, 1970.

The Random Finite Element Method (RFEM) in Settlement Analyses

Gordon A. Fenton*, D.V. Griffiths†, and W. Cavers‡

* Department of Engineering Mathematics, Dalhousie University, Canada

† Division of Engineering, Colorado School of Mines, U.S.A.

‡ Golder Associates Ltd., Ottawa, Canada

Abstract In order to control serviceability problems arising from excessive settlement of shallow footings, geotechnical design codes generally include specifications regarding maximum settlement which often govern the footing design. Once the footing has been designed and constructed, the actual settlement it experiences on a real three-dimensional soil mass can be quite different than expected, due to the soil's spatial variability. Because of this generally large variability (compared to other engineering materials such as concrete and steel) and because this particular serviceability limit state often governs the design, it makes sense to consider a reliability-based approach to settlement design. This chapter looks in some detail at a Load and Resistance Factor Design (LRFD) approach to limiting footing settlement. In particular, the resistance factors required to achieve a certain level of settlement reliability as a function of soil variability and site investigation intensity are determined analytically using random field theory. Simplified approximate relationships are proposed and tested using simulation via the Random Finite Element Method. It is found that the simplified relationships are validated both by theory and simulation and so can be used to augment the calibration of geotechnical LRFD code provisions with respect to shallow foundation settlement.

1 Reliability-Based Geotechnical Design Issues

In an effort to harmonize with structural codes, geotechnical design codes around the world are beginning to migrate towards some form of reliability-based design (RBD). Significant steps in this direction can be found in, for example, Eurocode 7, 2004, Australian Standard AS 4678, 2002, NCHRP Report 507, 2004, and the Canadian Foundation Engineering Manual, 1992. These RBD provisions are most often presented in the form of a Limit States Design (LSD), to define critical failure states, combined with load and resistance factors calibrated to achieve the target reliabilities associated with the various limit states. The use of load and resistance factors is generally referred to as Load and Resistance Factor Design (LRFD).

By and large, the random characteristics of loads, or "actions", in civil engineering projects, are fairly well known and so load factors are reasonably well established. On the resistance side, for most common structural materials representative tests can easily be performed, and have been, to establish material property distributions that apply with reasonable accuracy anywhere that the material is used. Thus, resistance factors for materials such as concrete, steel, and wood have been known for decades.

Unfortunately, the development of resistance factors for use in geotechnical engineering is much more difficult than for quality-controlled engineering materials, such as concrete, steel, or wood. For example, while the mean strength of a batch of 30 MPa concrete delivered to a site in one city, might differ by 5 to 10% from a batch delivered to a site in a second city, the soil strengths at the two sites may easily differ by orders of magnitude. A significant advantage of designing using quality-controlled materials is that the general form and, in particular, the variance of the material property distribution is well enough accepted by the engineering profession that only a few samples of the material are deemed necessary to ensure design requirements are met. That is, engineers rely on an *a priori* estimate of the material variance which means that testing can be aimed at only ensuring that the mean material resistance is sufficiently high (the number of samples taken is usually far too few to accurately estimate the variance). This is essentially a hypothesis test on the mean with variance known. Using this test to ensure that design requirements are met, combined with the known distributions and resulting codified load and resistance factors, is sufficient to provide a reasonably accurate reliability-based design.

Contrast the knowledge regarding the distribution of, say, concrete with that of soils. In analogy to the above discussion, it would be nice to have a reasonably accurate *a priori* estimate of soil property variance, so that only the mean soil property would have to be determined via a site investigation. Such an a priori estimate of variance would involve sampling many sites across the world – some in gravel, some in swamps, some in rock, some in clay, sand, and so on – and then estimating the variance across these samples. This a priori variance would be very large and this has two implications; first, more samples would be required to accurately estimate the mean at a site and secondly, and probably more importantly, the resulting reliability-based designs will often be overly conservative and expensive. That is, this 'worst case' *a priori* variance for soils would generally be much larger than the actual variance at a single site, which would typically lead to overdesign in order to achieve a certain reliability. Nevertheless, an a priori variance for soils would be of some value, particularly in situations where the site investigation is insufficient to estimate the variance, or for preliminary designs. In addition, it is better to start out on the safe side, and refine the design as sufficient information is gathered.

The above argument suggests that in order to achieve efficient reliability-based geotechnical designs, the site investigation must be intensive enough to allow the estimation of both the soil mean and variance – this level of site investigation intensity is typically what is aimed at in modern geotechnical codes, with varying degrees of success (for example, Australian Standard AS 4678, 2002, specifies three different investigation levels, associated with three different reliability levels). To date, however, little guidance is provided on how to determine "characteristic" design values for the soil on the basis of the gathered data, nor on how to use the estimated variance to adjust the design.

Another complicating factor, which is more of a concern in soils than in other quality-controlled materials, is that of spatial variability and its effect on design reliability. Soil properties often vary markedly from point to point and this variability can have quite different importance for different geotechnical issues. For example, footing settlement, which depends on an average property under the footing, is only moderately affected by spatial variability, while slope stability, which involves the path of least resistance, is more strongly affected by spatial variability. In this chapter, spatial variability will be simply characterized by a parameter referred to here as the correlation length – small correlation lengths imply more rapidly varying properties, and so on. In order to adequately characterize the probabilistic nature of a soil and arrive at reasonable reliability-based designs, then, three parameters need to be estimated at each site; the mean, variance, and correlation length.

Fortunately, evidence compiled by the authors in the past indicates that a 'worst case' correlation length typically exists – this means that, in the absence of sufficient data, this worst case can be used in reliability calculations. It will generally be true that insufficient data are collected at a site to reasonably estimate the correlation length, so the worst case value is appropriate to use (despite the fact that this is somewhat analogous to using the worst case a priori variance discussed above).

Once the random soil at a site has been characterized in some way, the question becomes how should this information be used in a reliability-based design? In this chapter, a Limit State Design approach will be considered, where a square footing is placed on a three-dimensional soil mass and the task is to design the footing to have a sufficiently high reliability against excessive settlement. Thus, the limit state in question is a serviceability limit state. In structural design, serviceability limit states are investigated using unfactored loads and resistances. In keeping with this, both the Eurocode 7 (2004) and Australian Standard AS 2159 (1995) specify unit resistance factors for serviceability limit states. The Australian Standard AS 5100.3 (2004) states that "a geotechnical reduction factor need not be applied" for serviceability limit states.

Due to the inherently large variability of soils, however, and because settlement often governs a design, it is the opinion of the authors that properly selected resistance factors should be used for both ultimate and serviceability limit states. The Australian Standard AS 4678 (2002), for example, agrees with this opinion and, in fact, distinguishes between resistance factors for ultimate limit states and serviceability limit states – the factors for the latter are closer to 1.0, reflecting the reduced reliability required for serviceability issues. Although the Canadian Foundation Engineering Manual (3rd Ed.,1992) suggests the use of a "performance factor" (foundation capacity reduction factor) of unity for settlement, it goes on to say "However, in view of the uncertainty and great variability in in situ soil-structure stiffnesses, Meyerhof (1982) has suggested that a performance factor of 0.7 should be used for an adequate reliability of serviceability estimates."

If resistance factors are to be used, how should they be selected so as to achieve a certain reliability? Statistical methods suggest that the resistance factors should be adjusted until a sufficiently small fraction of possible realizations of the soil enter the limit state being designed against. Unfortunately, there is only one realization of each site and, since all sites are different, it is difficult to apply statistical methods to this problem. For this reason geotechnical reliability-based code development has largely been accomplished by calibration with past experience as captured in previous codes. This is quite acceptable, since design methodologies have evolved over many years to produce a socially acceptable reliability, and this encapsulated information is very valuable – see, for example, Vick's (2002) discussion of the value of judgement in engineering.

On the other hand, a reliability-based design code derived purely from deterministic codes cannot be expected to provide the additional economies that a true reliability-based design code could provide, eg. by allowing the specification of the target reliability (lower reliability for less important structures, etc.), or by improving the design as uncertainty is reduced, and so on. To attain this level of control in a reliability-based design code, probabilistic modeling and/or simulation of many possible soil regimes should also be employed to allow the investigation of the effect that certain design parameters have on system reliability. This is an important issue – it means that probabilistic modeling is necessary if reliability-based design codes are to evolve beyond being mirror images of the deterministic codes they derive from. The randomness of soils must be acknowledged and propertly accounted for.

This chapter presents the results of a study in which a reliability-based settlement design approach is proposed and investigated via simulation using the Random Finite Element Method (RFEM). In particular, the effect of a soil's spatial variability and site investigation intensity on the resistance factors is quantified. The results of the chapter can and should be used to improve and generalize "calibrated" code provisions based purely on past experience.

2 Random Finite Element Method (RFEM)

A specific settlement design problem will be considered here in order to investigate the settlement probability distribution of footings designed against excessive settlement. The problem considered is that of a rigid rough square pad footing founded on the surface of a three-dimensional linearly elastic soil mass underlain by bedrock at depth H. Although only elastic settlement is specifically considered, the results can include consolidation settlement so long as the combined settlement can be adequately represented using an effective elastic modulus field. To the extent that the elastic modulus itself is a simplified representation of a soil's inverse compressibility, which is strain level dependent, the extension of the approximation to include consolidation settlement is certainly reasonable, and is as recommended in the Canadian Highway Bridge Design Code Commentary (2000).

The settlement of a rigid footing on a three-dimensional soil mass is estimated using a linear finite element analysis. The mesh selected is 64 elements by 64 elements in plan by 32 elements in depth. Eight-node hexahedral elements, each cubic with side length 0.15 m are used (note that metric units are used in this chapter, rather than making it non-dimensional, since footing design will be based on a maximum tolerable settlement which is specified in m) yielding a soil domain of size 9.6×9.6 m in plan by 4.8 m in depth. Because the stiffness matrix corresponding to a mesh of size $64 \times 64 \times 32$ occupies about 4 Gbytes of memory, a preconditioned conjugate gradient iterative solver, which avoids the need to assemble the global stiffness matrix, is employed in the finite element code. A max-norm relative error tolerance of 0.005 is used to determine when the iterative solver has converged to a solution.

The finite element model was tested (see also Griffiths and Fenton) in the deterministic case (uniform elastic soil properties) to validate its accuracy and was found to be about 20% stiffer (smaller settlements) than that derived analytically (see, eg, Milovic 1992). Using other techniques such as selectively reduced integration, non-conforming elements, and 20-node elements did not significantly affect the discrepancy between these results and Milovic's. The 'problem' is that the finite elements truncate the singular stresses that occur along the edge of a rigid footing, leading to smaller settlements than predicted by theory. In this respect, Seyček (1991) compares real settlements to those predicted by theory and concluded that predicted settlements are usually considerably higher than real settlements. This is because the true stresses measured in the soil near the footing edge are finite and significantly less than the singular stresses predicted by theory. Seyček improves the settlement calculations by reducing the stresses below the footing. Thus, the finite element results included here are apparently closer to actual settlements than those derived analytically, although a detailed comparison to Seyček's has yet to be performed by the authors. However, it is not believed that these possible discrepancies will make a significant difference to the probabilistic results of this chapter since the probability of failure (excessive settlement) involves a comparison between deterministic and random predictions arising from the same finite element model, thus cancelling out possible bias.

The rigid footing is assumed to have a rough interface with the underlying soil – no relative slip is permitted – and rotation of the footing is not permitted. Only square footings, of dimension $B \times B$ are considered, where the required footing width B is determined during the design phase, to be discussed in the next section. Once the required footing width has been found, the design footing width must be increased to the next larger element boundary – this is because the finite element mesh is fixed and footings must span an integer number of elements. For example, if the required footing width is 2.34 m, and elements have dimension $\Delta x = \Delta y = 0.15$ m square, then the design footing width must be increased to 2.4 m (since this corresponds to 16 elements, rather than the 15.6 elements that 2.34 m would entail). This corresponds roughly to common design practice, where element dimensions are increased to an easily measured quantity.

Once the design footing width has been found, it must be checked to ensure that it is physically reasonable, both economically and within the finite element model. First of all, there will be some minimum footing size. In this study the footings cannot be less than 4 x 4 elements in size – for one thing loaded areas smaller than this tend to have significant finite element errors, for another, they tend to be too small to construct. For example, if an element size of 0.15 m is used, then the minimum footing size is 0.6 x 0.6 m, which is not very big. French (1999) recommends a lower bound on footing size of 0.6 m and an upper economical bound of 3.7 m. If the design footing width is less than the minimum footing width, it is set equal to the minimum footing width. Secondly, there will be some maximum footing size. A spread footing bigger than about 4 m square would likely be replaced by some other foundation system (piles, mat, or raft). In this program, the maximum footing size is taken to be equal to 2/3 of the finite element mesh width. This limit has been found to result in less than a 1% error relative to the same footing founded on a mesh twice as wide, so boundary conditions are not significantly influencing the results. If the design footing width exceeds the maximum footing width then the probabilistic interpretation becomes somewhat complicated, since a different design solution would presumably be implemented. From the point of view of assessing the reliability of the 'designed' spread footing, it is necessary to decide if this excessively large footing design would correspond to a success, or to a failure. It is assumed in this study that the subsequent design of the alternative foundation would be a success, since it would have its own (high) reliability.

In all the simulations performed in this study, the lower limit on the footing size was never encountered, implying that for the choices of parameters selected in this study, the probability of a design footing being less than 0.6×0.6 in dimension was very remote. Similarly, the maximum footing size was not exceeded in any but the most severe parameter case considered (minimum sampling, lowest resistance factor, highest coefficient of variation), where it was only exceeded in 2% of the possible realizations. Thus, the authors were satisfied that the finite element analysis would give reasonably accurate settlement predictions over the entire study.

The soil property of primary interest to settlement is elastic modulus, E, which is taken to be spatially random and may represent both the initial elastic and consolidation behaviour. Its distribution is assumed to be lognormal for two reasons: the first is that a geometric average tends to a lognormal distribution by the central limit theorem and the effective elastic modulus, as 'seen' by a footing, was found to be closely represented by a geometric average in Fenton and Griffiths (2002), and second is that the lognormal distribution is strictly non-negative which is physically reasonable for elastic modulus. The lognormal distribution has two parameters, $\mu_{\ln E}$ and $\sigma_{\ln E}$ which can be estimated by the sample mean and sample standard deviation of observations of $\ln(E)$. They can also be obtained from the mean and standard deviation of E using the transformations

$$\sigma^2_{\ln E} = \ln(1 + V_E{}^2) \tag{1a}$$

$$\mu_{\ln E} = \ln(\mu_E) - \tfrac{1}{2}\sigma^2_{\ln E} \tag{1b}$$

where $V_E = \sigma_E/\mu_E$ is the coefficient of variation of the elastic modulus field. A Markovian spatial correlation function, which gives the correlation coefficient between log-elastic modulus values at points separated by the lag vector, $\underset{\sim}{\tau}$, is used in this study;

$$\rho_{\ln E}(\underset{\sim}{\tau}) = \exp\left\{-\frac{2|\underset{\sim}{\tau}|}{\theta_{\ln E}}\right\} \tag{2}$$

in which $\underset{\sim}{\tau} = \underset{\sim}{x} - \underset{\sim}{x}'$ is the vector between spatial points $\underset{\sim}{x}$ and $\underset{\sim}{x}'$, and $|\underset{\sim}{\tau}|$ is the absolute length of this vector (the lag distance). In this chapter, the word 'correlation' refers to the correlation coefficient. The results presented here are not particularly sensitive to the choice in functional form of the correlation – the Markov model is popular because of its simplicity. The correlation function decay rate is governed by the so-called correlation length, $\theta_{\ln E}$, which, loosely speaking, is the distance over which log-elastic moduli are significantly correlated (when the separation distance $|\underset{\sim}{\tau}|$ is greater than $\theta_{\ln E}$, the correlation between $\ln E(\underset{\sim}{x})$ and $\ln E(\underset{\sim}{x}')$ is less than 14%). The correlation structure is assumed to be isotropic in this study which is appropriate for investigating the fundamental stochastic behaviour of settlement. Anisotropic studies are more appropriate for site-specific analyses and for refinements to this study. In any case, anisotropy is not expected to have a large influence on the results of this chapter due to the averaging effect of the rigid footing on the properties it 'sees' beneath it.

Poisson's ratio, having only a relatively minor influence on settlement, is assumed to be deterministic and is set at 0.3 in this study.

Realizations of the random elastic modulus field are produced using the Local Average Subdivision (LAS) method (Fenton and Vanmarcke 1990). Specifically, LAS produces a discrete grid of local averages, $G(\underset{\sim}{x}_i)$, of a standard Gaussian random field, having correlation structure given by Eq. (2), where $\underset{\sim}{x}_i$ are the coordinates of the centroid of the ith grid cell. These local averages are then mapped to finite element properties according to

$$E(\underset{\sim}{x}_i) = \exp\{\mu_{\ln E} + \sigma_{\ln E} G(\underset{\sim}{x}_i)\} \tag{3}$$

(which assumes that the centroids of the random field cells and the finite elements coincide, as they do in this study).

Much discussion of the relative merits of various methods of representing random fields in finite element analysis has been carried out in recent years (see, for example, Li and Der Kiureghian, 1993). While the spatial averaging discretization of the random field used in this study is just one approach to the problem, it is appealing in the sense that it reflects the simplest idea of the finite element representation of a continuum as well as the way that soil samples are typically taken and tested in practice, ie. as local averages. Regarding the discretization of random fields for use in finite element analysis, Matthies et al. (1997) makes the comment that "One way of making sure that the stochastic field has the required structure is to assume that it is a local averaging process.", refering to the conversion of a nondifferentiable to a differentiable (smooth) stochastic process. Matthie further goes on to say that the advantage of the local average representation of a random field is that it yields accurate results even for rather coarse meshes.

Figure 1 illustrates the finite element mesh used in the study and Figure 2 shows a cross-section through the soil mass under the footing for a typical realization of the soil's elastic modulus field. Figure 2 also illustrates the boundary conditions.

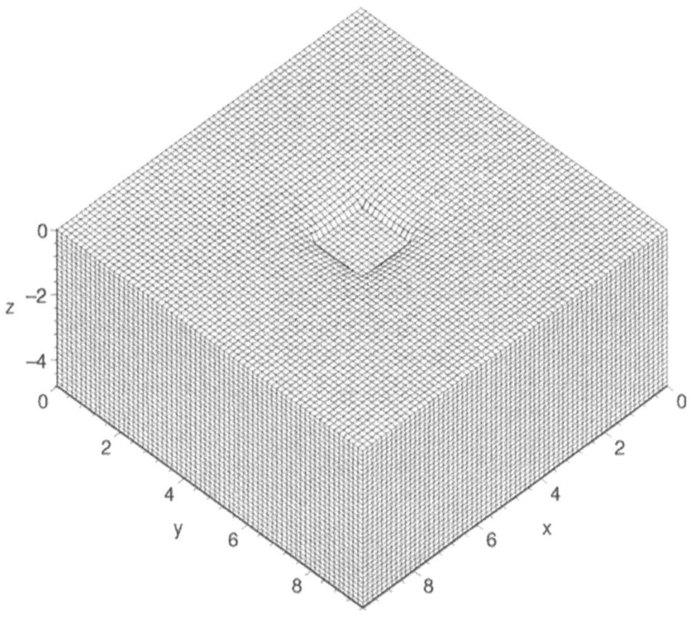

Figure 1. Finite element mesh with one square footing.

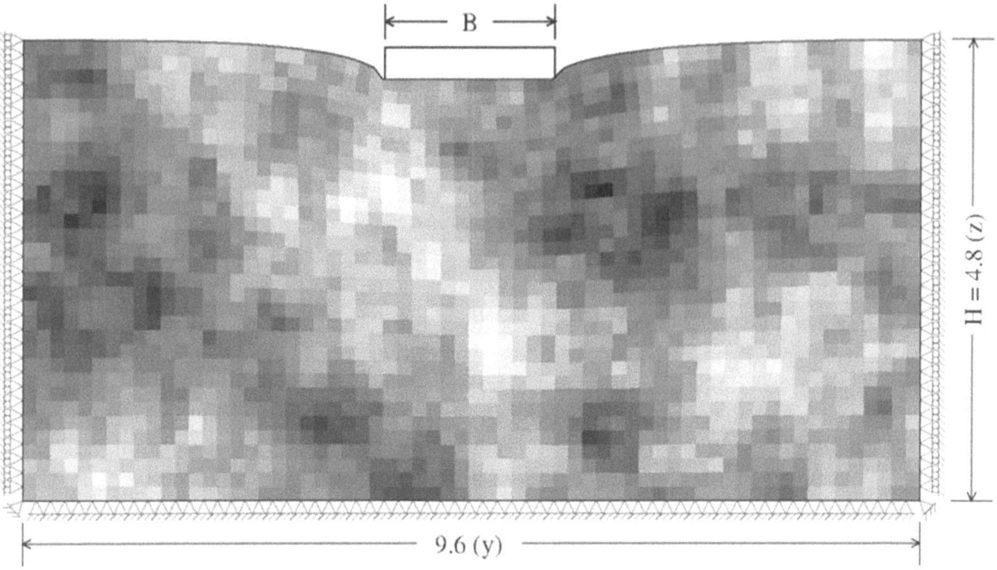

Figure 2. Cross-section through a realization of the random soil underlying the footing. Lighter soils are softer.

3 Reliability-Based Settlement Design

The goal of this chapter is to propose and investigate a reliability-based design methodology for the serviceability limit state of footing settlement. Footing settlement is predicted here using a modified Janbu (1956) relationship, and this is the basis of design used in this chapter;

$$\delta_p = u_1 \frac{\hat{q}B}{\hat{E}} \tag{4}$$

where δ_p is the predicted footing settlement, $\hat{q} = \hat{P}/B^2$ is the estimated stress applied to the soil by the estimated load, $\hat{P}$, acting over footing area $B \times B$, $\hat{E}$ is the (possibly drained) estimate of elastic modulus underlying the footing, u_1 is an influence factor which includes the effect of Poisson's ratio ($\nu = 0.3$ in this study). The estimated load, $\hat{P}$, is often a nominal load computed from the supported live and dead loads, while the estimated elastic modulus, $\hat{E}$, is usually a cautious estimate of the mean elastic modulus under the footing obtained by taking laboratory samples or by in-situ tests, such as CPT. In terms of the footing load, $\hat{P}$, the settlement predictor thus becomes

$$\delta_p = u_1 \frac{\hat{P}}{B\hat{E}} \tag{5}$$

The relationship above is somewhat modified from that given by Janbu (1956) and Christian and Carrier (1978) in that the influence factor, u_1, is calibrated specifically for a square rough rigid footing founded on the surface of an elastic soil using the same finite element model which is later used in the Monte Carlo simulations. This is done to remove bias (model) errors and concentrate specifically on the effect of spatial soil variability on required resistance factors. In practice, this means that the resistance factors proposed in this chapter are *upper bounds*, appropriate for use when bias and measurement errors are known to be minimal.

The calibration of u_1 is done by computing the deterministic (non-random) settlement of an elastic soil with elastic modulus $\hat{E}$ and Poisson's ratio ν under a square rigid rough footing supporting load $\hat{P}$ using the finite element program. Once the settlement is obtained, Eq. (5) can be solved for u_1. Repeating this over a range of H/B ratios leads to the curve shown in Figure **3**. (Note that this deterministic calibration was carried out over a larger range of mesh dimensions than indicated by Figure 1.) A very close approximation to the finite element results is given by the fitted relationship (obtained by consideration of the correct limiting form and by trial-and-error for the coefficients)

$$u_1 = 0.61\left(1 - e^{-1.18H/B}\right) \tag{6}$$

which is also shown on Figure **3**.

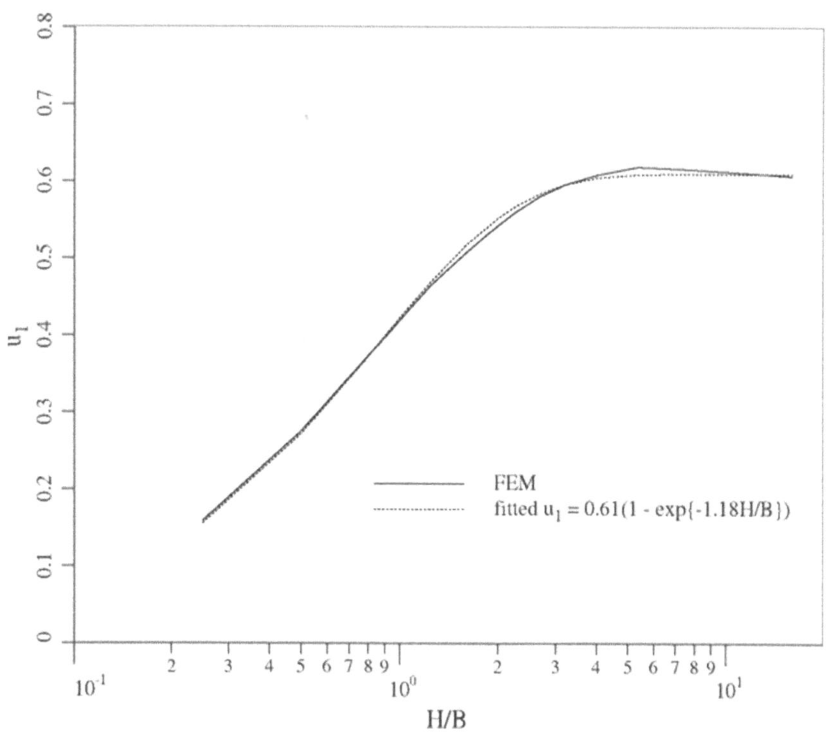

Figure 3. Calibration of u_1 using finite element model.

Using Eq. (**6**) in Eq. (5) gives the following settlement prediction

$$\delta_p = 0.61\left(1 - e^{-1.18H/B}\right)\left(\frac{\hat{P}}{B\hat{E}}\right) \tag{7}$$

The reliability-based design goal is to determine the footing width, B, such that the probability of exceeding a specified tolerable settlement, δ_{max}, is acceptably small. That is, to find B such that

$$\mathrm{P}\left[\delta > \delta_{max}\right] = p_f = p_{max} \tag{8}$$

where δ is the actual settlement of the footing as placed (which will be considered here to be the same as designed). Design failure is assumed to have occurred if the actual footing settlement, δ, exceeds the maximum tolerable settlement, δ_{max}. The probability of design failure is p_f and p_{max} is the maximum acceptable risk of design failure.

A realization of the footing settlement, δ, is determined here using a finite element analysis of a realization of the random soil. For u_1 calibrated to the finite element results, δ can also be computed from

$$\delta = u_1 \frac{P}{BE_{eff}} \tag{9}$$

where P is the actual footing load and E_{eff} is the effective elastic modulus as seen by the footing (ie, the uniform value of elastic modulus which would produce a settlement identical to the actual footing settlement). Both P and E_{eff} are random variables.

One way of achieving the desired design reliability is to introduce a load factor, $\alpha \geq 1$, and a resistance factor, $\phi \leq 1$, and then finding B, α and ϕ which satisfy both Eq. (8) and Eq. (5) with $\delta = \delta_{max}$. In other words, find B and α/ϕ such that

$$\delta_{max} = u_1 \left(\frac{\alpha \hat{P}}{B\phi\hat{E}}\right) \tag{10}$$

and

$$\mathrm{P}\left[u_1 \frac{P}{BE_{eff}} > u_1 \left(\frac{\alpha \hat{P}}{B\phi\hat{E}}\right)\right] = p_{max} \tag{11}$$

From these two equations, at most two unknowns can be found uniquely. For serviceability limit states, a load factor of 1.0 is commonly used, and $\alpha = 1$ will be used here. (Note: only the ratio α/ϕ need actually be determined for the settlement problem.)

Given α/ϕ, $\hat{P}$, $\hat{E}$, and H, Eq. (10) is relatively efficiently solved for B using 1-pt iteration;

$$B_{i+1} = 0.61\left(1 - e^{-1.18H/B_i}\right)\left(\frac{\alpha\hat{P}}{\delta_{max}\phi\hat{E}}\right) \tag{12}$$

for $i = 1, 2, \ldots$ until successive estimates of B are sufficiently similar. A reasonable starting guess is $B_1 = 0.4(\alpha\hat{P})/(\delta_{max}\phi\hat{E})$.

In Eq. (11), the random variables u_1 and B are common to both sides of the inequality and so can be canceled. It will also be assumed that the footing load is lognormally distributed and that the estimated load, $\hat{P}$, equals the (non-random) *median load*, that is

$$\hat{P} = \exp\{\mu_{\ln P}\} \tag{13}$$

Setting the value of $\hat{P}$ to the median load considerably simplifies the theory in the sequel, but it should be noted that the definition of $\hat{P}$ will directly affect the magnitude of the estimated resistance factors. The lognormal distribution was selected because it results in loads which are strictly non-negative (uplift problems should be dealt with separately, and not handled via the tail end of a normal distribution assumption). The results to follow should be similar for any reasonable load distribution (e.g. Gamma, Chi-Square, etc) having the same mean and variance.

Collecting all remaining random quantities leads to the simplified design probability

$$\mathrm{P}\left[P\frac{\hat{E}}{E_{eff}} > \frac{\alpha}{\phi}e^{\mu_{\ln P}}\right] = p_{max} \tag{14}$$

The estimated modulus, $\hat{E}$, and the effective elastic modulus, E_{eff}, will also be assumed to be lognormally distributed. Under these assumptions, if W is defined as

$$W = P\frac{\hat{E}}{E_{eff}} \tag{15}$$

then W is also lognormally distributed, so that

$$\ln W = \ln P + \ln \hat{E} - \ln E_{eff} \tag{16}$$

is normally distributed with mean

$$\mu_{\ln W} = \mu_{\ln P} + \mu_{\ln \hat{E}} - \mu_{\ln E_{eff}} \tag{17}$$

It is assumed that the load distribution is known, so that $\mu_{\ln P}$ and $\sigma^2_{\ln P}$ are known. The nature of the other two terms on the right hand side will now be investigated.

Assume that $\hat{E}$ is estimated from a series of m soil samples that yield the observations $E^o_1, E^o_2, \ldots, E^o_m$. To investigate the nature of this estimate, it is constructive to first consider the effective elastic modulus, E_{eff}, as seen by the footing. Analogous to the estimate for $\hat{E}$, it can be imagined that the soil volume under the footing is partitioned into a large number of soil 'samples' (although most of them, if not all, will remain unsampled), $E_1, E_2, \ldots, E_n$. Investigations by Fenton and Griffiths (2002) suggest that the effective elastic modulus, as seen by the footing, E_{eff}, is a geometric average of the soil properties in the block under the footing, that is

$$E_{eff} = \left(\prod_{i=1}^{n} E_i\right)^{1/n} = \exp\left\{\frac{1}{n}\sum_{i=1}^{n}\ln E_i\right\} \tag{18}$$

If $\hat{E}$ is to be a good estimate of E_{eff}, which is desirable, then it should be similarly determined as a geometric average of the observed samples $E^o_1, E^o_2, \ldots, E^o_m$,

$$\hat{E} = \left(\prod_{j=1}^{m} E^o_j\right)^{1/m} = \exp\left\{\frac{1}{m}\sum_{j=1}^{m}\ln E^o_j\right\} \tag{19}$$

since this estimate of E_{eff} is unbiased in the median, ie. the median of $\hat{E}$ is equal to the median of E_{eff}. This is a fairly simple estimator, and no attempt is made here to account for the location of samples relative to the footing. Note that if the soil is layered horizontally and it is desired to specifically capture the layer information, then Eqs. 18 and 19 can be applied to each layer individually – the final $\hat{E}$ and E_{eff} values are then computed as harmonic averages of the layer values. Although the distribution of a harmonic average is not simply defined, a lognormal approximation is often reasonable.

Under these definitions, the means of $\mu_{\ln \hat{E}}$ and $\mu_{\ln E_{eff}}$ are identical,

$$\mu_{\ln E_{eff}} = \mathrm{E}\left[\ln E_{eff}\right] = \mu_{\ln E} \tag{20}$$

$$\mu_{\ln \hat{E}} = \mathrm{E}\left[\ln \hat{E}\right] = \mu_{\ln E} \tag{21}$$

where $\mu_{\ln E}$ is the mean of the logarithm of elastic moduli of any sample. Thus, as long as Equations (18) and (19) hold, the mean of $\ln W$ simplifies to

$$\mu_{\ln W} = \mu_{\ln P} \tag{22}$$

Now, attention can be turned to the variance of $\ln W$. If the variability in the load P is independent of the soil's elastic modulus field then the variance of $\ln W$ is

$$\sigma^2_{\ln W} = \sigma^2_{\ln P} + \sigma^2_{\ln \hat{E}} + \sigma^2_{\ln E_{eff}} - 2\mathrm{Cov}\left[\ln \hat{E}, \ln E_{eff}\right] \tag{23}$$

The variances of $\ln \hat{E}$ and $\ln E_{eff}$ can be expressed in terms of the variance of $\ln E$ using two variance reduction functions, γ^o and γ, defined as follows

$$\gamma^o(m) = \frac{1}{m^2}\sum_{i=1}^{m}\sum_{j=1}^{m}\rho^o_{ij} \tag{24a}$$

$$\gamma(n) = \frac{1}{n^2}\sum_{i=1}^{n}\sum_{j=1}^{n}\rho_{ij} \tag{24b}$$

where ρ^o_{ij} is the correlation coefficient between $\ln E^o_i$ and $\ln E^o_j$ and ρ_{ij} is the correlation coefficient between $\ln E_i$ and $\ln E_j$. These functions can be computed numerically once the locations of all soil 'samples' are known. Both $\gamma^o(1)$ and $\gamma(1)$ have value 1.0 when only 1 sample is used to specify $\hat{E}$ or E_{eff}, respectively (when samples are 'point' samples then 1 sample corresponds to zero volume – however, in this chapter, it is assumed that there is some representative sample volume from which the mean and variance of the elastic modulus field are estimated and this corresponds to the 'point' measure). As the number of samples increases the variance reduction function decreases towards zero at a rate inversely proportional to the total sample volume (see Vanmarcke 1984). If the volume of the soil under the footing is $B \times B \times H$ then a reasonable approximation to $\gamma(n)$ is obtained by assuming a separable form;

$$\gamma(n) \simeq \gamma_1(2B/\theta_{\ln E})\gamma_1(2B/\theta_{\ln E})\gamma_1(2H/\theta_{\ln E}) \tag{25}$$

where $\gamma_1(a)$ is the 1-D variance function corresponding to a Markov correlation;

$$\gamma_1(a) = \frac{1}{a^2}\left[a + e^{-a} - 1\right] \tag{26}$$

As an aside, Fenton and Griffiths, 2002, suggest that the depth to the bedrock, H, be limited to no more than about $10B$ in the calculation of the soil volume under the footing. However, the effective strain zone is generally quite a bit shallower, so a maximum depth of $2B$ to $4B$ might be more appropriate as a limitation on H.

An approximation to $\gamma^o(m)$ is somewhat complicated by the fact that samples for $\hat{E}$ are likely to be collected at separate locations. If the observations are sufficiently separated that they can be considered independent (eg. separated by more than $\theta_{\ln E}$), then $\gamma^o(m) = 1/m$. If they are collected from within a contiguous volume, V^o, then

$$\gamma^o(m) \simeq \gamma_1(2R/\theta_{\ln E})\gamma_1(2R/\theta_{\ln E})\gamma_1(2H/\theta_{\ln E}) \tag{27}$$

where the total plan area of soil sampled is $R \times R$ (for example, a CPT sounding can probably be assumed to be sampling an effective area equal to about 0.2×0.2 m^2, so that $R = 0.2$ m). The true variance reduction function will be somewhere in between. In this chapter, the soil is sampled by examining one or more columns of the finite element model, and so for an individual column, $R \times R$ becomes replaced by, $\Delta x \times \Delta y$, which are the plan dimensions of the finite elements and Eq. (27) can be used to obtain the variance reduction function for a single column. If more than one column is sampled, then

$$\gamma^o(m) \simeq \frac{\gamma_1(2\Delta x/\theta_{\ln E})\gamma_1(2\Delta y/\theta_{\ln E})\gamma_1(2H/\theta_{\ln E})}{n_{eff}} \tag{28}$$

where n_{eff} is the effective number of independent columns sampled. If the sampled columns are well separated (ie, by more than the correlation length), then they could be considered independent, and n_{eff} would be equal to the number of columns sampled. If the columns are closely clustered (relative to the correlation length), then n_{eff} would decrease towards 1. The actual number is somewhere in between and can be estimated by judgement.

With these results,

$$\sigma^2_{\ln \hat{E}} = \gamma^o(m)\sigma^2_{\ln E} \tag{29a}$$

$$\sigma^2_{\ln E_{eff}} = \gamma(n)\sigma^2_{\ln E} \tag{29b}$$

The covariance term in Eq. (23) is computed from

$$\begin{aligned}
\text{Cov}\left[\ln \hat{E}, \ln E_{eff}\right] &= \frac{1}{mn}\sum_{j=1}^{m}\sum_{i=1}^{n}\text{Cov}\left[\ln E^o_j, \ln E_i\right] \\
&= \sigma^2_{\ln E}\left[\frac{1}{mn}\sum_{j=1}^{m}\sum_{i=1}^{n}\rho'_{ij}\right] \\
&= \sigma^2_{\ln E}\rho'_{ave}
\end{aligned} \tag{30}$$

where ρ'_{ij} is the correlation coefficient between $\ln E^o_j$ and $\ln E_i$ and ρ'_{ave} is the average of all these correlations. If the estimate, $\ln \hat{E}$, is to be at all useful in a design, the value of ρ'_{ave} should be reasonably high. However, its magnitude depends on the degree of spatial correlation (measured by $\theta_{\ln E}$) and the distance between the observations E^o_i and the soil volume under the footing. The correlation function of Eq. (2) captures both of these effects. That is, there will exist an 'average' distance τ'_{ave} such that

$$\rho'_{ave} = \exp\left\{\frac{-2\tau'_{ave}}{\theta_{\ln E}}\right\} \tag{31}$$

and the problem is to find a reasonable approximation to τ'_{ave} if the numerical calculation of Eq. (30) is to be avoided. The approximation considered in this study is that τ'_{ave} is defined as the average absolute distance between the E^o_i samples and a vertical line below the center of the footing, with a sample taken anywhere under the footing to be considered to be taken at the footing corner (eg, at a distance $B/\sqrt{2}$ from the centerline) – this latter restriction is taken to avoid a perfect correlation when a sample is taken directly at the footing centerline, which would be incorrect. In addition, a side study performed by the authors, which is not reported here, indicated that for all moderate correlation lengths ($\theta_{\ln E}$ of the order of the footing width) the true τ'_{ave} differed by less than about 10% from the approximation $B/\sqrt{2}$ for any sample taken under the footing.

Using these definitions, the variance of $\ln W$ can be written

$$\begin{aligned}\sigma^2_{\ln W} &= \sigma^2_{\ln P} + \sigma^2_{\ln E}\left[\gamma^o(m) + \gamma(n) - 2\rho'_{ave}\right] \\ &\geq \sigma^2_{\ln P}\end{aligned} \tag{32}$$

The limitation $\sigma^2_{\ln W} \geq \sigma^2_{\ln P}$ is introduced because it is possible, using the approximations suggested above, for the quantity inside the square brackets to become negative, which is physically inadmissable. It is assumed that if this happens that the sampling has reduced the uncertainty in the elastic modulus field essentially to zero.

With these results in mind the design probability becomes

$$\begin{aligned}\mathrm{P}\left[P\frac{\hat{E}}{E_{eff}} > \frac{\alpha}{\phi}e^{\mu_{\ln P}}\right] &= \mathrm{P}\left[W > \frac{\alpha}{\phi}e^{\mu_{\ln P}}\right] \\ &= \mathrm{P}\left[\ln W > \ln\alpha - \ln\phi + \mu_{\ln P}\right] \\ &= 1 - \Phi\left(\frac{-\ln\phi}{\sigma_{\ln W}}\right) \qquad \text{(assuming } \alpha = 1\text{)} \\ &= p_{max}\end{aligned} \tag{33}$$

from which the required resistance factor, ϕ, can be found as

$$\phi = \exp\{-z_{p_{max}}\sigma_{\ln W}\} \tag{34}$$

where $z_{p_{max}}$ is the point on the standard normal distribution having exceedance probability p_{max}. For $p_{max} = 0.05$, which will be assumed in this chapter, $z_{p_{max}} = 1.645$.

It is instructive at this point to consider a limiting case, namely where $\hat{E}$ is a perfect estimate of E_{eff}. In this case, $\hat{E} = E_{eff}$, which implies that $m = n$ and the observations $E_1^o, \ldots$ coincide identically with the 'samples' $E_1, \ldots$. In this case, $\gamma^o = \gamma$ and $\rho = 1$, so that

$$\sigma^2_{\ln W} = \sigma^2_{\ln P} \tag{35}$$

from which the required resistance factor can be calculated as

$$\phi = \exp\{-z_{p_{max}} \cdot \sigma_{\ln P}\} \tag{36}$$

For example, if $p_{max} = 0.05$ and the coefficient of variation of the load is $V_P = 0.1$, then $\phi = 0.85$. Alternatively, for the same maximum acceptable failure probability, if $V_P = 0.3$, then ϕ decreases to 0.62.

One difficulty with the computation of $\sigma^2_{\ln E_{eff}}$, that is apparent in the approximation of Eq. (25), is that it depends on the footing dimension B. From the point of view of the design probability, Eq. (14), this means that B does not entirely disappear, and the equation is still interpreted as the probability that a footing of a certain size will fail to stay within the serviceability limit state. The major implication of this interpretation is that if Eq. (14) is used conditionally to determine ϕ, then the design resistance factor, ϕ, will have some dependence on the footing size – this is not convenient for a design code (imagine designing a concrete beam if ϕ_c varied with the beam dimension). Thus, as is, Eq. (14) should be used conditionally to determine the reliability of a footing against settlement failure once it has been designed. The determination of ϕ must then proceed by using the total probability theorem; that is, find ϕ such that

$$p_{max} = \int_0^\infty \mathrm{P}\left[W > \frac{\alpha}{\phi}\hat{P} \mid B\right] f_B(b)db \tag{37}$$

where f_B is the probability distribution of the footing width B. The distribution of B is not easily obtained – it is a function of H, $\hat{P}$, δ_{max}, the parameters of $\hat{E}$, and the load and resistance factors, α and ϕ, see Eq. (12) – and so the value of ϕ is not easily determined using Eq. (37). One possible solution is to assume that changes in B do not have a great influence on the computed value of ϕ and to take $B = B_{med}$, where B_{med} is the (non-random) footing width required by the median elastic modulus using a moderate resistance factor of $\phi = 0.5$ in Eq. (12). This approach will be adopted in this chapter, and will be validated by the simulation to be discussed next.

4 Design Simulations

As mentioned above, the resistance factor ϕ cannot be directly obtained by solving Eq. (14), for given B, simultaneously with Eq. (10) since this would result in a resistance factor which depends on the footing dimension. To find the value of ϕ to be used for any footing size involves solving Eq. (37). Unfortunately, this is not feasible since the distribution of B is unknown (or, at least very difficult to compute). A simple solution is to use Monte Carlo simulation to estimate the probability on the right hand side of Eq. (37) and then use the simulation results to assess the validity of the simplifying assumption that B_{med} can be used to find ϕ using Eq. (14). In this chapter, the Random

Finite Element Method (RFEM) will be employed within a design context to perform the desired simulation. The approach is described as follows;

1) decide on a maximum tolerable settlement, δ_{max}. In this chapter, $\delta_{max} = 0.025$ m.
2) estimate the nominal footing load, $\hat{P}$, to be the median load applied to the footing by the supported structure (it is assumed that the load distribution is known well enough to know its median, $\hat{P} = e^{\mu_{\ln P}}$).
3) simulate an elastic modulus field, $E(\underset{\sim}{x})$, for the soil from a lognormal distribution with specified mean, μ_E, variance, σ_E^2, and correlation structure (Eq. 2) with correlation length $\theta_{\ln E}$. The field is simulated using the Local Average Subdivision (LAS) method (Fenton, 1990) whose local average values are assigned to corresponding finite elements.
4) 'virtually' sample the soil to obtain an estimate, $\hat{E}$, of its elastic modulus. In a real site investigation, the geotechnical engineer may estimate the soil's elastic modulus and depth to firm stratum by performing one or more CPT or SPT soundings. In this simulation, one or more vertical columns of the soil model are selected to yield the elastic modulus samples. That is, $\hat{E}$ is estimated using a geometric average, Eq. (19), where E_1^o is the elastic modulus of the top element of a column, E_2^o is the elastic modulus of the 2nd to top element of the same column, and so on to the base of the column. One or more columns may be included in the estimate, as will be discussed shortly, and measurement and model errors are not included in the estimate – the measurements are assumed precise.
5) letting $\delta_p = \delta_{max}$, and for given factors α and ϕ solve Eq. (12) for B. This constitutes the footing design. Note that design widths are normally rounded up to the next most easily measured dimension (eg 1684 mm would probably be rounded up to 1700 mm). In the same way, the design value of B is rounded up to the next larger element boundary, since the finite element model assumes footings are a whole number of elements wide. (The finite element model uses elements which are 0.15 m wide, so B is rounded up to the next larger multiple of 0.15 m.)
6) simulate a lognormally distributed footing load, P, having median $\hat{P}$ and variance σ_P^2.
7) compute the 'actual' settlement, δ, of a footing of width B under load P on a random elastic modulus field using the finite element model. In this step, the virtually sampled random field generated in step (3) above is mapped to the finite element mesh, the footing of width B (suitably rounded up to a whole number of elements wide) is placed on the surface and the settlement computed by finite element analysis.
8) if $\delta > \delta_{max}$, the footing design is assumed to have failed.
9) repeat from step (3) a large number of times ($n = 1000$, in this chapter), counting the number of footings, n_f, which experienced a design failure. The failure probability is then estimated as $\hat{p}_f = n_f/n$.

By repeating the entire process over a range of possible values of ϕ the resistance factor which leads to an acceptable probability of failure, $p_f = p_{max}$, can be selected. This 'optimal' resistance factor will also depend on;

1) the number and locations of sampled columns (analogous to the number and locations of CPT/SPT soundings),
2) the coefficient of variation of the soil's elastic modulus, V_E,

3) the correlation length, $\theta_{\ln E}$;
and the simulation will be repeated over a range of values of these parameters to see how they affect ϕ.

Five different sampling schemes will be considered in this study, as illustrated in Figure 4. The outer solid line denotes the edge of the soil model, and the interior dashed line the location of the footing. The small black squares show the plan locations where the site is virtually sampled. It is expected that the quality of the estimate of E_{eff} will improve for higher numbered sampling schemes. That is, the probability of design failure will decrease for higher numbered sampling schemes, everything else being held constant.

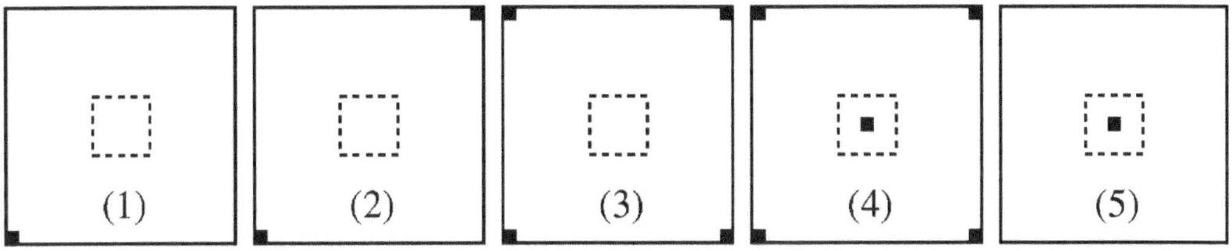

Figure 4. Sampling schemes considered in this study.

Table 1 lists the other parameters, aside from sampling schemes, varied in this study. In total 300 RFEM runs, each involving 1000 realizations were performed. Based on 1000 independent realizations, the estimated failure probability, $\hat{p}_f$, has standard error $\sqrt{\hat{p}_f(1-\hat{p}_f)/1000}$, which for a probability level of 5% is 0.7%.

Table 1. Input parameters varied in the study while holding $H = 4.8$ m, $D = 9.6$ m, $\mu_P = 1200$ kN , $V_P = 0.25$, $\mu_E = 20$ MPa, and $\nu = 0.3$ constant.

Parameter	Values Considered
V_E	0.1, 0.2, 0.5
$\theta_{\ln E}(m)$	0.1, 1.0 10.0, 100.0
ϕ	0.4, 0.5, 0.6, 0.7, 0.8

5 Simulation Results

Figure 5 shows the effect of the correlation length on the probability of failure for sampling scheme #1 (a single sampled column at the corner of site) and for $V_E = 0.5$. The other sampling schemes and values of V_E displayed similarly shaped curves. Of particular note in Figure 5 is the fact that the probability of failure reaches a maximum for an intermediate correlation length, in this case when $\theta_{\ln E} \simeq 10$ m. This is as expected, since for stationary random fields the values of $\hat{E}$ and E_{eff} will coincide for both vanishingly small correlation lengths (where local averaging results in both becoming equal to the median) and for very large correlation lengths (where $\hat{E}$ and E_{eff} become perfectly correlated) and so the largest differences between $\hat{E}$ and E_{eff} will occur at intermediate correlation lengths. The true maximum could lie somewhere between $\theta_{\ln E} = 1$ m and $\theta_{\ln E} = 100$ m in this particular study.

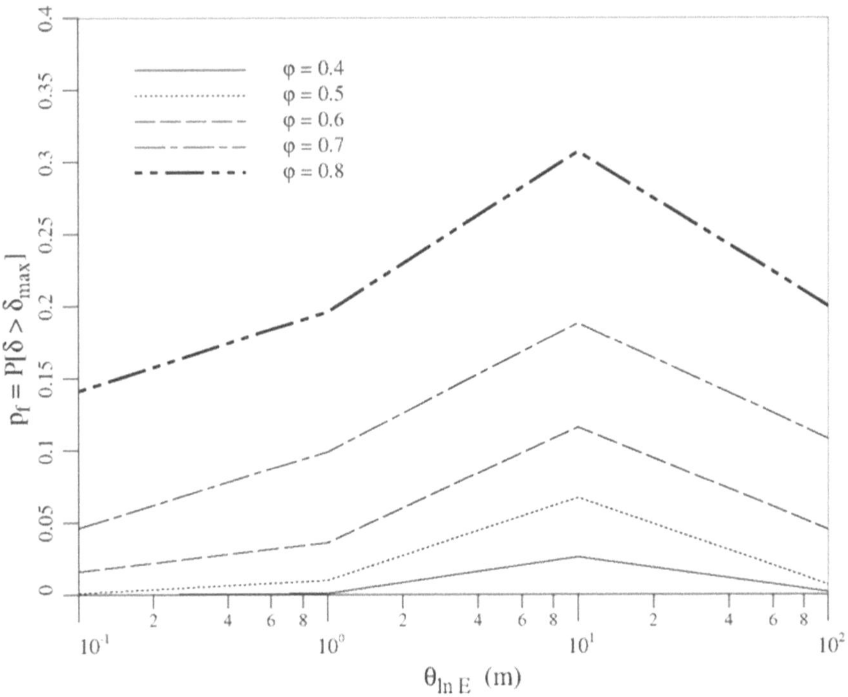

Figure 5. Effect of correlation length, $\theta_{\ln E}$, on probability of failure, $p_f = \mathrm{P}[\delta > \delta_{max}]$.

Where the maximum lies for arbitrary sampling patterns is still unknown, but the authors expect that it is probably safe to say that taking $\theta_{\ln E}$ approximately equal to the average distance between sample locations and the footing center (but not less than approximately the footing size) would yield suitably conservative failure probabilities. In this chapter, the $\theta_{\ln E} = 10$ m results will be concentrated on since these yielded the most conservative designs in this study.

Figure 6 shows how the estimated probability of failure varies with resistance factor for the five sampling schemes considered with $V_E = 0.2$ and $\theta_{\ln E} = 10$ m. This Figure can be used for design by drawing a horizontal line across at the target probability, p_{max} – to illustrate this, a light line has been drawn across at $p_{max} = 0.05$ – and then reading off the required resistance factor for a given sampling scheme. For $p_{max} = 0.05$, it can be seen that $\phi \simeq 0.62$ for the 'worst case' sampling scheme #1. For all the other sampling schemes considered, the required resistance factor is between about 0.67 and 0.69. Because the standard error of the estimated p_f values is 0.7% at this level, the relative positions of the lines tends to be somewhat erratic. What Figure 6 is saying, essentially, is that at low levels of variability, increasing the number of samples does not greatly affect the probability of failure.

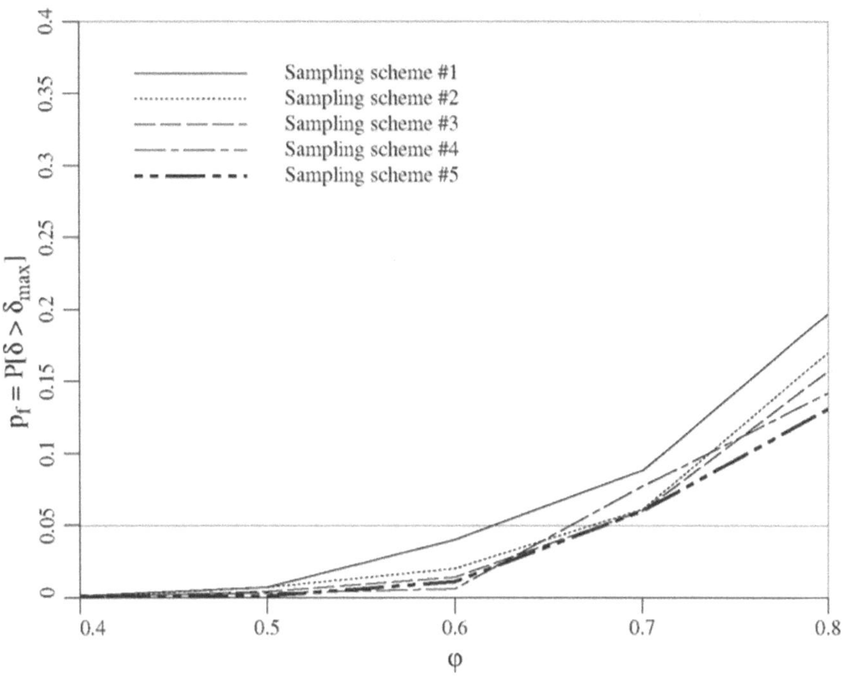

Figure 6. Effect of resistance factor, ϕ, on probability of failure, $p_f = \mathrm{P}\left[\delta > \delta_{max}\right]$ for $V_E = 0.2$ and $\theta_{\ln E} = 10$ m.

When the coefficient of variation, V_E, increases the distinction between sampling schemes becomes more pronounced. Figure 7 shows the failure probability for the various sampling schemes at $V_E = 0.5$ and $\theta_{\ln E} = 10$ m. Improved sampling now makes a significant difference to the required value of ϕ, which ranges from $\phi \simeq 0.46$ for sampling scheme #1 to $\phi \simeq 0.65$ for sampling scheme #5, assuming a target probability of $p_{max} = 0.05$. The implications of Figure 7 are that when soil variability is significant, considerable design/construction savings can be achieved when the sampling scheme is improved.

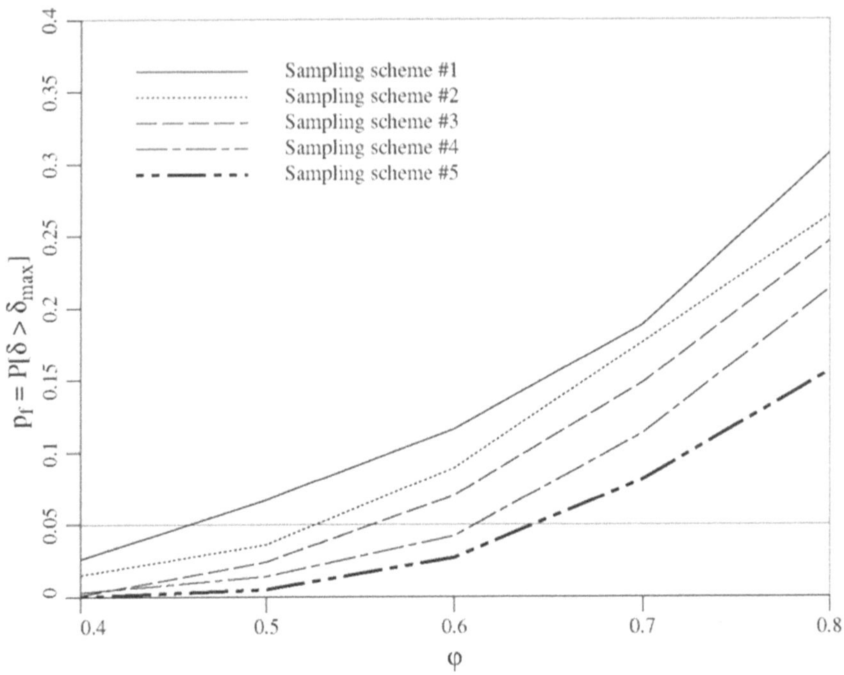

Figure 7. Effect of resistance factor, ϕ, on probability of failure, $p_f = \mathrm{P}\left[\delta > \delta_{max}\right]$ for $V_E = 0.5$ and $\theta_{\ln E} = 10$ m.

The approximation to the analytical expression for the failure probability can now be evaluated. For the case considered in Figure 7, $V_E = 0.5$ and $V_P = 0.25$ so that

$$\sigma^2_{\ln E} = \ln(1 + {V_E}^2) = 0.2231$$
$$\sigma^2_{\ln P} = \ln(1 + {V_P}^2) = 0.0606$$

To compute the variance reduction function, $\gamma(n)$, the footing width corresponding to the median elastic modulus is needed. For this calculation, an initial value of ϕ is also needed, and the moderate value of $\phi = 0.5$ is recommended. For $\mu_E = 20000$ kPa, the median elastic modulus, $\tilde{E}$, is

$$\tilde{E} = \frac{\mu_E}{\sqrt{1 + {V_E}^2}} = \frac{20000}{\sqrt{1 + 0.5^2}} = 17889 \ \ \text{kPa}$$

and for $\mu_P = 1200$ kN, the median footing load is

$$\hat{P} = \frac{\mu_P}{\sqrt{1 + {V_P}^2}} = \frac{1200}{\sqrt{1 + 0.25^2}} = 1164.2 \ \ \text{kN}$$

Solving Eq. (12) iteratively gives $B_{med} = 2.766$ m. The corresponding variance reduction factors are

$$\gamma_1\left(\frac{2(4.8)}{10}\right) = \frac{1}{0.96^2}\left[0.96 + e^{-0.96} - 1\right] = 0.74413$$

$$\gamma_1\left(\frac{2(2.766)}{10}\right) = \frac{1}{0.5532^2}\left[0.5532 + e^{-0.5532} - 1\right] = 0.83852$$

which gives

$$\gamma(n) \simeq (0.83852)^2(0.74413) = 0.5232$$

Now consider sampling scheme #1 which involve a single vertical sample with $R = \Delta x = 0.15$ m and corresponding variance reduction factor,

$$\gamma_1\left(\frac{2(0.15)}{10}\right) = \frac{1}{0.03^2}\left[0.03 + e^{-0.03} - 1\right] = 0.99007$$

$$\gamma^o(m) \simeq (0.99007)^2(0.74413) = 0.7294$$

For sampling scheme #1, $\tau'_{ave} \simeq \sqrt{2}(9.6/2) = 6.79$ m is the (approximate) distance from the sample point to the center of the footing. In this case,

$$\rho'_{ave} = \exp\left\{-\frac{2(6.79)}{10}\right\} = 0.2572$$

which gives us, using Eq. (32),

$$\sigma^2_{\ln W} = 0.0606 + 0.2231\left[0.7294 + 0.5232 - 2(0.2572)\right] = 0.2253$$

so that $\sigma_{\ln W} = 0.4746$. For $z_{0.05} = 1.645$, the required resistance factor is determined by Eq. (34) to be

$$\phi = \exp\{-1.645(0.4746)\} = 0.46$$

The corresponding value on Figure 7 is also 0.46. Although this agreement is excellent, it must be remembered that this is an approximation, and the precise agreement may be due somewhat to mutually cancelling errors and to chance, since the simulation estimates are themselves somewhat random. For example, if the more precise formulas of Eq's (24a), (24b), and (30) are used then $\gamma^o(m) = 0.7432$, $\gamma(n) = 0.6392$, and $\rho'_{ave} = 0.2498$, which gives

$$\sigma^2_{\ln W} = 0.0606 + 0.2231\left[0.7432 + 0.6392 - 2(0.2498)\right] = 0.2576$$

so that the 'more precise' required resistance factor actually has poorer agreement with simulation;

$$\phi = \exp\{-1.645\sqrt{0.2576}\} = 0.43$$

It is also to be remembered that the 'more precise' result above is still conditioned on $B = B_{med}$ and $\phi = 0.5$, whereas the simulation results are unconditional. Nevertheless, these results suggest that the approximations are insensitive to variations in B and ϕ, and are thus reasonably general.

Sampling scheme #2 involves two sampled columns separated by more than $\theta_{\ln E} = 10$ m so that n_{eff} can be taken as 2. This means that $\gamma^o(m) \simeq 0.7294/2 = 0.3647$. The average distance from the footing centerline to the sampled columns is still about 6.79 m, so that $\rho'_{ave} = 0.2572$. Now

$$\sigma^2_{\ln W} = 0.0606 + 0.2231\left[0.3647 + 0.5232 - 2(0.2572)\right] = 0.1439$$

and the required resistance factor is

$$\phi = \exp\{-1.645\sqrt{0.1439}\} = 0.54$$

The corresponding value on Figure 7 is about 0.53.

Sampling scheme #3 involves four sampled columns, separated by somewhat less than $\theta_{\ln E} = 10$ m. Due to the resulting correlation between columns, $n_{eff} \simeq 3$ is selected (ie. somewhat less than the 'independent' value of 4). This gives $\gamma^o(m) \simeq 0.7294/3 = 0.2431$. Since the average distance from the footing centerline to the sample columns is still about 6.79 m,

$$\sigma^2_{\ln w} = 0.0606 + 0.2231\,[0.2431 + 0.5232 - 2(0.2572)] = 0.1268$$

The required resistance factor is

$$\phi = \exp\{-1.645\sqrt{0.1268}\} = 0.57$$

The corresponding value on Figure 7 is about 0.56.

Sampling scheme #4 involves 5 sampled columns, also separated by somewhat less than $\theta_{\ln E} = 10$ m and $n_{eff} \simeq 4$ is selected to give $\gamma^o(m) \simeq 0.7294/4 = 0.1824$. One of the sampled columns lies below the footing, and so its 'distance' to the footing centerline is taken to be $B_{med}/\sqrt{2} = 2.766/\sqrt{2} = 1.96$ m to avoid complete correlation. The average distance to sampling points is thus

$$\tau'_{ave} = \frac{4}{5}(6.79) + \frac{1}{5}(1.96) = 5.82$$

so that $\rho'_{ave} = 0.3120$. This gives

$$\sigma^2_{\ln w} = 0.0606 + 0.2231\,[0.1824 + 0.5232 - 2(0.3120)] = 0.0788$$

The required resistance factor is

$$\phi = \exp\{-1.645\sqrt{0.0788}\} = 0.63$$

The corresponding value on Figure 7 is about 0.62.

For sampling scheme #5, the distance from the sample point to the center of the footing is zero, so τ'_{ave} is taken to equal the distance to the footing corner, $\tau'_{ave} = (2.766)/\sqrt{2} = 1.96$ m, as recommended earlier. This gives $\rho'_{ave} = 0.676$ and

$$\sigma^2_{\ln w} = 0.0606{+}0.2231\,[0.7294 + 0.5232 - 2(0.676)] = 0.0606{+}0.2231\,[-0.0994] \;\rightarrow\; 0.0606$$

where approximation errors led to a negative variance contribution from the elastic modulus field which was ignored (ie. set to zero). In this case, the sampled information is deemed sufficient to render uncertainties in the elastic modulus negligible, so that $\hat{E} \simeq E_{eff}$ and

$$\phi = \exp\{-1.645\sqrt{0.0606}\} = 0.67$$

The value of ϕ read from Figure 7 is about 0.65. If the more precise formulas for the variance reduction functions and covariance terms are used, then $\gamma^o(m) = 0.7432$, $\gamma(n) = 0.6392$, and $\rho'_{ave} = 0.6748$, which gives

$$\sigma^2_{\ln w} = 0.0606 + 0.2231\,[0.7432 + 0.6392 - 2(0.6748)] = 0.0679$$

Notice that this is very similar to the approximate result obtained above, which suggests that the assumption that samples taken below the footing largely eliminate uncertainty in the effective elastic modulus is reasonable. For this more accurate result,

$$\phi = \exp\{-1.645\sqrt{0.0679}\} = 0.65$$

which is the same as the simulation results.

Perhaps surprisingly, sampling scheme #5 outperforms, in terms of failure probability and resistance factor, sampling scheme #4, even though sampling scheme #4 involves considerably more information. The reason for this is that the good information taken below the footing is diluted by poorer information taken from farther away. This implies that when a sample is taken below the footing, other samples taken from farther away should be downweighted.

The computations illustrated above for all five sampling schemes can be summarized as follows:

1) Decide on an acceptable maximum settlement, δ_{max}. Since serviceability problems in a structure usually arise as a result of differential settlement, rather than settlement itself, the choice of an acceptable maximum settlement is usually made assuming that differential settlement will be less than the total settlement of any single footing (see, eg. D'Appolonia, 1968).
2) Choose statistical parameters of the elastic modulus field, μ_E, σ_E, and $\theta_{\ln E}$. The last can be the 'worst case' correlation length, suggested here to approximately equal the average distance between sample locations and the footing center, but not to be taken less than the median footing dimension. The values of μ_E and σ_E can be estimated from site samples (although the effect of using estimated values of μ_E and σ_E in these computations has not been investigated) or from the literature.
3) Use Eqs.(1) to compute the statistical parameters of $\ln E$ and then compute the median $\tilde{E} = \exp\{\mu_{\ln E}\}$.
4) Choose statistical parameters for the load, μ_P and σ_P, and use these to compute the mean and variance of $\ln P$. Set $\hat{P} = \exp\{\mu_{\ln P}\}$.
5) Using a moderate resistance factor, $\phi = 0.5$, and the median elastic modulus, $\tilde{E}$, compute the 'median' value of B using the 1-pt iteration of Eq. (12). Call this B_{med}.
6) Compute $\gamma(n)$ using Eq. (25) (or Eq. 24b) with $B = B_{med}$.
7) Compute $\gamma^o(m)$ using Eq. (28) (or Eq. 24a).
8) Compute ρ'_{ave} using Eq. (31) (or Eq. 30) after selecting a suitable value for τ'_{ave} as the average absolute distance between the sample columns and the footing center (where distances are taken to be no less than the distance to the footing corner, $B_{med}/\sqrt{2}$).
9) Compute $\sigma_{\ln w}$ using Eq. (32).
10) Compute the required resistance factor, ϕ, using Eq. (34).

6 Conclusions

The chapter presents approximate relationships based on random field theory which can be used to estimate resistance factors for appropriate for the LRFD settlement design of shallow foundations. Some specific comments arising from this research are as follows;

1) Two assumptions deemed to have the most influence on the resistance factors estimated in this study are 1) that the nominal load used for design, $\hat{P}$, is the median load and 2) that the load factor, α, is equal to 1.0. Changes in α result in a linear change in the resistance factor, e.g. $\phi' = \alpha\phi$, where ϕ is the resistance factor found in this study and ϕ' is the resistance factor corresponding to an α which is not equal to 1.0. Changes in $\hat{P}$ (for example, if $\hat{P}$ were taken as some other load exceedance percentile) would result in first order linear changes to ϕ, but further study would be required to specify the actual effect on the resistance factor.
2) The resistance factors obtained in this study should be considered to be *upper bounds* since the additional uncertainties arising from measurement and model errors have not been considered. To some extent, these additional error sources can be accommodated here simply by using a value of V_E greater than would actually be true at a site. For example, if $V_E = 0.35$ at a site, the effects of measurement and model error might be accommodated by using $V_E = 0.5$ in the relationships presented here. This issue needs additional study, but Meyerhof's (1982) comment that a "performance factor of 0.7 should be used for adequate reliability of serviceability estimates" suggests that the results presented here are reasonable (possibly a little conservative at the $V_E = 0.5$ level) for all sources of error.
3) The use of a 'median' footing width, B_{med}, derived using a median elastic modulus and moderate $\phi = 0.5$ value, rather than by using the full B distribution in the computation of $\gamma(n)$ appears to be quite reasonable. This is validated by the agreement between the simulation results (where B varies with each realization) and the results obtained using the approximate relationships (see previous Section).
4) The computation of a required resistance factor assumes that the uncertainty (eg, V_E) is known. In fact, at a given site, all three parameters μ_E, V_E, and $\theta_{\ln E}$ will be unknown and only estimated to various levels of precision by sampled data. To establish a LRFD code, at least V_E and $\theta_{\ln E}$ need to be known *a priori*. One of the significant results of this research is that a worst-case correlation length exists, which can be used in the development of a design code. While, the value of σ_E^2 remains an outstanding issue, calibration with existing codes may very well allow its 'practical' estimation.
5) At low uncertainty levels, that is when $V_E \leq 0.2$ or so, there is not much advantage to be gained by taking more than two sampled columns (eg. SPT or CPT borings) *in the vicinity of the footing*, as seen in Figure 6. This statement assumes that the soil is *stationary*. The assumption of stationarity implies that samples taken in one location are as good an estimator of the mean, variance, etc., as samples taken elsewhere. Since this is rarely true of soils, the qualifier "in the vicinity" was added to the above statement.
6) Although sampling scheme #4 involved five sampled columns and sampling scheme #5 involved only one sampled column, sampling scheme #5 outperformed #4. This is because the distance to the samples was not considered in the calculation of $\hat{E}$.

Thus, in sampling scheme #4 the good estimate taken under the footing was diluted by four poorer estimates taken some distance away. Whenever a soil is sampled directly under a footing, the sample results should be given much higher precedence than soil samples taken elsewhere. That is, the concepts of Best Linear Unbiased Estimation (BLUE), which takes into account the correlation between estimate and observation, should be used. In this chapter a straightforward geometric average was used (arithmetic average of logarithms in log-space) for simplicity. Further work on the effect of the form of the estimator on the required resistance factor is needed.

Notation

The following symbols are used in this chapter:

B= footing width, as designed

B_{med}= footing width required on median elastic modulus using moderate resistance factor

D= plan width of soil model (= 9.6 m in this study)

E= elastic modulus

$E(\underset{\sim}{x}_i)$= elastic modulus at the spatial location $\underset{\sim}{x}_i$

$\hat{E}$= estimate of effective elastic modulus, derived from soil samples

E_{eff}= effective uniform elastic modulus that, if underlying the footing, would yield the same settlement as actually observed

E_i= one of n elastic modulus 'samples' forming a partition in the region under the footing

E_j^o= one of m elastic modulus soil samples actually observed

f_B= footing width probability density function

$G(\underset{\sim}{x})$= standard normal (Gaussian) random field

H= overall depth of soil layer

n= number of simulations

n_f= number of simulations resulting in failure ($\delta > \delta_{max}$)

n_{eff}= effective number of *independent* sampled soil columns

P= actual applied footing load

$\hat{P}$= median applied footing load

p_f= probability of failure ($\delta > \delta_{max}$)

$\hat{p}_f$= estimated probability of failure

p_{max}= maximum acceptable probability of failure

$\hat{q}$= estimated soil stress applied by footing

R= sample plan dimension

u_1= settlement influence factor

V_E= elastic modulus coefficient of variation (μ_E/σ_E)

V_P = load coefficient of variation (μ_P/σ_P)

$W = P\hat{E}/E_{eff}$

$\underset{\sim}{x}$ = spatial coordinate or position

y = horizontal component of spatial position

z = vertical component of spatial position

$z_{p_{max}}$ = point on standard normal distribution with exceedance probability p_{max}

α = load factor

γ = variance reduction function (due to local averaging)

γ^o = variance reduction function for observed samples

γ_1 = 1-dimensional variance reduction function for Markov correlation

δ = footing settlement, positive downwards

δ_p = predicted footing settlement

δ_{max} = maximum acceptable footing settlement

$\theta_{\ln E}$ = isotropic correlation length of the log-elastic modulus field

μ_E = mean elastic modulus

$\mu_{\ln E}$ = mean of log-elastic modulus

$\mu_{\ln \hat{E}}$ = mean of the logarithm of the estimated effective elastic modulus

$\mu_{\ln E_{eff}}$ = mean of the logarithm of the effective elastic modulus underlying the footing

μ_P = mean footing load

$\mu_{\ln P}$ = mean of the log-footing load

$\mu_{\ln W}$ = mean of $\ln W$

Φ = standard normal cumulative distribution function

ϕ = resistance factor

ν = Poisson's ratio

ρ = correlation coefficient

ρ_{ij} = correlation coefficient between $\ln E_i$ and $\ln E_j$

ρ^o_{ij} = correlation coefficient between $\ln E^o_i$ and $\ln E^o_j$

ρ'_{ij} = correlation coefficient between $\ln E_i$ and $\ln E^o_j$

ρ'_{ave} = average correlation coefficient between $\ln E_i$ and $\ln E^o_j$

σ_E = standard deviation of elastic modulus

$\sigma_{\ln E}$ = standard deviation of log-elastic modulus

$\sigma_{\ln \hat{E}}$ = standard deviation of the logarithm of the estimated effective elastic modulus

$\sigma_{\ln E_{eff}}$ = standard deviation of the logarithm of the effective elastic modulus underlying the footing

$\sigma_{\ln P}$= standard deviation of the log-footing load

$\underset{\sim}{\tau}$= spatial lag vector

τ= lag distance, equal to $|\underset{\sim}{\tau}|$

τ'_{ave}= average distance between the footing center and the sampled soil columns

Bibliography

Australian Standard . Bridge Design, Part 3: Foundations and Soil-Supporting Structures, AS 5100.3–2004, Sydney, Australia, 2004.

Australian Standard . Earth-Retaining Structures, AS 4678–2002, Sydney, Australia, 2003.

Australian Standard . Piling – Design and Installation, AS 2159–1995, Sydney, Australia, 1995.

CEN . EN 1997-1:2004 (E) Geotechnical Design – Part 1: General Rules, European Committee for Standardization, Brussels, 2004.

Canadian Geotechnical Society . Canadian Foundation Engineering Manual, 3rd Ed., Montreal, Quebec, 1992.

Canadian Standards Association . Commentary on CAN/CSA-S6-00 Canadian Highway Bridge Design Code, CAN/CSA-S6.1-00, Mississauga, Ontario, Canada, 2000.

Christian, J.T. and Carrier, W.D. . Janbu, Bjerrum and Kjaernsli's chart reinterpreted. Canadian Geotechnical Journal, 15: 123–128, 1978.

D'Appolonia, D.J., D'Appolonia, E. and Brissette, R.F. . Settlement of spread footings on sand. ASCE Journal of Soil Mechanics, Foundations Division, 94(SM3): 735–760, 1968.

Fenton, G.A. and Griffiths, D.V. . Probabilistic foundation settlement on spatially random soil. ASCE Journal of Geotechnical and Geoenvironmental Engineering, 128(5): 381–390, 2002.

Fenton, G.A. and Griffiths, D.V. . Reliability of traditional retaining wall design. Géotechnique, 55(1): 55–62, 2005.

Fenton, G.A. and Vanmarcke, E.H. . Simulation of random fields via Local Average Subdivision. ASCE Journal of Engineering Mechanics, 116(8): 1733–1749, 1990.

Fenton, G.A. . Simulation and Analysis of Random Fields, Ph.D. Thesis, Princeton University , Department of Civil Engineering and Operations Research, Princeton, New Jersey, 1990.

French, S.E. . Design of Shallow Foundations. ASCE Press, Reston, Virginia, 1999.

Janbu, N., Bjerrum, L. and Kjaernsli, B. . Veiledning ved losning av fundamenteringsoppgaver. Norwegian Geotechnical Institute Publication 16, Oslo, pp. 30–32, 1956.

Li, C.-C. and Der Kiureghian, A. . Optimal discretization of random fields. ASCE Journal of Engineering Mechanics, 119(6): 1136–1154. 1993.

Matthies, H.G., Brenner, C.E., Bucher, C.G. and Soares, C.G. . Uncertainties in probabilistic numerical analysis of structures and solids – stochastic finite elements. Structural Safety, 19(3): 283–336, 1997.

Meyerhof, G.G. . Limit States Design in Geotechnical Engineering. Structural Safety Journal, 1: 67–71, 1982.

Milovic, D. . Stresses and Displacements for Shallow Foundations. *In* Developments in Geotechnical Engineering, **70**. Elsevier, Amsterdam, 1992.

NCHRP . Load and Resistance Factor Design (LRFD) for Deep Foundations, Report 507, National Cooperative Highway Research Program, Transportation Research Board, NRC, Washington, DC, 2004.

Seycek, J. . Settlement calculation limited to actual deformation zone. *In* Deformations of Soils and Displacements of Structures, Proceedings of the 10th European Conference on Soil Mechanics and Foundation Engineering, Florence, Italy, May 1991. pp. 543–548, 1991.

Vanmarcke, E.H. . Random Fields: Analysis and Synthesis. The MIT Press, Cambridge, Massachusetts, 1984.

Vick, S.G. . Degrees of Belief: Subjective Probability and Engineering Judgement. American Society of Civil Engineers, Reston, Virginia, 2002.

The Random Finite Element Method RFEM) in Mine Pillar Stability Analysis

D.V. Griffiths, Gordon A. Fenton and Carisa B. Lemons
Division of Engineering, Colorado School of Mines, U.S.A.
Department of Engineering Mathematics, Dalhousie University, Canada

bstract The majority of geotechnical analyses are deterministic, in that the inherent variability of the materials is not modeled directly, rather some "factor of safety" is applied to results computed using "average" properties. In the present study, the influence of spatially varying strength is assessed via numerical experiments involving the compressive strength and stability of pillars typically used in underground construction and mining operations. The model combines random eld theory with an elasto-plastic nite element algorithm in a Monte-Carlo framework. It is found that the average strength of the rock is not a good indicator of the overall strength of the pillar. The results of this study enable traditional approaches involving factors of safety to be re-interpreted as a "probability of failure" in the context of reliability based design.

1 Introduction

A review and assessment of existing design methods for estimating the factor of safety of coal pillars based on statistical approaches was covered recently by Salamon (1999). This paper follows this philosophy by investigating in a rigorous way the influence of rock strength variability on the overall compressive strength of rock pillars typically used in mining and underground construction. The technique merges elasto-plastic finite element analysis (e.g. Smith and Griffiths 1998) with random field theory (e.g. Vanmarcke 1984, Fenton 1990) within a Monte-Carlo framework. The rock strength is characterized by its unconfined compressive strength or "cohesion" c using an elastic-perfectly plastic Tresca failure criterion. The variable c, is defined by a lognormal distribution with three parameters as shown in Table 1.

Table 1. Input parameters for rock strength c

		Units
Mean	μ_c	kN/m^2
Standard Deviation	σ_c	kN/m^2
Spatial Correlation Length	$_{\ln c}$	m

The Spatial Correlation Length describes the distance over which the spatially random values will tend to be correlated in the underlying Gaussian field. Thus, a large value will imply a smoothly varying field, while a small value will imply a ragged field. Initial studies on a similar problem were reported by Paice and Griffiths (1999).

In order to non-dimensionalize the input, the rock strength variability is expressed in terms of the Coefficient of Variation $C.O.V._c = \sigma_c/\mu_c$, and a normalized spatial correlation length $_{\ln c} = {}_{\ln c}/B$ where B is the height (and width) of the pillar.

The spatially varying rock strength field is simulated using the Local Average Subdivision method (Fenton, 1994, Fenton and Vanmarcke, 1990) which produces local arithmetic averages of the $\ln c$ field over each element. Thus, each element is assigned a random value of $\ln c$ as a local average, over the element size, of the continuously varying random field having point statistics derived from Table 1. The element values thus correctly reflect the variance reduction due to arithmetic averaging over the element as well as the cross-correlation structure dictated by the spatial correlation length, $_{\ln c}$. In this study, an exponentially decaying (Markovian) correlation function is assumed,

$$\rho(\tau) = \exp\left(\frac{2\tau}{_{\ln c}} \right) \tag{1}$$

where τ is the absolute distance between any two points in the rock mass. Notice that the above correlation function is isotropic, which is to say two points separated by 0.2 m vertically have the same correlation coefficient as two points separated by 0.2 m horizontally. While it is unlikely that actual rock properties will have an isotropic correlation structure (due to layering, etc.), the basic probabilistic behaviour of pillar failure can be established in the isotropic case and anisotropic 'site-specific' refinements left for future work. The methodologies and general trends will be similar to the results presented here.

The present study is confined to plane strain pillars with square dimensions in the plane of the analysis. A typical finite element mesh is shown in Figure 1 and consists of 400 8-node plane strain quadrilateral elements. Each element is assigned a different c-value based on the underlying lognormal distribution, as discussed above. For each Monte-Carlo simulation, the block is compressed by incrementally displacing the top surface vertically downwards. At convergence following each displacement increment, the nodal reaction loads are summed and divided by the width of the block B to give the average axial stress. When this axial stress levels out to quite strict tolerances, it is then defined as the compressive strength of the block, q_f.

This study focuses on the dimensionless "bearing capacity factor" N_c defined for each of the n_{sim} Monte-Carlo simulations as:

$$N_c^i = q_f^i/\mu_c, \quad i = 1, 2, ..., n_{sim} \tag{2}$$

It should be noted that N_c^i, for each simulation, is normalized by dividing q_f by the *mean* compressive strength μ_c. The N_c^i values are then analyzed statistically leading to a sample mean m , and sample standard deviation, s . These, in turn, can be used to estimate probabilities concerning the compressive strength of the pillar.

A uniform rock, having spatially constant strength given by c has an unconfined compressive strength from Mohr's circle given by $N_c = 2$, hence,

$$q_f = 2c \tag{3}$$

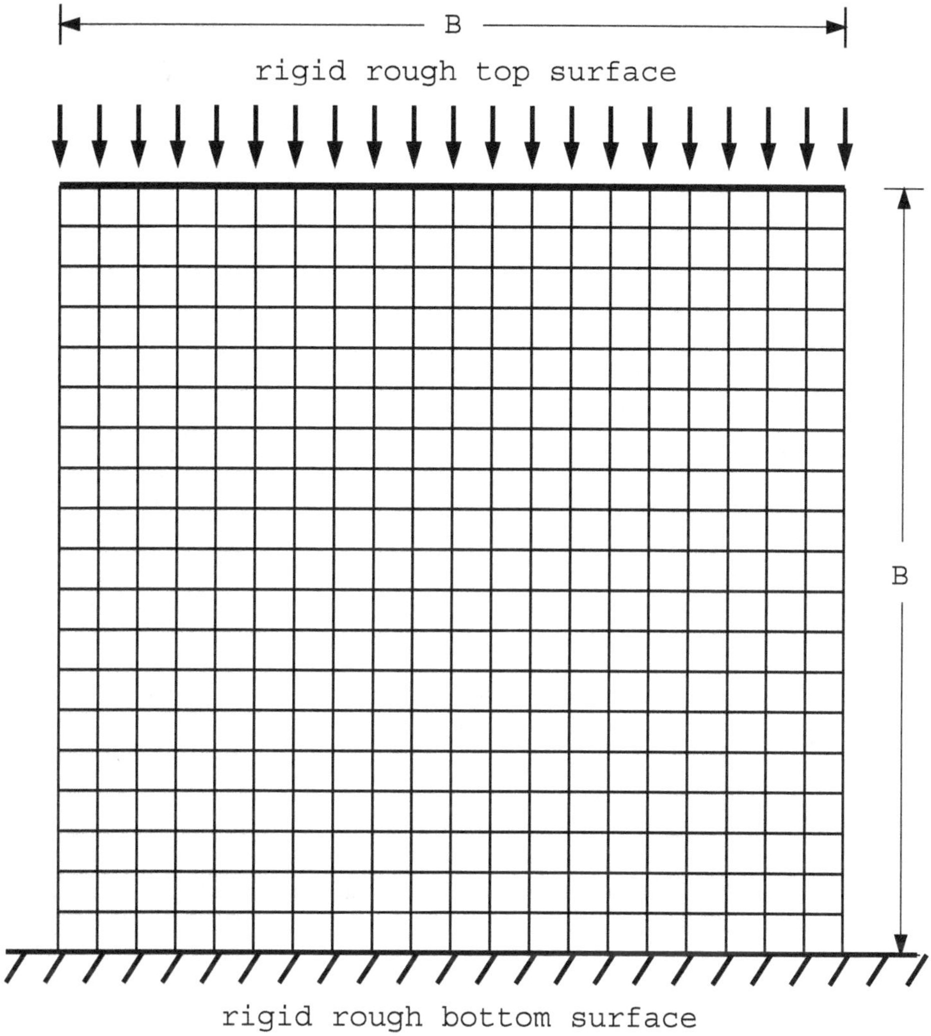

Figure 1. Mesh used for finite element pillar analysis

Of particular interest in this study therefore, is to compare this deterministic value of 2, with m from the Monte-Carlo simulations.

2 Literature

Although reliability based approaches have not yet been widely implemented by geotechnical engineers in routine design, there has been a significant growth in interest in this area as an alternative to the more traditional factor of safety. A valid criticism of the factor of safety, is that it does not give as much physical insight into the likelihood of design failure as a probabilistic measure. Even though a reliability based analysis tells more about the safety of a design, engineers have tended to prefer the factor of safety approach since there is a perception that it takes less time to compute . This perception is no doubt well based, since factor of safety approaches are generally fairly simple, but the old addage 'you get what you pay for' applies here. The understanding of the basic failure mechanism afforded by the consideration of spatial variation is well worth the effort. In addition to increasing understanding and safety, reliability based design can also maximize cost efficiency (e.g. Call 1985).

Both variability and spatial correlation lengths of material properties can affect the reliability of geotechnical systems. While the variability of geotechnical properties are hard to determine, since soil and rock properties can vary widely (e.g. Phoon and Kulhawy 1999, Harr 1987, Lumb 1970, Lee et al 1983), there is some consensus that $C.O.V._c$ values for rock strength range from 0.30 to 0.50 (e.g. Hoek 1998, Savely 1987, Hoek and Brown 1997). This variability has been represented in the present study by a lognormal distribution that ensures non-negative strength values. The spatial correlation length can also affect system reliability, although it is not often accounted for properly (e.g. Mostyn and Li 1993, Lacasse and Nadim 1996, DeGroot 1996, Wickremesinghe and Campanella 1993, Cherubini 2000).

In mining applications, material variability is not usually accounted for directly, however empirical formulas have been developed to make adjustments to the factors of safety (e.g Salamon 1999, Peng and Dutta 1992, Scovazzo 1992).

Finite element analysis has been used in the past to account for varying properties of geotechnical problems including pillar design (See e.g. Park 1992, Tan et al 1993, Mellah et al 2000, Dai et al 1993). In this paper, elasto-plastic finite element analysis has been combined with random field theory to investigate the influence of material variability and spatial correlation lengths on mine pillar stability. By using multiple simulations, the Monte-Carlo technique can be used to predict pillar reliability involving materials with high variances and spatial variability that would not be amenable to analysis by first order second moment methods.

3 Introduction

Parametric studies

Analyses were performed with input parameters within the following ranges:

$$0.01 < {}_{\ln c} < 10$$

$$0.05 < C.O.V._c < 1.6$$

For each pair of values of $C.O.V._c$ and ${}_{\ln c}$, 2500 Monte-Carlo simulations were performed, and from these, the *estimated* statistics of the bearing capacity factor N_c were computed leading to the sample mean, m , and sample standard deviation, s .

In order to maintain reasonable accuracy and run-time efficiency, the sensitivity of results to mesh density and the number of Monte-Carlo simulations was examined. Figure 2 shows the effect of varying the mesh size with all other variables held constant. Since there is little change from the 20x20 element mesh to the 40x40 element mesh, the 20x20

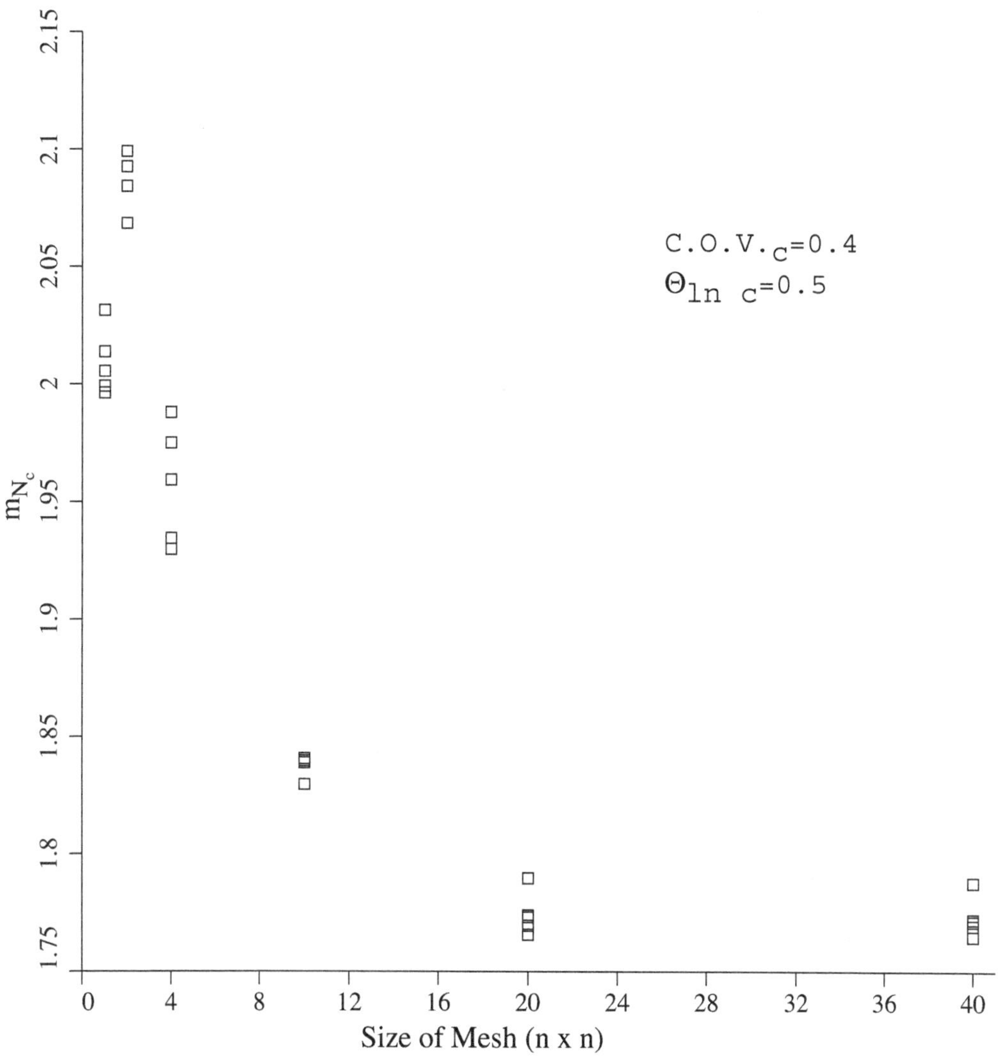

Figure 2. Influence of mesh density on the computer m $_c$

element mesh is deemed to give reasonable precision for the analysis. Figure 3 shows the convergence of m as the number of simulations increases. The figure displays five repeated analyses with identical properties and indicates that 2500 simulations gives reasonable precision and reproducibility. Although higher precision could be achieved with greater mesh density and simulation counts, the use of a 20x20 element mesh with n_{sim} = 2500 simulations is considered to be accurate enough in view of the inherent variability of the input data.

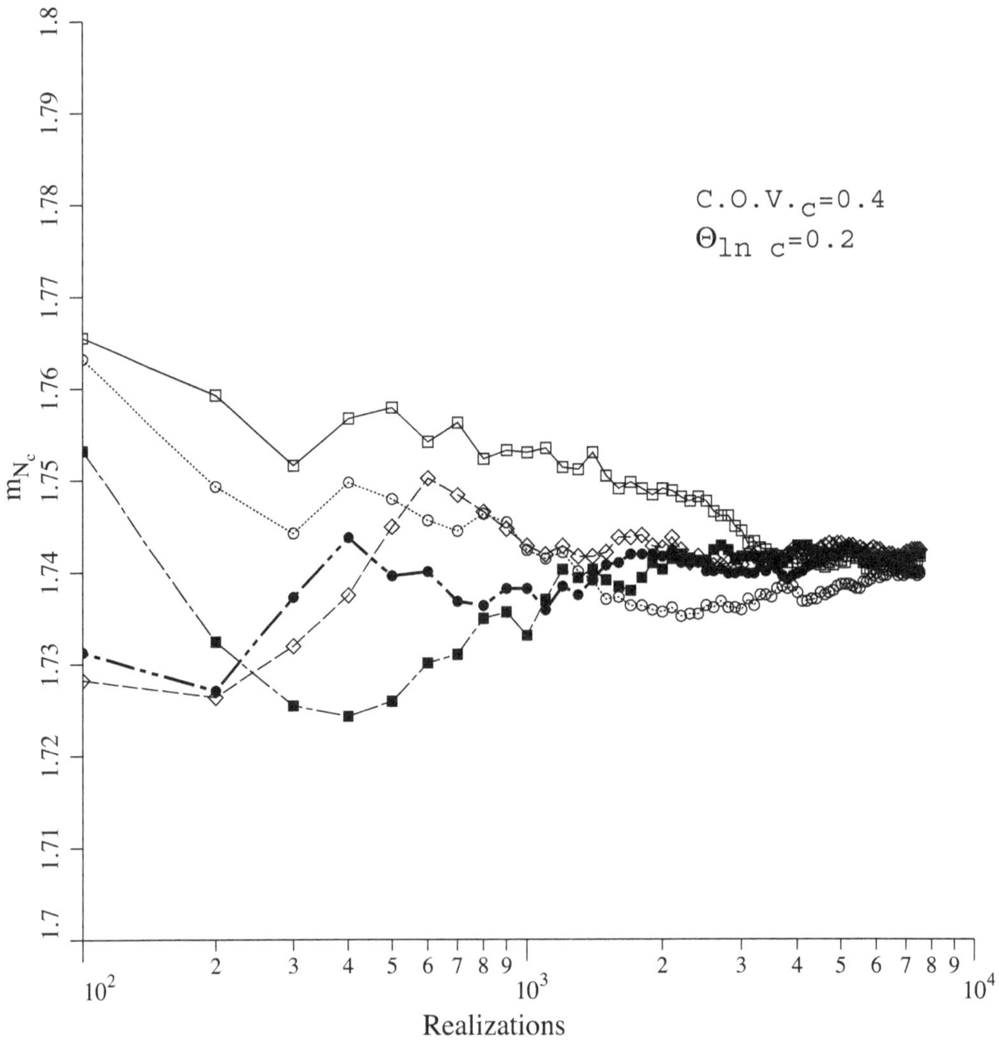

Figure 3. Influence of the number of simulations on $m \quad _c$

The accuracy of results obtained from Monte-Carlo analyses can also be directly computed from the number of simulations. Estimated mean bearing capacities will have a standard error ($\pm$ one standard deviation) equal to the sample standard deviation times $1/\sqrt{n_{sim}} = 1/\sqrt{2500} = 0.020$ or about 2% of the sample standard deviation. Similarly, the estimated variance will have a standard error equal to the sample variance times $\sqrt{(2/(n_{sim} \quad 1))} = \sqrt{(2/2499)} = 0.028$, or about 3% of the sample variance. This means that estimated quantities will generally be within about 4% of the true (ie. finite element) quantities, statistically speaking.

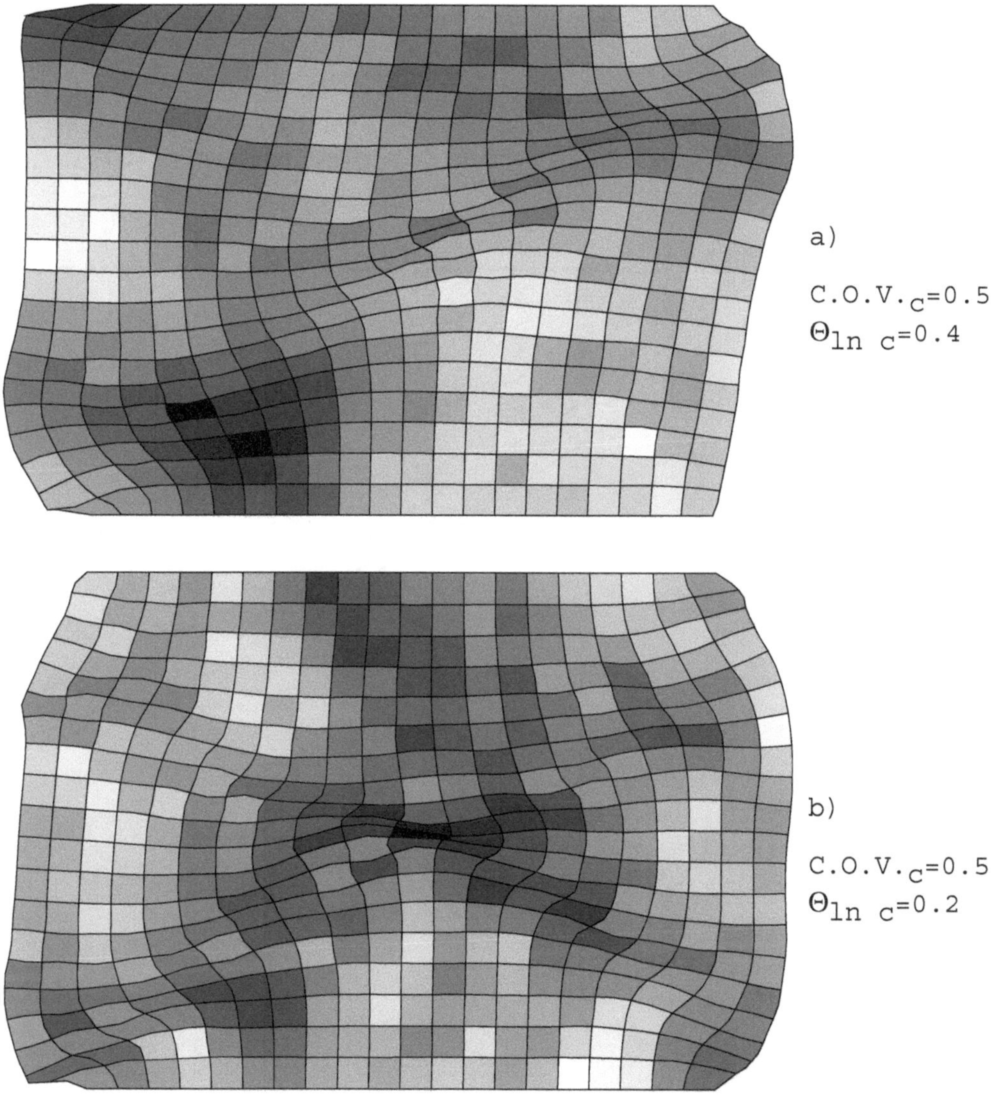

Figure 4. Typical deformed meshes and grey scales at failure. Darker zones signify weaker rock

Figures 4a and 4b show two typical deformed meshes at failure, corresponding to $_{\ln c} = 0.4$ and $_{\ln c} = 0.2$, respectively. Lighter regions in the plots indicate stronger rock and darker regions indicate weaker rock. It is clear that the weak (dark) regions have triggered quite irregular failure mechanisms. In general, the mechanism is attracted to the weak zones and "avoids" the strong zones. This suggests that failure is not simply a function of the arithmetic average of rock strength – it is somewhat reduced due to the failure path preferentially selecting weak materials.

3.1 Mean of N_c

A summary of the sample mean bearing capacity factor (m), computed using the values provided by equation (2), for each simulation is shown in Figures 5 and 6.

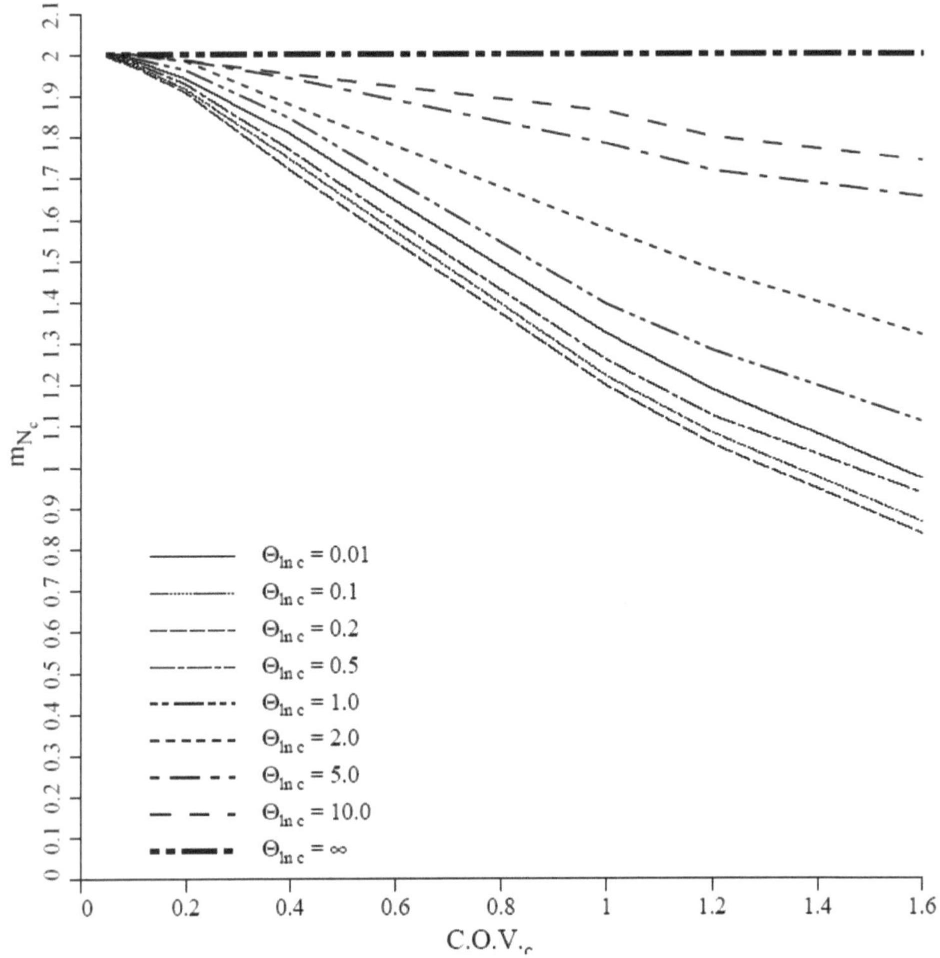

Figure 5. Variation of $m\ _c$ with COV_c

The plots confirm that for low values of $C.O.V._c$, m tends to the deterministic value of 2. As the $C.O.V._c$ of the rock increases, the mean bearing capacity factor falls quite rapidly, especially for smaller values of $_{\ln c}$. As shown in Figure 6, however, m reaches a minimum at about $_{\ln c} = 0.2$ and starts to climb again.

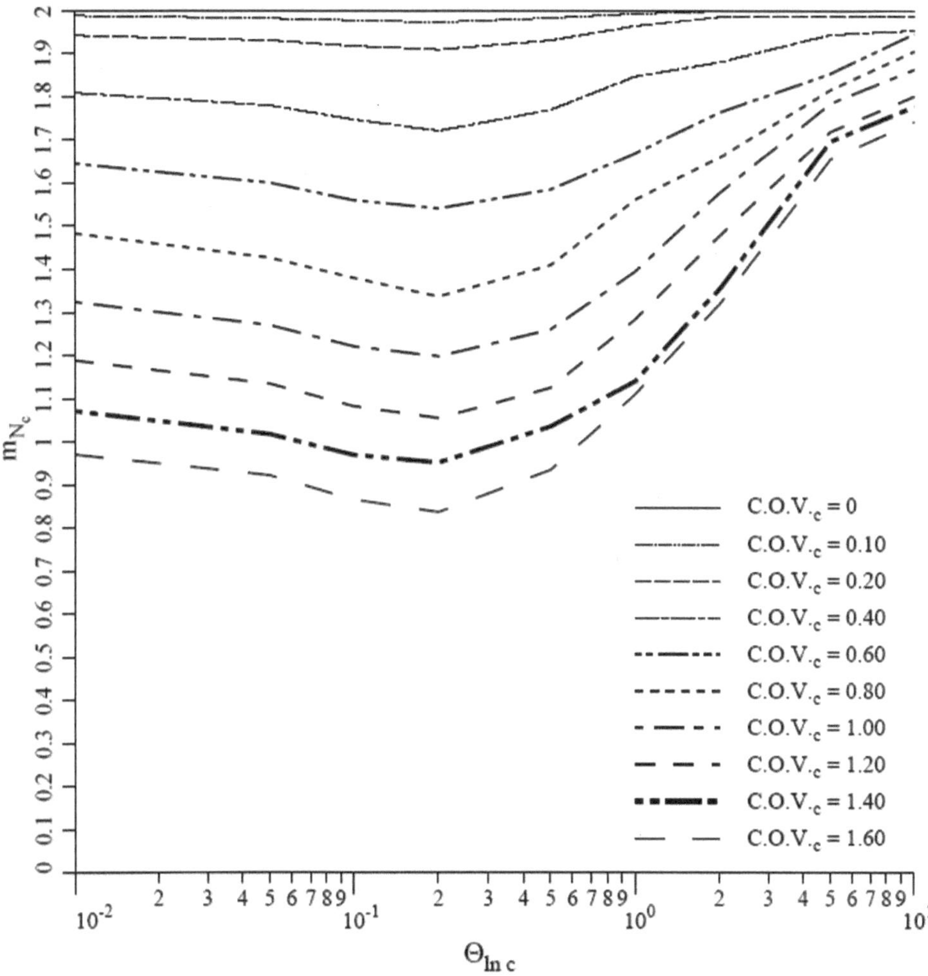

Figure 6. Variation of $m\ \ _c$ with $\ _{\ln c}$

It is speculated that in the limit of $\ _{\ln c} \to 0$, there are no "preferential" weak paths the failure mechanism can follow, and the mean bearing capacity factor will return once more to the deterministic value of 2. This is as suggested by Figure 6. In principle, the $\ _{\ln c} = 0$ case is somewhat delicate to investigate. Strictly speaking, any local average of a (finite variance) random $\ln c$ field having $\ _{\ln c} = 0$ will have zero variance (since the local average will involve an infinite number of independent points). Thus, in the $\ _{\ln c} = 0$ case the 'local average' representation, ie. the finite element method (as interpreted here), will necessarily return to the deterministic case. The detailed investigation of this trend is also complicated by the fact that rock properties are never determined at the 'point' level – they are based on a local average over the rock sample volume. Thus, while recognizing the apparent trend with small $_{\ln c}$ in this study, the theoretical and numerical verification of the limiting trend is left for further research.

Also included on Figure 5 is the horizontal line corresponding to the solution that would be obtained for $_{\ln c} = \infty$. This hypothetical case implies that each simulation of the Monte-Carlo process involves an essentially uniform soil, albeit with properties varying from one simulation to the next. In this case, the distribution of q_f will be statistically similar to the lognormal distribution of c but magnified by 2, thus $m \quad = 2$ for all values of $C.O.V._c$.

3.2 Coe cient of Variation of N_c

Figure 7 shows the influence of $_{\ln c}$ and $C.O.V._c$ on the sample coefficient of variation of the estimated bearing capacity factor, $C.O.V. \quad = s \quad /m$. The plots indicate that $C.O.V.$ is positively correlated with both $C.O.V._c$ and $_{\ln c}$, with the limiting value of $_{\ln c} = \infty$ giving the straight line $C.O.V. \quad = C.O.V._c$.

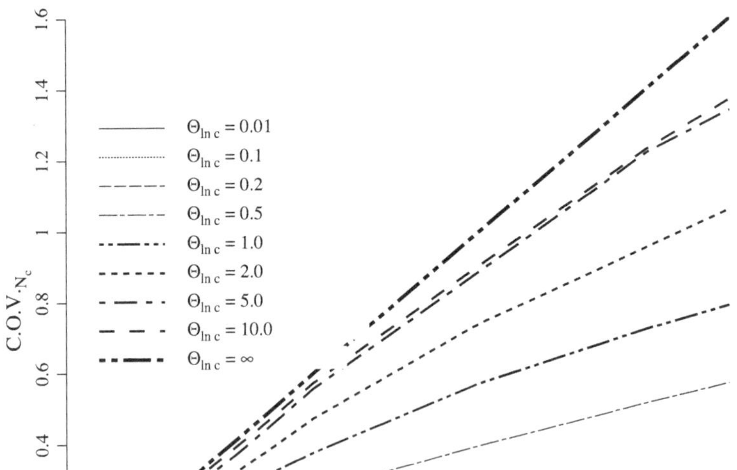

Figure 7. Variation of $COV \quad _c$ with COV_c

4 Probabilistic interpretation

Following Monte-Carlo simulations for each parametric combination of input parameters ($_{\ln c}$ and $C.O.V._c$), the suite of computed bearing capacity factor values from equation (1) was plotted in the form of a histogram, and a "best-fit" lognormal distribution superimposed. An example of such a plot is shown in Figure 8 for the case where $_{\ln c} = 0.2$ and $C.O.V._c = 0.4$.

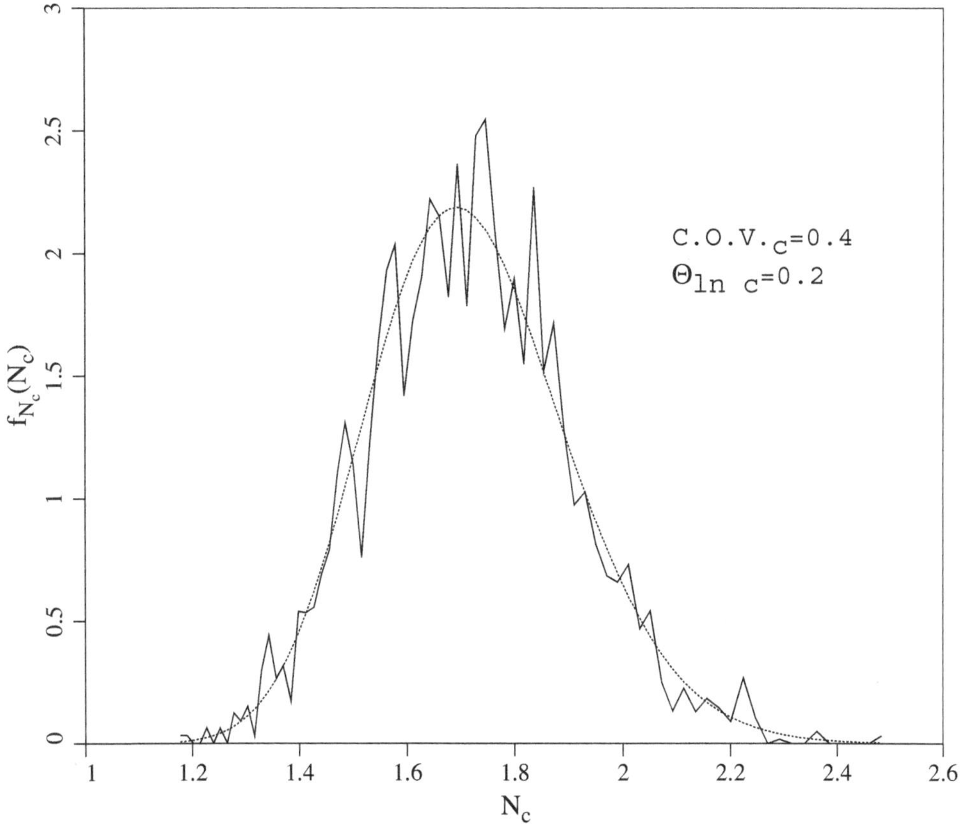

Figure 8. Histogram and lognormal fit for a typical set of computed N_c values

Since the lognormal fit has been normalized to enclose an area of unity, areas under the curve can be directly related to probabilities. From a practical viewpoint, it would be of interest to estimate the probability of "design failure", defined here as occurring when the computed compressive strength is less than the deterministic value based on the mean strength divided by a "factor of safety" F, i.e.

$$\text{"Design failure" if } q_f < 2\mu_c/F \tag{4}$$

or alternatively,

$$\text{“Design failure” if } N_c < 2/F \tag{5}$$

The probability of failure as defined in equation 4 can be expressed as the area under the probability density function to the left of a “target” design value $2/F$, hence from the properties of the underlying normal distribution we get:

$$p(N_c < 2/F) = \Phi \quad \frac{\ln 2/F \quad m_{\ln}}{s_{\ln}} \Big) \tag{6}$$

where Φ is the cumulative standard normal distribution function.

For the particular case shown in Figure 8, the fitted lognormal distribution has the properties m = 1.721 and s = 0.185. These values indicate a median given by Median = 1.711 and a mode given by Mode = 1.692. Furthermore, the underlying normal distribution (see Appendix) is easily shown to have the properties $m_{\ln}$ = 0.537 and $s_{\ln}$ = 0.107. For the particular case of $F = 1.5$, equation (4) gives $p(N_c < 2/1.5) = 0.01$, indicating a 1% probability of “design failure” as defined above. This implies a 99% reliability that the pillar will remain stable. It should be noted that for the relatively low variance indicated in Figure 8, the lognormal distribution looks very similar to a normal distribution.

4.1 General observation on the lognormal distribution

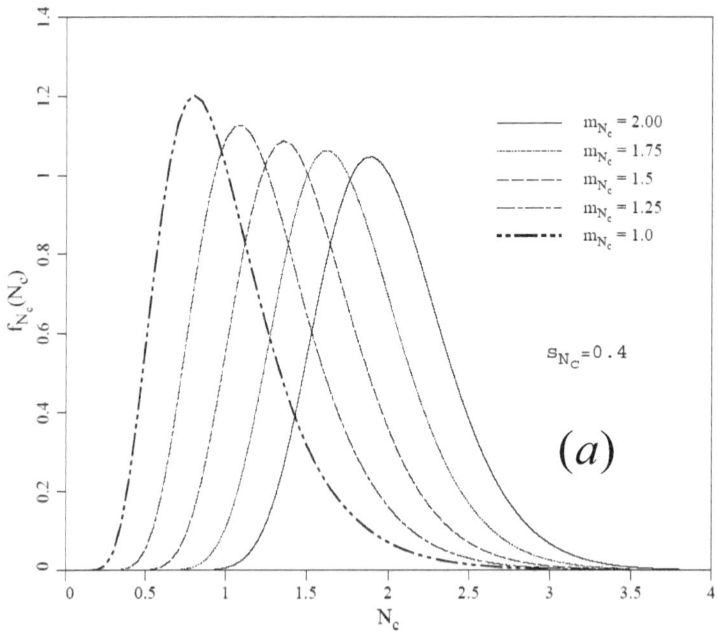

Figure 9. Lognormal plots with constant s $_c$ and varying m $_c$

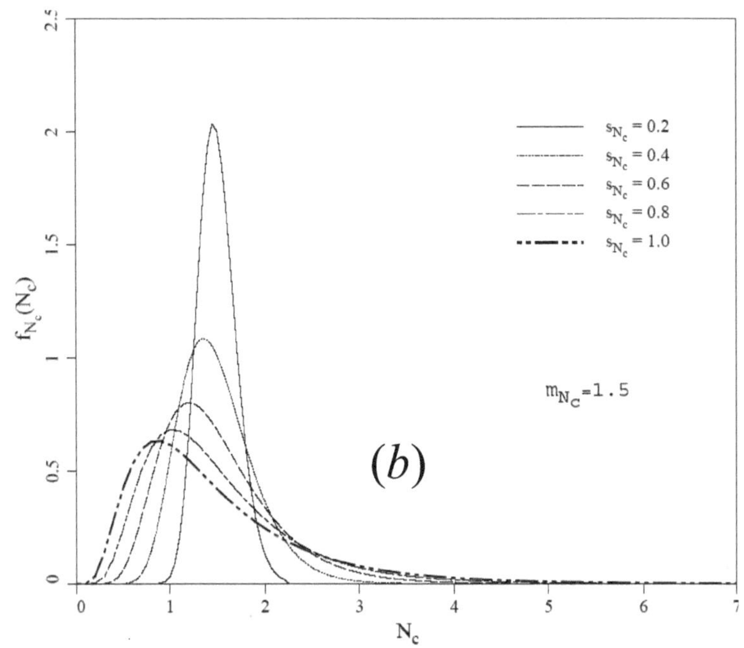

Figure 10. Lognormal plots with constant $m_{\;c}$ and varying $s_{\;c}$

While the probability of design failure is directly related to the estimated values of m and s, it is of interest to observe the separate influences of m and s.

If s is held constant, increasing m clearly decreases the probability of failure as shown in Figure 9, since the curves move consistently to the right and the area to the left of any stationary "target" decreases. The situation is less clear if m is held constant and s is varied as shown in Figure 10.

Figure 11 shows how the probability of design failure as defined in equation 4, varies as a function of *C.O.V.* and the ratio of the target value $2/F$ to the *mean* of the lognormal distribution m. If the target values is less than or equal to the mean, the probability of failure always increases as *C.O.V.* is increased. If the target values is larger than the mean however, the probability of failure initially falls and then gradually rises.

A more fundamental parameter when estimating probabilities of lognormal distributions is the median, which represents the 50% probability location. Figure 12 shows how the probability of design failure varies as a function of *C.O.V.* and the ratio of the target value $2/F$ to the *median*. In this case the probabilistic interpretation is clearly defined. If the target is less than the median, the probability always increases as *C.O.V.* is increased, whereas if the target is greater than the median, the probability always decreases. If the target equals the median, the probability of failure is 50% irrespective of the value of *C.O.V.*. It might also be noted in Figure 12, that while the rate of change of probability is quite high at low values of $C.O.V.$, the curves tend to flatten out quite rapidly as $C.O.V.$ is increased.

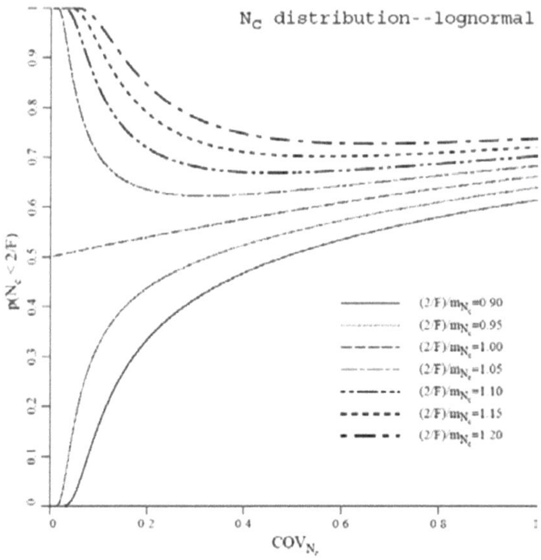

Figure 11. Probability of N_c being less than $2/F$ as a function of $COV\ _c$ for different $2/F/m\ _c$ values

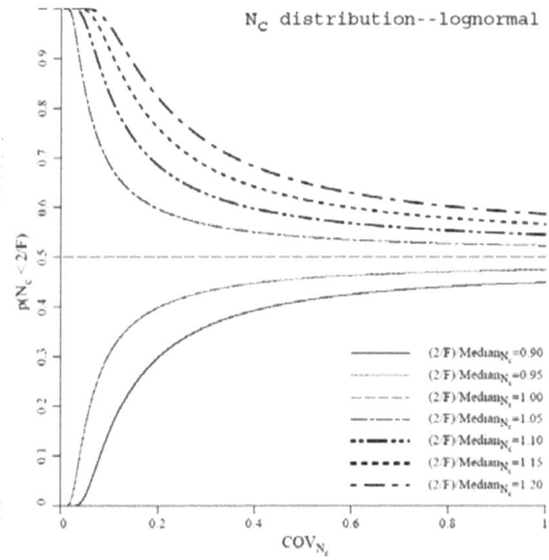

Figure 12. Probability of N_c being less than $2/F$ as a function of $COV\ _c$ for different $2/F/\text{median}\ _c$ values

4.2 Results from pillar analyses

The influence of these rather complex interactions on the pillar stability analyses can be seen in Figures 13-16 where the probability of design failure is shown as a function of the correlation length $\ln c$ for different values of $C.O.V._c$.

Each of the four plots corresponds to a different value of the factor of safety, where $F = 1.5$, 2.0, 2.5, and 3.0 respectively. Consider in more detail the results shown in Figure 13 for the case of F=1.5, where the target value is $2/F = 1.33$. To help with the interpretation tabulated values of the statistics of N_c corresponding to different values of $C.O.V._c$ are presented.

Table 2. Probability of Design Failure, $F = 1.5$, $C.O.V._c = 0.2$

$\ln c$	m	$C.O.V.$	1.33/Median	$p(N_c < 1.33)$
0.01	1.943	0.008	0.686	0.000
0.1	1.917	0.031	0.696	0.000
0.2	1.909	0.056	0.670	0.000
0.5	1.930	0.099	0.694	0.000
1.0	1.964	0.134	0.685	0.002
2.0	1.985	0.164	0.681	0.009
5.0	1.987	0.180	0.682	0.016
10.0	1.987	0.190	0.683	0.021
∞	2.000	0.200	0.680	0.026

Small values of $C.O.V._c \leq 0.20$, result in correspondingly small values of $C.O.V.$ and high values of $m \approx 2$ as shown in Table 2, leading to low probabilities of design failure for all $\ln c$.

Table 3. Probability of Design Failure, $F = 1.5$, $C.O.V._c = 0.4$

$\ln c$	m	$C.O.V.$	1.33/Median	$p(N_c < 1.33)$
0.01	1.809	0.014	0.737	0.000
0.1	1.747	0.058	0.764	0.000
0.2	1.721	0.107	0.779	0.010
0.5	1.770	0.193	0.767	0.083
1.0	1.847	0.264	0.747	0.130
2.0	1.880	0.310	0.743	0.163
5.0	1.944	0.358	0.728	0.181
10.0	1.953	0.380	0.730	0.196
∞	2.000	0.400	0.718	0.195

For larger values of $C.O.V._c$, e.g. $C.O.V._c = 0.4$, the mean m has fallen but is still always higher than the target value of 1.33 as shown in Table 3. With $1.33/m < 1$, the Table indicates that that the increasing values of $C.O.V.$ result in a gradually increasing probability of design failure. This trend is also confirmed by Figure 11.

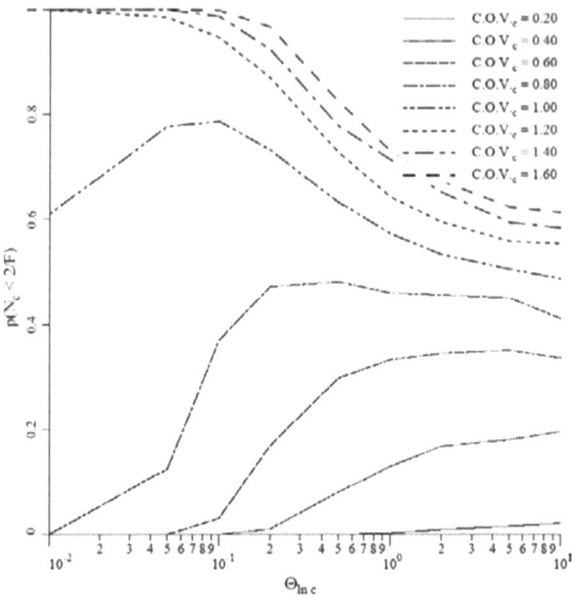

Figure 13. Probability of design failure as a function of COV_c and $_{\ln c}$ with $F = 1.5$

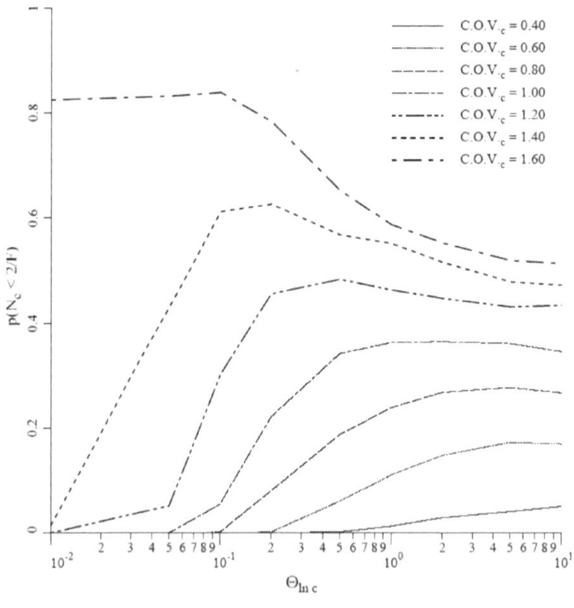

Figure 14. Probability of design failure as a function of COV_c and $_{\ln c}$ with $F = 2,0$

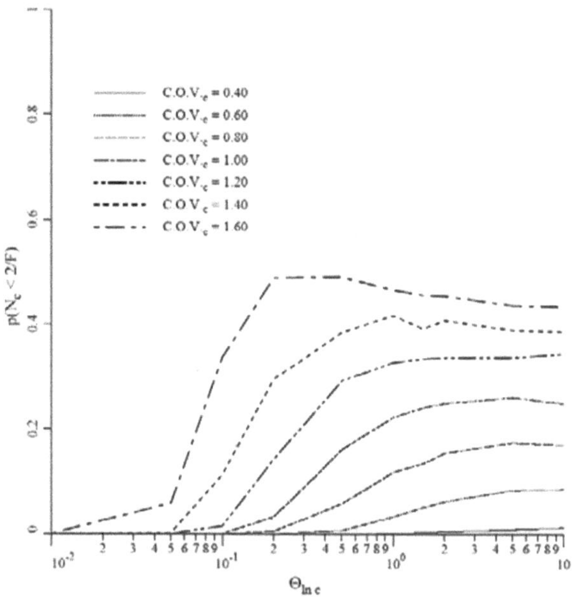

Figure 15. Probability of design failure as a function of COV_c and $_{\ln c}$ with $F = 2.5$

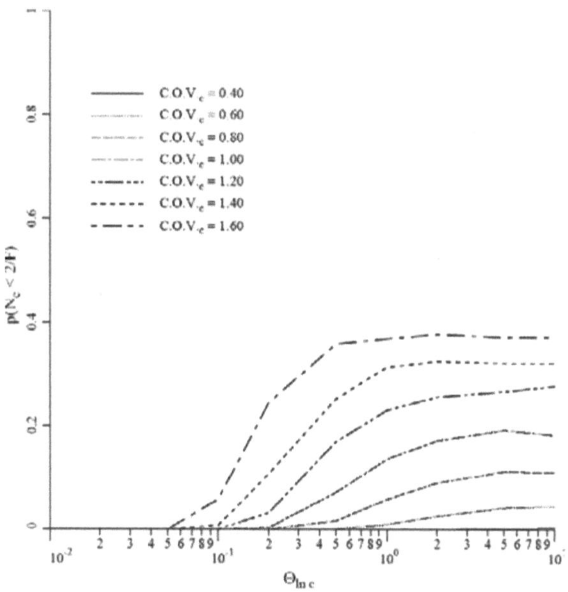

Figure 16. Probability of design failure as a function of COV_c and $_{\ln c}$ with $F = 3.0$

Table 4. Probability of Design Failure, $F = 1.5$, $C.O.V._c = 1.2$

ln c	m	$C.O.V.$	1.33/Median	$p(N_c < 1.33)$
0.01	1.189	0.028	1.122	1.000
0.1	1.083	0.136	1.242	0.946
0.2	1.055	0.239	1.299	0.867
0.5	1.125	0.468	1.309	0.727
1.0	1.283	0.662	1.246	0.643
2.0	1.479	0.838	1.176	0.588
5.0	1.719	1.003	1.099	0.545
10.0	1.801	1.108	1.105	0.545
∞	2.000	1.200	1.041	0.517

Consider now the behavior of the probabilities for rather high values of $C.O.V._c$, such as $C.O.V._c = 1.2$. From Table 4, the mean values of m have fallen quite significantly, and are often smaller than the target value of 1.33. More significantly in this case, the median of N_c is *always* smaller than the target of 1.33. Small values of ln c imply small values of $C.O.V.$ and an almost certain probability of design failure (≈ 1). With 1.33/Median > 1, the Table indicates that that the increasing values of $C.O.V.$ result in a falling probability of design failure. This trend is also confirmed by Figure 12.

Table 5. Probability of Design Failure, $F = 1.5$, $C.O.V._c = 0.8$

ln c	m	$C.O.V.$	1.33/Median	$p(N_c < 1.33)$
0.01	1.478	0.022	0.902	0.000
0.1	1.387	0.103	0.966	0.370
0.2	1.371	0.178	0.988	0.472
0.5	1.429	0.336	0.984	0.481
1.0	1.542	0.472	0.956	0.460
2.0	1.659	0.607	0.940	0.456
5.0	1.816	0.754	0.920	0.450
10.0	1.905	0.738	0.870	0.416
∞	2.000	0.800	0.854	0.411

For intermediate values of $C.O.V._c$, such as $C.O.V._c = 0.8$, the probability of design failure from Figure 13 is seen to rise and then fall. This interesting result implies a "worst case" combination of $C.O.V._c$ and ln c which would give a maximum probability of design failure.

The results tabulated in Table 5 indicate that at low values of ln c, the Median is slightly larger than the target and this, combined with the low value of $C.O.V.$, gives a negligible probability of failure. As ln c is increased, $C.O.V.$ increases and the Median decreases. Both of these effect cause the probability of failure to rise as confirmed by Figure 12.

At approximately ln c $= 0.5$, the Median approaches the target, giving a maximum probability of design failure close to 0.5. As indicated in Table 5, further increase in ln c causes the 1.33/Median ratio to fall quite consistently. Although $C.O.V.$ is still rising, the overall behavior is dominated by the falling 1.33/Median ratio and the probability of failure falls as implied in Figure 12.

Figures 13-16 corresponding to higher factors of safety, display similar maxima in their probabilities, however there is an overall trend that shows the expected reduction in the probability of failure as the Factor of Safety is increased. Figure 16, corresponding to $F = 3$, indicates that for a reasonable upper-bound value of $C.O.V. = 0.6$, the probability of design failure will be negligible for $_{\ln c} < 1$.

The program that was used to produce the results in this paper enables the reliability of rock pillars with varying compressive strength and spatial correlation to be assessed. In particular, a direct comparison can be made between the probability of failure and the more traditional Factor of Safety.

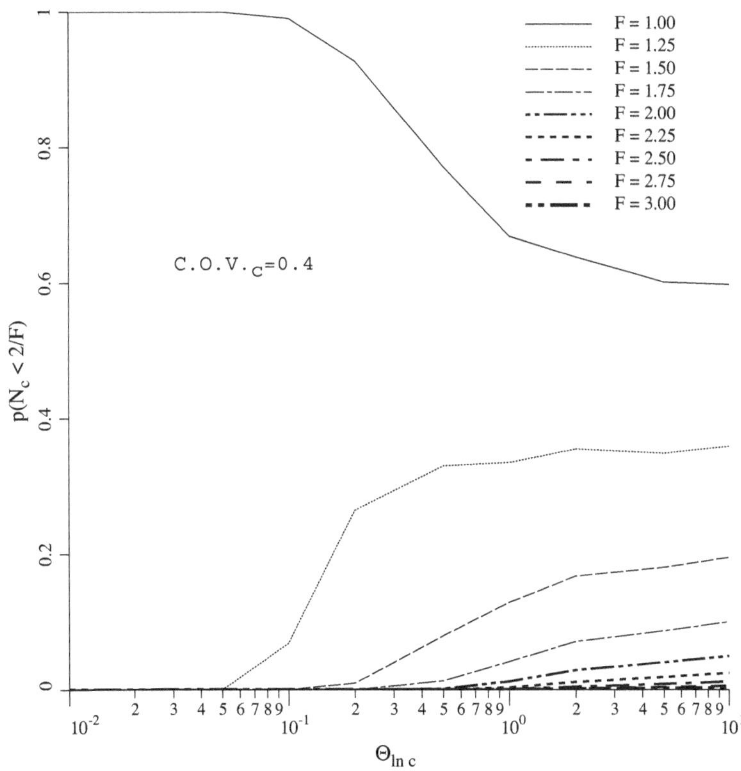

Figure 17. Probability of design failure as a function of $_{\ln c}$ for different factors of safety F

Table 6 shows the factor of safety and probability of failure for pillar strength as a function of $_{\ln c}$ for the particular case of $C.O.V._c = 0.4$. When $C.O.V._c$ and $_{\ln c}$ are known, a Factor of Safety can be chosen to meet the desired probability of failure or acceptable risk. For instance, if a target probability of failure of 1% is desired for $C.O.V._c = 0.4$ and $_{\ln c} = 0.2$, a Factor of Safety of at least $F = 1.5$ should be applied to the mean shear strength value. When $_{\ln c}$ is not known, a conservative estimate should made that would lead to the most conservative prediction. For instance, if a 1% probability of failure is acceptable for $C.O.V._c = 0.4$ with unknown $_{\ln c}$, a factor of safety of at least $F = 2.75$ is called for.

Table 6. Probability of Pillar Failure (%) for $C.O.V._c = 0.4$

			ln c		
F	0.10	0.20	1.00	2.00	10.0
1.0	99%	93%	67%	64%	60%
1.25	7%	27%	34%	36%	36%
1.50	0%	1%	13%	13%	20%
1.75	0%	0%	4%	7%	10%
2.00	0%	0%	1%	3%	5%
2.25	0%	0%	0%	1%	2%
2.50	0%	0%	0%	0%	1%
2.75	0%	0%	0%	0%	1%
3.00	0%	0%	0%	0%	0%

Figure 17 shows a plot of the results from Table 6.

5 Concluding remarks

The paper has shown that rock strength variability in the form of a spatially varying lognormal distribution can significantly reduce the compressive strength of an axially loaded rock pillar.

The following more specific conclusions can be made:

1. As the coefficient of variation of the rock strength increases, the expected compressive strength decreases. The decrease in compressive strength is greatest for small correlation lengths, however there appears to be a critical value of the spatial correlation length for which the reduction in mean compressive strength is greatest. It is speculated that as the correlation length becomes vanishingly small, the compressive strength rises again (slowly) towards the deterministic value.

2. The coefficient of variation of the compressive strength is observed to be positively correlated with both the spatial correlation length and the coefficient of variation of the rock strength.

3. The probability of failure is a function of m , s and the "target" design value $2/F$. The paper has shown that the interpretation of the probability of failure is most conveniently explained by comparing the target design value with the median of the lognormal distribution.

4. By interpreting the Monte-Carlo simulations in a probabilistic context, a direct relationship between the Factors of Safety and Probability of Failure can be established.

6 Appendix

A lognormal distribution for the rock shear strength c has been adopted in this study, meaning that $\ln c$ is normally distributed. If the mean and standard deviation of the shear strength are μ_c and σ_c respectively, then the standard deviation and mean of the underlying normal distribution of $\ln c$ are given by:

$$\sigma_{\ln c} = \sqrt{\ln \left\{ 1 + \quad \frac{\sigma_c}{\mu_c} \right)^2 \right\}} \qquad \text{(i)}$$

$$\mu_{\ln c} = \ln \mu_c \quad \frac{1}{2}\sigma^2_{\ln c} \qquad \text{(ii)}$$

and the probability density function of the lognormal distribution by:

$$f(c) = \frac{1}{c\, \sigma_{\ln c} \sqrt{2\pi}} \exp \left\{ \quad \frac{1}{2} \quad \frac{\ln c \quad \mu_{\ln c}}{\sigma_{\ln c}} \right)^2 \right\} \qquad \text{(iii)}$$

In terms of the properties of the underlying normal distribution, the properties of the lognormal distribution can therefore be summarized as:

$$\begin{aligned} \mu_c &= \exp \quad \mu_{\ln c} + \frac{1}{2}\sigma^2_{\ln c} \Big) && \text{(iv)} \\ \sigma_c &= \mu_c \sqrt{\exp(\sigma^2_{\ln c}) \quad 1} && \text{(v)} \\ \text{Median}_c &= \exp(\mu_{\ln c}) && \text{(vi)} \\ \text{Mode}_c &= \exp(\mu_{\ln c} \quad \sigma^2_{\ln c}) && \text{(vii)} \end{aligned}$$

7 Acknowledgement

The writers acknowledge the support of NSF Grant No. CMS-9877189.
This is a modified version of a paper that first appeared as, D.V. Griffiths, Gordon A. Fenton and Carisa B. Lemons (2002). "Probabilistic analysis of underground pillar stability" *Int. J. Numer. Anal. Meth. Geomech.* **26** , No. 8, 775-791

8 Notation

B	height (and width) of pillar
c	shear strength
$C.O.V$	coefficient of variation, $s\quad/m$
$C.O.V._{\ln c}$	coefficient of variation, σ_c/μ_c
F	factor of safety
$f(x)$	probability density function of x
m	sample mean of N_c
$m_{\ln}$	sample mean of $\ln N_c$
n	number of elements across the mesh
n_{sim}	number of Monte-Carlo simulations
N_c, N_c^i	bearing capacity factor, at the ith iteration
$p(x)$	probability of x
q_f, q_f^i	bearing capacity, at the ith iteration
s	sample standard deviation of N_c
$s_{\ln}$	sample standard deviation of $\ln N_c$
$_{\ln c}$	spatial correlation length of $\ln c$
$_{\ln c}$	non-dimensional spatial correlation $_{\ln c}/B$
μ_c	mean shear strength c
$\mu_{\ln c}$	mean of $\ln c$
ρ	correlation coefficient
σ_c	standard deviation of shear strength c
$\sigma_{\ln c}$	standard deviation of $\ln c$
τ	absolute distance between points
Φ	cumulative standard normal function

Bibliography

R.D. Call. Probability of stability design of open pit slopes. In *Proc Symp Geotech Div, Denver, Colorado*, pages 56–71. ASCE, 1985.

C. Cherubini. Reliability evaluation of shallow foundation bearing capacity on c , ϕ soils. *Can Geotech J*, 37(1):264–269, 2000.

Y. Dai, D.G. Fredlund, and W.J. Stolte. A probabilistic slope stability analysis using deterministic computer software. In K.S. Li and S-C. R. Lo, editors, *Probabilistic Methods in Geotechnical Engineering*, pages 267–274. Balkema, Rotterdam, 1993.

D.J. DeGroot. Analyzing spatial variability of in situ properties. In C.D. Shackelford *et al*, editor, *Geotechnical Special Publication No 58, Proceedings of Uncertainity '96 held in Madison, Wisconson, July 31 - August 3, 1996*, pages 210–238. ASCE, 1996.

G.A. Fenton. Error evaluation of three random field generators. *J Eng Mech, ASCE*, 120 (12):2478, 1994.

G.A. Fenton. *Simulation and analysis of random elds.* PhD thesis, Department of Civil Engineering and Operations Research, Princeton University, 1990.

G.A. Fenton and E.H. Vanmarcke. Simulation of random fields via local average subdivision. *J Eng Mech, ASCE*, 116(8):1733–1749, 1990.

M.E. Harr. *Reliability based design in civil engineering.* McGraw Hill, London, New York, 1987a.

M.E. Harr. *Reliability-based design in civil engineering.* McGraw-Hill Book Company, New York, 1987b.

E. Hoek. Reliability of Hoek-Brown estimates of rock mass properties and their impact on design. *International journal of Rock Mechanics, Mineral Science, and Geomechanics Abstr.*, 34(5):63–68, 1998.

E. Hoek and E. T. Brown. Practical estimates of rock mass strength. *International journal of rock mechanics and mining science*, 34(8):1165, 1997.

S. Lacasse and F. Nadim. Uncertainties in characterising soil properties. In C.D. Shackelford *et al*, editor, *Geotechnical Special Publication No 58, Proceedings of Uncertainity '96 held in Madison, Wisconson, July 31 - August 3, 1996*, pages 49–75. GSP 58, ASCE, 1996.

I. K. Lee, W. White, and O. G. Ingles. *Geotechnical Engineering.* Pitman, London, 1983.

P. Lumb. The variability of natural soils. *Can Geotech J*, 3(2):74–97, 1966.

P. Lumb. Safety factors and the probability distribution of soil strength. *Can Geotech J*, 7(3):225–241, 1970.

R. Mellah, G. Auvinet, and F. Masrouri. Stochastic finite element method applied to non-linear analysis of embankments. *Probablistic Engineering Mechanics*, 15:251–259, 2000.

G.R. Mostyn and K.S. Li. Probabilistic slope stability – State of play. In K.S. Li and S-C.R. Lo, editors, *Proc. Conf. Probabilistic Meths. Geotech. Eng.*, pages 89–110. A. A. Balkema, Rotterdam, 1993.

G. M. Paice and D. V. Griffiths. Bearing capacity reliability of an undrained clay block formed from spatially random soil. In P. Bettess, editor, *Proc. 7th Conf. Assoc. Comp. Mech. Eng. (ACME)*, pages 203–206. Penshaw Press, 1999.

D. Park. Numerical modeling as a tool for mine design. In *Prodeedings of the Workshop on Coal Pillar Mechanics and Design, 33rd U.S. Symposium on on Rock Mechanics*, pages 250–268. U.S. Bureau of Mines, Sante Fe, New Mexico, 1992.

S.S. Peng and D. Dutta. Evaluation of various pillar design methods: a case study. In *Prodeedings of the Workshop on Coal Pillar Mechanics and Design, 33rd U.S. Symposium on on Rock Mechanics*, pages 269–276. U.S. Bureau of Mines, Sante Fe, New Mexico, 1992.

K. Phoon and F.H. Kulhawy. Characterization of geotechnical variability. *Can Geotech J*, 36:612–624, 1999.

M.D.G. Salamon. Strength of coal pillars from back-calculation. In B. Amadei *et al*, editor, *Rock Mechanics for Industry*, pages 29–36. 1999.

J.P. Savely. Probabilistic analysis of intensely fractured rock masses. In *Proc. 6th Int. Cong. Rock Mech.*, volume 1, pages 509–514. 1987.

V.A. Scovazzo. A practitioner's approach to pillar design. In *Prodeedings of the Workshop on Coal Pillar Mechanics and Design, 33rd U.S. Symposium on on Rock Mechanics*, pages 277–282. U.S. Bureau of Mines, Sante Fe, New Mexico, 1992.

I. M. Smith and D. V. Griffiths. *Programming the Finite Element Method.* John Wiley and Sons, Chichester, New York, 3rd edition, 1998.

C.P. Tan, I.B. Donald, and R.E. Melchers. Probabilistic slip circle analysis of earth and rock fill dams. In K.S. Li and S-C. R. Lo, editors, *Probabilistic Methods in Geotechnical Engineering*, pages 281–288. Balkema, Rotterdam, 1993.

E.H. Vanmarcke. *Random elds: Analysis and synthesis.* The MIT Press, Cambridge, Mass., 1984.

D. Wickremesinghe and R.G. Campanella. Scale of fluctuation as a descriptor of soil variability. In K.S. Li and S-C. R. Lo, editors, *Probabilistic Methods in Geotechnical Engineering*, pages 233–239. Balkema, Rotterdam, 1993.

The Random Finite Element Method (RFEM) in Bearing Capacity Analyses

Gordon A. Fenton* and D.V. Griffiths†

* Department of Engineering Mathematics, Dalhousie University, Canada

† Division of Engineering, Colorado School of Mines, U.S.A.

Abstract In this chapter, soils with spatially varying shear strengths are modeled using random field theory and elasto-plastic finite element analysis to evaluate the extent to which spatial variability and cross-correlation in soil properties (c and ϕ) affect bearing capacity. The analysis is two dimensional, corresponding to a strip footing with infinite correlation length in the out-of-plane direction, and the soil is assumed to be weightless with footing placed on the soil surface. Theoretical predictions of the mean and standard deviation of bearing capacity, for the case where c and ϕ are independent, are derived using a geometric averaging model and then verified via Monte Carlo simulation. The standard deviation prediction is found to be quite accurate, while the mean prediction is found to require some additional semi-empirical adjustment to give accurate results for 'worst case' correlation lengths. Combined, the theory can be used to estimate the probability of bearing capacity failure, but also sheds light on the stochastic behaviour of foundation bearing failure.

1 Introduction

The design of a footing involves two limit states; a serviceability limit state, which generally translates into a maximum settlement or differential settlement, and an ultimate limit state. The latter is concerned with the maximum load which can be placed on the footing just prior to a bearing capacity failure. This chapter looks at the ultimate bearing capacity of a smooth strip footing founded on a soil having spatially random properties.

Most modern bearing capacity predictions involve a relationship of the form (Terzaghi, 1943)

$$q_f = cN_c + \bar{q}N_q + \tfrac{1}{2}\gamma BN_\gamma \tag{1}$$

where q_f is the ultimate bearing stress, c is the cohesion, $\bar{q}$ is the overburden stress, γ is the unit soil weight, B is the footing width, and N_c, N_q, and N_γ are the bearing capacity factors. To simplify the analysis in this chapter, and to concentrate on the stochastic behaviour of the most important term (at least as far as spatial variation is concerned), the soil is assumed weightless. Under this assumption, the bearing capacity equation simplifies to

$$q_f = cN_c \tag{2}$$

Bearing capacity predictions, involving specification of the N factors, are often based on plasticity theory (see, e.g., Prandtl, 1921, Terzaghi, 1943, and Sokolovski, 1965) of a rigid base punching into a softer material. These theories assume a *homogeneous* soil

The Random Finite Element Method (RFEM) in Bearing Capacity Analyses

Gordon A. Fenton* and D.V. Griffiths†

* Department of Engineering Mathematics, Dalhousie University, Canada

† Division of Engineering, Colorado School of Mines, U.S.A.

Abstract In this chapter, soils with spatially varying shear strengths are modeled using random field theory and elasto-plastic finite element analysis to evaluate the extent to which spatial variability and cross-correlation in soil properties (c and ϕ) affect bearing capacity. The analysis is two dimensional, corresponding to a strip footing with infinite correlation length in the out-of-plane direction, and the soil is assumed to be weightless with footing placed on the soil surface. Theoretical predictions of the mean and standard deviation of bearing capacity, for the case where c and ϕ are independent, are derived using a geometric averaging model and then verified via Monte Carlo simulation. The standard deviation prediction is found to be quite accurate, while the mean prediction is found to require some additional semi-empirical adjustment to give accurate results for 'worst case' correlation lengths. Combined, the theory can be used to estimate the probability of bearing capacity failure, but also sheds light on the stochastic behaviour of foundation bearing failure.

1 Introduction

The design of a footing involves two limit states; a serviceability limit state, which generally translates into a maximum settlement or differential settlement, and an ultimate limit state. The latter is concerned with the maximum load which can be placed on the footing just prior to a bearing capacity failure. This chapter looks at the ultimate bearing capacity of a smooth strip footing founded on a soil having spatially random properties.

Most modern bearing capacity predictions involve a relationship of the form (Terzaghi, 1943)

$$q_f = cN_c + \bar{q}N_q + \tfrac{1}{2}\gamma BN_\gamma \tag{1}$$

where q_f is the ultimate bearing stress, c is the cohesion, $\bar{q}$ is the overburden stress, γ is the unit soil weight, B is the footing width, and N_c, N_q, and N_γ are the bearing capacity factors. To simplify the analysis in this chapter, and to concentrate on the stochastic behaviour of the most important term (at least as far as spatial variation is concerned), the soil is assumed weightless. Under this assumption, the bearing capacity equation simplifies to

$$q_f = cN_c \tag{2}$$

Bearing capacity predictions, involving specification of the N factors, are often based on plasticity theory (see, e.g., Prandtl, 1921, Terzaghi, 1943, and Sokolovski, 1965) of a rigid base punching into a softer material. These theories assume a *homogeneous* soil underlying the footing – that is, the soil is assumed to have properties which are spatially constant. Under this assumption, most bearing capacity theories (e.g., Prandtl, 1921, and Meyerhof, 1951, 1963) assume that the failure slip surface takes on a logarithmic spiral shape to give

$$N_c = \frac{e^{\pi \tan\phi} \tan^2\left(\frac{\pi}{4} + \frac{\phi}{2}\right) - 1}{\tan\phi} \tag{3}$$

This relationship has been found to give reasonable agreement with test results (Bowles, 1996) under ideal conditions. In practice, however, it is well known that the actual failure conditions will be somewhat more complicated than a simple logarithmic spiral. Due to spatial variation in soil properties the failure surface under the footing will follow the weakest path through the soil, constrained by the stress field. For example, Figure 1 illustrates the bearing failure of a realistic soil with spatially varying properties. It can be seen that the failure surface only approximately follows a log-spiral on the right side and is certainly not symmetric. In this plot lighter regions represent stronger soil and darker regions indicate weaker soil. The weak (dark) region near the ground surface to the right of the footing has triggered a non-symmetric failure mechanism that is typically at a lower bearing load than predicted by traditional homogeneous and symmetric failure analysis.

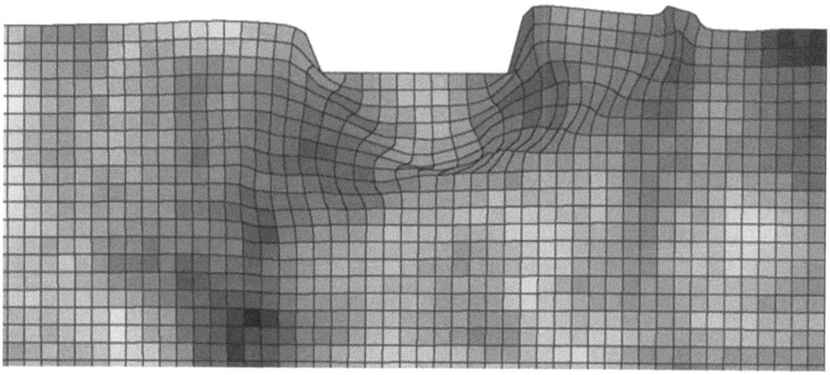

Figure 1. Typical deformed mesh at failure, where the darker regions indicate weaker soil.

The problem of finding the minimum strength failure slip surface through a soil mass is very similar in nature to the slope stability problem, and one which currently lacks a closed form stochastic solution, so far as the authors are aware. In this chapter the traditional relationships shown above will be used as a starting point to this problem.

For a realistic soil, both c and ϕ are random, so that both quantities in the right hand side of Eq. (2) are random. This equation can be non-dimensionalized by dividing through by the cohesion mean,

$$M_c = \frac{q_f}{\mu_c} = \frac{c}{\mu_c} N_c \tag{4}$$

where μ_c is the mean cohesion and M_c is the stochastic equivalent of N_c, ie., $q_f = \mu_c M_c$. The stochastic problem is now boiled down to finding the distribution of M_c. A theoretical model for the first two moments (mean and variance) of M_c, based on geometric averaging, are given in the next section. Monte Carlo simulations are then performed to assess the quality of the predictions and determine the approximate form of the distribution of M_c. This is followed by an example illustrating how the results can be used to compute the probability of a bearing capacity failure. Finally, an overview of the results is given, including their limitations.

2 The Random Soil Model

In this study, the soil cohesion, c, is assumed to be lognormally distributed with mean μ_c, standard deviation σ_c, and spatial correlation length $\theta_{\ln c}$. The lognormal distribution is selected because it is commonly used to represent non-negative soil properties and since it has a simple relationship with the normal. A lognormally distributed random field is obtained from a normally distributed random field, $G_{\ln c}(\underset{\sim}{x})$, having zero mean, unit variance, and spatial correlation length $\theta_{\ln c}$ through the transformation

$$c(\underset{\sim}{x}) = \exp\{\mu_{\ln c} + \sigma_{\ln c} G_{\ln c}(\underset{\sim}{x})\} \tag{5}$$

where $\underset{\sim}{x}$ is the spatial position at which c is desired. The parameters $\mu_{\ln c}$ and $\sigma_{\ln c}$ are obtained from the specified cohesion mean and variance using the lognormal distribution transformations,

$$\sigma^2_{\ln c} = \ln\left(1 + \frac{\sigma_c^2}{\mu_c^2}\right) \tag{6a}$$

$$\mu_{\ln c} = \ln \mu_c - \tfrac{1}{2}\sigma^2_{\ln c} \tag{6b}$$

The correlation coefficient between the log-cohesion at a point $\underset{\sim}{x}_1$ and a second point $\underset{\sim}{x}_2$ is specified by a correlation function, $\rho_{\ln c}(\tau)$, where $\tau = |\underset{\sim}{x}_1 - \underset{\sim}{x}_2|$ is the absolute distance between the two points. In this chapter, a simple exponentially decaying (Markovian) correlation function will be assumed, having the form

$$\rho_{\ln c}(\tau) = \exp\left(-\frac{2|\tau|}{\theta_{\ln c}}\right) \tag{7}$$

The spatial correlation length, $\theta_{\ln c}$, is loosely defined as the separation distance within which two values of $\ln c$ are significantly correlated. Mathematically, $\theta_{\ln c}$ is defined as the area under the correlation function, $\rho_{\ln c}(\tau)$ (Vanmarcke, 1984). (Note that geostatisticians often define the correlation length as the area under the non-negative half of the correlation function so that there is a factor of two difference between the two lengths – under their definition, the factor of 2 appearing in Eq. (7) is absent. The more general definition is retained here since it can be used also in higher dimensions where the correlation function is not necessarily symmetric in all directions about the origin.)

It should also be noted that the correlation function selected above acts between values of $\ln c$. This is because $\ln c$ is normally distributed and a normally distributed random field is simply defined by its mean and covariance structure. In practice, the correlation length $\theta_{\ln c}$ can be estimated by evaluating spatial statistics of the log-cohesion data directly (see, e.g., Fenton, 1999). Unfortunately, such studies are scarce so that little is currently known about the spatial correlation structure of natural soils. For the problem considered here, it turns out that a worst case correlation length exists which should be assumed in the absence of improved information.

The random field is also assumed here to be statistically isotropic (the same correlation length in any direction through the soil). Although the horizontal correlation length is often greater than the vertical, due to soil layering, taking this into account was deemed to be a refinement beyond the scope of this study. The main aspects of the stochastic behaviour of bearing capacity needs to be understood for the simplest case first and more complex variations on the theme, such as site specific anisotropy, left for later work.

The friction angle, ϕ, is assumed to be bounded both above and below, so that neither normal nor lognormal distributions are appropriate. A beta distribution is often used for bounded random variables. Unfortunately, a beta distributed random field has a very complex joint distribution and simulation is cumbersome and numerically difficult. To keep things simple, a bounded distribution is selected which resembles a beta distribution but which arises as a simple transformation of a standard normal random field, $G_\phi(\underset{\sim}{x})$, according to

$$\phi(\underset{\sim}{x}) = \phi_{min} + \tfrac{1}{2}(\phi_{max} - \phi_{min})\left\{1 + \tanh\left(\frac{sG_\phi(\underset{\sim}{x})}{2\pi}\right)\right\} \tag{8}$$

where ϕ_{min} and ϕ_{max} are the minimum and maximum friction angles, respectively, and s is a scale factor which governs the friction angle variability between its two bounds. Figure 2 shows how the distribution of ϕ (normalized to the interval [0, 1]) changes as s changes, going from an almost uniform distribution at $s = 5$ to a very normal looking distribution for smaller s. In all cases, the distribution is symmetric so that the midpoint between ϕ_{min} and ϕ_{max} is the mean. Values of s greater than about 5 lead to a U-shaped distribution (higher at the boundaries), which is not deemed realistic. Thus, varying s between about 0.1 and 5.0 leads to a wide range in the stochastic behaviour of ϕ.

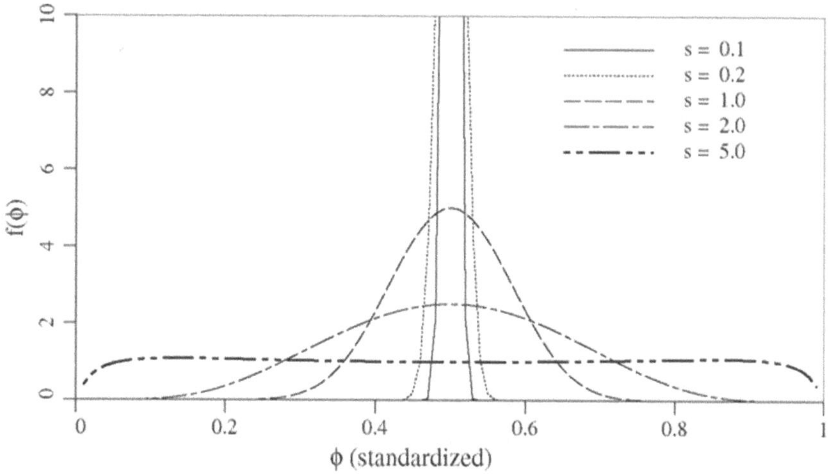

Figure 2. Bounded distribution of friction angle normalized to the interval [0, 1].

The random field, $G_\phi(\underset{\sim}{x})$, has zero mean and unit variance, as does $G_{\ln c}(\underset{\sim}{x})$. Conceivably, $G_\phi(\underset{\sim}{x})$ could also have its own correlation length θ_ϕ distinct from $\theta_{\ln c}$. However, it seems reasonable to assume that if the spatial correlation structure is caused by changes in the constitutive nature of the soil over space, then both cohesion and friction angle would have similar correlation lengths. Thus, θ_ϕ is taken to be equal to $\theta_{\ln c}$ in this study. Both lengths will be referred to generically from now on simply as θ, remembering that this length reflects correlation between points in the underlying normally distributed random fields, $G_{\ln c}(\underset{\sim}{x})$ and $G_\phi(\underset{\sim}{x})$, and not directly between points in the cohesion and friction fields. As mentioned above, both lengths can be estimated from data sets obtained over some spatial domain by statistically analyzing the suitably transformed data (inverses of Eq's 5 and 8). After transforming to the c and ϕ fields, the transformed correlation lengths will no longer be the same, but since both transformations are monotonic (ie. larger values of $G_{\ln c}$ give larger values of c, etc.), the correlation lengths will be similar (for $s = \text{C.O.V.} = 1.0$, the difference is less than 15% from each other and from the original correlation length). In that all engineering soil properties are derived through various transformations of the physical soil behaviour (eg. cohesion is a complex function of electrostatic forces between soil particles), the final correlation lengths between engineering properties cannot be expected to be identical, only similar. For the purposes of a generic non-site specific study, the above assumptions are believed reasonable.

The question as to whether the two parameters c and ϕ are correlated is still not clearly decided in the literature, and no doubt depends very much on the soil being studied. Cherubini (2000) quotes values of ρ ranging from -0.24 to -0.70, as does Wolff (1985) (see also Yuceman et al., 1973, Lumb, 1970, and Cherubini, 1997). As Wolff says (private correspondence, 2000),

> *The practical meaning of this [negative correlation] is that we are more certain of the undrained strength at a certain confining pressure than the values of the two parameters we use to define it.*

This observation arises from the fact that the variance of the shear strength is reduced if there is a negative correlation between c and ϕ.

In that the correlation between c and ϕ is not certain, this chapter investigates the correlation extremes to determine if cross-correlation makes a significant difference. As will be seen, under the given assumptions regarding the distributions of c (lognormal) and ϕ (bounded), varying the cross-correlation ρ from -1 to $+1$ was found to have only a minor influence on the stochastic behaviour of the bearing capacity.

3 Bearing Capacity Mean and Variance

The determination of the first two moments of the bearing capacity (mean and variance) requires first a failure model. Equations 2 and 3 assume that the soil properties are spatially uniform. When the soil properties are spatially varying, the slip surface no longer follows a smooth log-spiral and the failure becomes unsymmetric. The problem of finding the constrained path having the lowest total shear strength through the soil is mathematically difficult, especially since the constraints are supplied by the stress field. A simpler approximate model will be considered here wherein geometric averages of c and ϕ, over some region under the footing, are used in Equations 2 and 3. The geometric average is proposed because it is dominated more by low strengths than is the arithmetic average. This is deemed reasonable since the failure slip surface preferentially travels through lower strength areas.

Consider a soil region of some size D discretized into a sequence of non-overlapping rectangles, each centered on $\underset{\sim}{x}_i$, $i = 1, 2, \ldots, n$. The geometric average of the cohesion, c, over the domain D may then be defined as

$$\begin{aligned} \bar{c} &= \left[\prod_{i=1}^{n} c(\underset{\sim}{x}_i) \right]^{1/n} = \exp\left\{ \frac{1}{n} \sum_{i=1}^{n} \ln c(\underset{\sim}{x}_i) \right\} \\ &= \exp\left\{ \mu_{\ln c} + \sigma_{\ln c} \bar{G}_{\ln c} \right\} \end{aligned} \tag{9}$$

where $\bar{G}_{\ln c}$ is the *arithmetic* average of $G_{\ln c}$ over the domain D. Note that an assumption is made in the above concerning $c(\underset{\sim}{x}_i)$ being constant over each rectangle. In that cohesion is generally measured using some representative volume (eg. a lab sample), the values of $c(\underset{\sim}{x}_i)$ used above are deemed to be such measures.

In a similar way, the exact expression for the geometric average of ϕ over the domain D is

$$\bar{\phi} = \exp\left\{ \frac{1}{n} \sum_{i=1}^{n} \ln \phi(\underset{\sim}{x}_i) \right\} \tag{10}$$

where $\phi(\underset{\sim}{x}_i)$ is evaluated using Eq. (8). A close approximation to the above geometric average, accurate for $s \leq 2.0$, is

$$\bar{\phi} \simeq \phi_{min} + \tfrac{1}{2}(\phi_{max} - \phi_{min}) \left\{ 1 + \tanh\left(\frac{s\bar{G}_\phi}{2\pi} \right) \right\} \tag{11}$$

where $\bar{G}_\phi$ is the *arithmetic* average of G_ϕ over the domain D. For $\phi_{min} = 5^o$, $\phi_{max} = 45^o$, this expression has relative error of less than 5% for $n = 20$ independent samples. While the relative error rises to about 12%, on average, for $s = 5.0$, this is an extreme case, corresponding to a uniformly distributed ϕ between the minimum and maximum values, which is felt to be unlikely to occur very often in practice. Thus, the above approximation is believed reasonable in most cases.

Using the latter result in Eq. (3) gives the 'effective' value of N_c, $\bar{N}_c$, where the log-spiral model is assumed to be valid using a geometric average of soil properties within the failed region,

$$\bar{N}_c = \frac{e^{\pi \tan \bar{\phi}} \tan^2\left(\frac{\pi}{4} + \frac{\bar{\phi}}{2} \right) - 1}{\tan \bar{\phi}} \tag{12}$$

so that, now

$$M_c = \frac{\bar{c}}{\mu_c} \bar{N}_c \tag{13}$$

If c is lognormally distributed, an inspection of Eq. (9) indicates that $\bar{c}$ is also lognormally distributed. If we can assume that $\bar{N}_c$ is at least approximately lognormally distributed, then M_c will also be at least approximately lognormally distributed (the Central Limit Theorem helps out somewhat here). In this case, taking logarithms of Eq. (13) gives

$$\ln M_c = \ln \bar{c} + \ln \bar{N}_c - \ln \mu_c \tag{14}$$

so that, under the given assumptions, $\ln M_c$ is at least approximately normally distributed.

The task now is to find the mean and variance of $\ln M_c$. The mean is obtained by taking expectations of Eq. (14),

$$\mu_{\ln M_c} = \mu_{\ln \bar{c}} + \mu_{\ln \bar{N}_c} - \ln \mu_c \tag{15}$$

where

$$\begin{aligned} \mu_{\ln \bar{c}} &= \mathrm{E}\left[\mu_{\ln c} + \sigma_{\ln c}\bar{G}_{\ln c}\right] \\ &= \mu_{\ln c} + \sigma_{\ln c}\mathrm{E}\left[\bar{G}_{\ln c}\right] \\ &= \mu_{\ln c} \\ &= \ln \mu_c - \tfrac{1}{2}\ln\left(1 + \frac{\sigma_c^2}{\mu_c^2}\right) \end{aligned} \tag{16}$$

which used the fact that since $\bar{G}_{\ln c}$ is normally distributed, its arithmetic average has the same mean as $G_{\ln c}$, that is $\mathrm{E}\left[\bar{G}_{\ln c}\right] = \mathrm{E}\left[G_{\ln c}\right] = 0$. The above result is as expected since the geometric average of a lognormally distributed random variable preserves the mean of the logarithm of the variable. Also Eq. (6b) was used to express the mean in terms of the prescribed statistics of c.

A second order approximation to the mean of the logarithm of Eq. (12), $\mu_{\ln \bar{N}_c}$, is

$$\mu_{\ln \bar{N}_c} \simeq \ln \bar{N}_c(\mu_{\bar{\phi}}) + \sigma_{\bar{\phi}}^2 \left(\left. \frac{d^2 \ln \bar{N}_c}{d\bar{\phi}^2} \right|_{\mu_{\bar{\phi}}} \right) \tag{17}$$

where $\mu_{\bar{\phi}}$ is the mean of the geometric average of ϕ. Since $\bar{G}_\phi$ is an arithmetic average, its mean is equal to the mean of G_ϕ, which is zero. Thus, since the assumed distribution of ϕ is symmetric about its mean, $\mu_{\bar{\phi}} = \mu_\phi$ so that $\ln \bar{N}_c(\mu_{\bar{\phi}}) = \ln N_c(\mu_\phi)$.

A first order approximation to $\sigma_{\bar{\phi}}^2$ is

$$\sigma_{\bar{\phi}}^2 = \left[\frac{s}{4\pi} (\phi_{max} - \phi_{min}) \sigma_{\bar{G}_\phi} \right]^2 \tag{18}$$

where, from local averaging theory (Vanmarcke, 1984), the variance of a local average over the domain D is given by (recalling that G_ϕ is normally distributed with zero mean and unit variance),

$$\sigma_{\bar{G}_\phi}^2 = \sigma_{G_\phi}^2 \gamma(D) = \gamma(D) \tag{19}$$

where $\gamma(D)$ is the 'variance function' which reflects the amount that the variance is reduced due to local arithmetic averaging. It can be obtained directly from the correlation function (see Appendix I).

The derivative in Eq. (17) is most easily obtained numerically using any reasonably accurate (N_c is quite smooth) approximation to the second derivative. See, for example, Press et. al. (1997). If $\mu_{\bar{\phi}} = \mu_\phi = 25^o = 0.436$ radians (note that in all mathematical expressions, ϕ is assumed to be in radians), then

$$\left. \frac{d^2 \ln \bar{N}_c}{d\bar{\phi}^2} \right|_{\mu_{\bar{\phi}}} = 5.2984 \ \ (\mathrm{rad})^{-2} \tag{20}$$

Using these results with $\phi_{max} = 45^o$ and $\phi_{min} = 5^o$ so that $\mu_\phi = 25^o$ gives

$$\mu_{\ln \bar{N}_c} = \ln(20.72) + 0.0164 s^2 \gamma(D) \tag{21}$$

Some comments need to be made about this result: First of all it increases with increasing variability in ϕ (increasing s). It seems doubtful that this increase would occur since increasing variability in ϕ would likely lead to more lower strength paths through the soil mass for moderate θ. Aside from ignoring the weakest path issue, some other sources of error in the above analysis are

1) the geometric average of ϕ given by Eq. (10) actually shows a slight decrease with s (about 12% less, relatively, when $s = 5$). Although the decrease is only slight, it at least is in the direction expected.
2) an error analysis of the second order approximation in Eq. (17) and the first order approximation in Eq. (18) has not been carried out. Given the rather arbitrary nature of the assumed distribution on ϕ, and the fact that this chapter is primarily aimed at establishing the approximate stochastic behaviour, such refinements have been left for later work.

In light of these observations, a first order approximation to $\mu_{\ln \bar{N}_c}$ may actually be more accurate. Namely,

$$\mu_{\ln \bar{N}_c} \simeq \ln \bar{N}_c(\mu_{\bar{\phi}}) \simeq \ln N_c(\mu_\phi) \tag{22}$$

Finally, combining Equations (16) and (22) into Eq. (15) gives

$$\mu_{\ln M_c} \simeq \ln N_c(\mu_\phi) - \tfrac{1}{2} \ln \left(1 + \frac{\sigma_c^2}{\mu_c^2}\right) \tag{23}$$

For independent c and ϕ, the variance of $\ln M_c$ is

$$\sigma^2_{\ln M_c} = \sigma^2_{\ln \bar{c}} + \sigma^2_{\ln \bar{N}_c} \tag{24}$$

where

$$\sigma^2_{\ln \bar{c}} = \gamma(D)\sigma^2_{\ln c} = \gamma(D) \ln \left(1 + \frac{\sigma_c^2}{\mu_c^2}\right) \tag{25}$$

and, to first order,

$$\sigma^2_{\ln \bar{N}_c} \simeq \sigma^2_{\bar{\phi}} \left(\left. \frac{d \ln \bar{N}_c}{d\bar{\phi}} \right|_{\mu_{\bar{\phi}}} \right)^2 \tag{26}$$

The derivative appearing in Eq. (26), which will be denoted as $\beta(\phi)$, is

$$\begin{aligned} \beta(\phi) &= \frac{d \ln \bar{N}_c}{d\bar{\phi}} = \frac{d \ln N_c}{d\phi} \\ &= \frac{bd}{bd^2 - 1}\left[\pi(1+a^2)d + 1 + d^2\right] - \frac{1+a^2}{a} \end{aligned} \tag{27}$$

where $a = \tan(\phi)$, $b = e^{\pi a}$, and $d = \tan\left(\frac{\pi}{4} + \frac{\phi}{2}\right)$.

The variance of $\ln M_c$ is thus

$$\sigma^2_{\ln M_c} \simeq \gamma(D)\left\{\ln\left(1+\frac{\sigma_c^2}{\mu_c^2}\right)+\left[\left(\frac{s}{4\pi}\right)(\phi_{max}-\phi_{min})\beta(\mu_\phi)\right]^2\right\} \tag{28}$$

where ϕ is measured in radians.

4 Monte Carlo Simulation

A finite element computer program was written to compute the bearing capacity of a smooth rigid strip footing (plane strain) founded on a weightless soil with shear strength parameters c and ϕ represented by spatially varying and cross-correlated (point-wise) random fields, as discussed above. The bearing capacity analysis uses an elastic-perfectly plastic stress-strain law with a classical Mohr-Coulomb failure criterion. Plastic stress redistribution is accomplished using a viscoplastic algorithm. The program uses 8-node quadrilateral elements and reduced integration in both the stiffness and stress redistribution parts of the algorithm. The theoretical basis of the method is described more fully in Chapter 6 of the text by Smith and Griffiths (1998). The finite element model incorporates five parameters; Young's modulus (E), Poisson's ratio (ν), dilation angle (ψ), shear strength (c), and friction angle (ϕ). The program allows for random distributions of all five parameters, however in the present study, E, ν and ψ are held constant (at 100000 kN/m^2, 0.3, and 0, respectively) while c and ϕ are randomized. The Young's modulus governs the initial elastic response of the soil, but does not affect bearing capacity. Setting the dilation angle to zero means that there is no plastic dilation during yield of the soil. The finite element mesh consists of 1000 elements, 50 elements wide by 20 elements deep. Each element is a square of side length 0.1m and the strip footing occupies 10 elements, giving it a width of $B = 1$ m.

The random fields used in this study are generated using the Local Average Subdivision (LAS) method (Fenton, 1994, Fenton and Vanmarcke 1990). Cross-correlation between the two soil property fields (c and ϕ) is implemented via Covariance Matrix Decomposition (Fenton, 1994). The algorithm is given in Appendix II.

In the parametric studies that follow, the mean cohesion (μ_c) and mean friction angle (μ_ϕ) have been held constant at 100 kN/m^2 and 25° (with $\phi_{min} = 5^o$ and $\phi_{max} = 45^o$), respectively, while the C.O.V. ($= \sigma_c/\mu_c$), spatial correlation length (θ), and correlation coefficient, ρ, between $G_{\ln c}$ and G_ϕ are varied systematically according to Table 1.

Table 1. Random field parameters used in the study.

θ	=	0.5	1.0	2.0	4.0	8.0	50.
C.O.V.	=	0.1	0.2	0.5	1.0	2.0	5.0
ρ	=	−1.0	0.0	1.0			

It will be noticed that C.O.V.'s up to 5.0 are considered in this study, which is an order of magnitude higher than generally reported in the literature (see, eg. Phoon and Kulhawy, 1999). There are two considerations which complicate the problem of defining typical C.O.V.'s for soils that have not yet been clearly considered in the literature (although Fenton, 1999, does introduce these issues). The first has to do with the level of information known about a site. Prior to any site investigation, there will be plenty of uncertainty about soil properties, and an appropriate C.O.V. comes by using a C.O.V. obtained from regional data over a much larger scale. Such a C.O.V. will typically be much greater than that found when soil properties are estimated over a much smaller scale, such as a specific site. As investigation proceeds at the site of interest, the C.O.V. drops. For example, a single sample at the site will reduce the C.O.V. slightly, but as the investigation intensifies, the C.O.V. drops towards zero, reaching zero when the entire site has been sampled (which, of course, is clearly impractical). The second consideration, which is actually closely tied to the first, has to do with scale. If one were to take soil samples every 10 km over 5000 km (macroscale), one will find that the C.O.V. of those samples will be very large. A C.O.V. of 5.0 would not be unreasonable. Alternatively, suppose one were to concentrate one's attention on a single cubic metre of soil. If several 50 mm cubed samples were taken and sent to the laboratory, one would expect a fairly small C.O.V. On the other hand, if samples of size 0.1 μm cubed were taken and tested (assuming this was possible), the resulting C.O.V. could be very large since some samples might consist of very hard rock particles, others of water, and others just of air (ie. the sample location falls in a void). In such a situation, a C.O.V. of 5.0 could easily be on the low side. While the last scenario is only conceptual, it does serve to illustrate that C.O.V. is highly dependent on the ratio between sample volume and sampling domain volume. This dependence is certainly pertinent to the study of bearing capacity since it is currently not known at what scale bearing capacity failure operates. Is the weakest path through a soil dependent on property variations at the micro-scale (having large C.O.V.), or does the weakest path 'smear' the small-scale variations and depend primarily on local average properties over, say, laboratory scales (small C.O.V.)? Since laboratory scales are merely convenient for us, it is unlikely that nature has selected that particular scale to accommodate us. From the point of view of reliability estimates, where the failure mechanism might depend on microscale variations for failure initiation, the small C.O.V.'s reported in the literature might very well be dangerously unconservative. Much work is still required to establish the relationship between C.O.V., site investigation intensity, and scale. In the meantime, this chapter considers C.O.V.'s over a fairly wide range, since it is entirely possible that the higher values more truly reflect failure variability.

In addition, it is assumed that when the variability in the cohesion is large, the variability in the friction angle will also be large. Under this reasoning, the scale factor, s, used in Eq. (8) is set to $s = \sigma_c/\mu_c =$ C.O.V.. This choice is arbitrary, but results in the friction angle varying from quite narrowly (when C.O.V. $= 0.1$ and $s = 0.1$) to very widely (when C.O.V. $= 5.0$ and $s = 5$) between its lower and upper bounds, 5° and 45°, as illustrated in Figure 2.

For each set of assumed statistical properties given by Table 1, Monte-Carlo simulations have been performed. These involve 1000 realizations of the soil property random fields and the subsequent finite element analysis of bearing capacity. Each realization, therefore, has a different value of the bearing capacity and, after normalization by the *mean* cohesion, a different value of the bearing capacity factor,

$$M_{c_i} = \frac{q_{f_i}}{\mu_c}, \quad i = 1, 2, \ldots, 1000, \quad \Longrightarrow \quad \hat{\mu}_{\ln M_c} = \frac{1}{1000}\sum_{i=1}^{1000} \ln M_{c_i} \tag{29}$$

where $\hat{\mu}_{\ln M_c}$ is the sample mean of $\ln M_c$ estimated over the ensemble of realizations. Because of the non-linear nature of the analysis, the computations are quite intensive. One run of 1000 realizations typically takes about 2 days on a dedicated 800 MHz Pentium III computer (which, by the time of printing, is likely obsolete). For the 108 cases considered in Table 1, the total single CPU time required is about 220 days (run time varies with the number of iterations required to analyze various realizations).

4.1 Simulation Results

Figure 3(a) shows how the sample mean log-bearing capacity factor, taken as the average over the 1000 realizations of $\ln M_{c_i}$, and referred to as $\hat{\mu}_{\ln M_c}$ in the Figure, varies with correlation length, soil variability, and cross-correlation between c and ϕ. For small soil variability, $\hat{\mu}_{\ln M_c}$ tends towards the deterministic value of $\ln(20.72) = 3.03$, which is found when the soil takes on its mean properties everywhere. For increasing soil variability, the mean bearing capacity factor becomes quite significantly reduced from the traditional case. What this implies from a design standpoint is that the bearing capacity of a spatially variable soil will, on average, be *less* than the Prandtl solution based on the mean values alone. The greatest reduction from the Prandtl solution is observed for perfectly correlated c and ϕ ($\rho = +1$), the least reduction when c and ϕ are negatively correlated ($\rho = -1$), and the independent case ($\rho = 0$) lies between these two extremes. However, the effect of cross-correlation is seen to be not particularly large. If the negative cross-correlation indicated by both Cherubini (2000) and Wolff (1985) is correct, then the independent, $\rho = 0$, case is conservative, having mean bearing capacities consistently somewhat less than the $\rho = -1$ case.

The cross-correlation between c and ϕ is seen to have minimal effect on the sample standard deviation, $\hat{\sigma}_{\ln M_c}$, as shown in Figure 3(b). The sample standard deviation is most strongly affected by the correlation length and somewhat less so by the soil property variability. A decreasing correlation length results in a decreasing $\hat{\sigma}_{\ln M_c}$. As suggested by Eq. (28), the function $\gamma(D)$ decays approximately with θ/D and so decreases with decreasing θ. This means that $\hat{\sigma}_{\ln M_c}$ should decrease as the correlation length decreases, which is as seen in Figure 3(b).

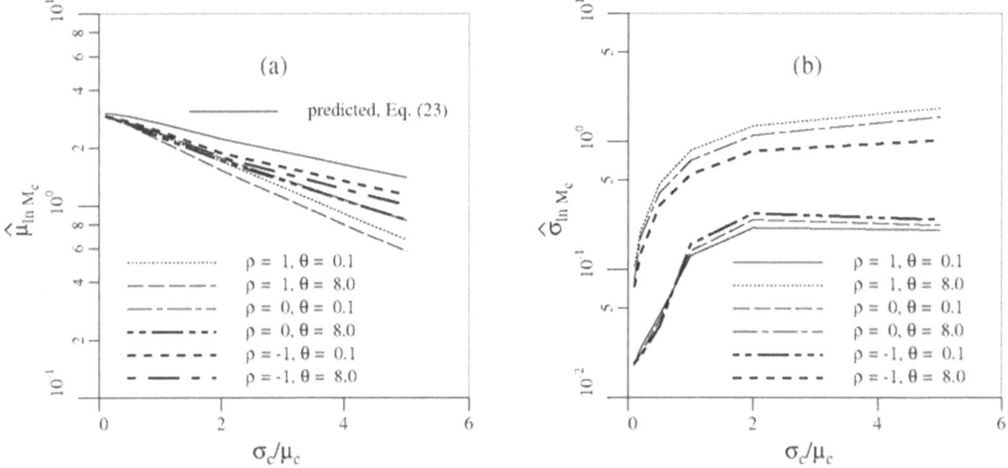

Figure 3. a) Sample mean of log-bearing capacity factor, $\ln M_c$, and b) its sample standard deviation.

Figure 3(a) also seems to show that the correlation length, θ, does not have a significant influence in that the $\theta = 0.1$ and $\theta = 8$ curves for $\rho = 0$ are virtually identical. However, the $\theta = 0.1$ and $\theta = 8$ curves are significantly lower than that predicted by Eq. (23) implying that the plot is somewhat misleading with respect to the dependence on θ. For example, when the correlation length goes to infinity, the soil properties become spatially constant, albeit still random from realization to realization. In this case, because the soil properties are spatially constant, the weakest path returns to the log-spiral and $\mu_{\ln M_c}$ will rise towards that given by Eq. (23), namely $\mu_{\ln M_c} = \ln(20.72) - \frac{1}{2}\ln(1 + \sigma_c^2/\mu_c^2)$, which is also shown on the plot. This limiting value holds because $\mu_{\ln N_c} \simeq \ln N_c(\mu_\phi)$, as discussed for Eq. (22), where for spatially constant properties $\bar{\phi} = \phi$.

Similarly, when $\theta \to 0$, the soil property field becomes infinitely "rough", in that all points in the field become independent. Any point at which the soil is weak will be surrounded by points where the soil is strong. A path through the weakest points in the soil might have very low average strength, but at the same time will become infinitely tortuous and thus infinitely long. This, combined with shear interlocking dictated by the stress field, implies that the weakest path should return to the traditional log-spiral with average shear strength along the spiral given by μ_ϕ and the median of c which is $\exp\{\mu_{\ln c}\}$. Again, in this case, $\mu_{\ln M_c}$ should rise to that given by Eq. (23).

The variation of $\mu_{\ln M_c}$ with respect to θ is more clearly seen in Figure 4. Over a range of values of σ_c/μ_c, the value of $\mu_{\ln M_c}$ rises towards that predicted by Eq. (23) at both high and low correlation lengths. At intermediate correlation lengths, the weakest path issue is seen to result in $\mu_{\ln M_c}$ being less than that predicted by Eq. (23) (see Figure 3a), the greatest reduction in $\mu_{\ln M_c}$ occurring when θ is of the same order as the footing width, B. It is hypothesized that $\theta \simeq B$ leads to the greatest reduction in $\mu_{\ln M_c}$ because it allows enough spatial variability for a failure surface which deviates somewhat from the log-spiral but which is not too long (as occurs when θ is too small) yet has significantly lower average strength than the $\theta \to \infty$ case. The apparent agreement between the $\theta = 0.1$ and $\theta = 8$ curves in Figure 3(a) is only because they are approximately equispaced on either side of the minimum at $\theta \simeq 1$.

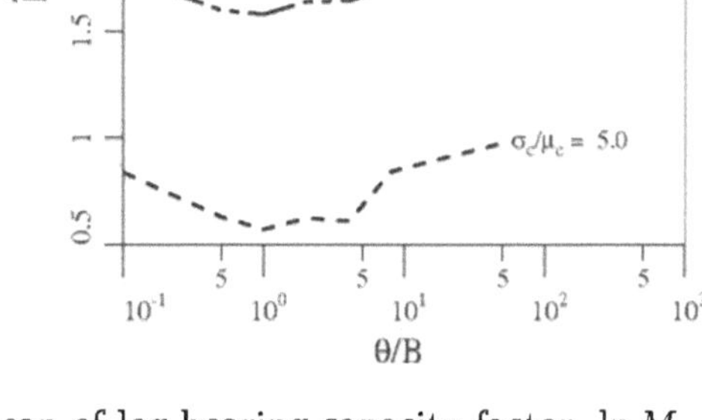

Figure 4. Sample mean of log-bearing capacity factor, $\ln M_c$, versus normalized correlation length.

As noted above, in the case where c and ϕ are independent ($\rho = 0$) the predicted mean, $\mu_{\ln M_c}$, given by Eq. (23) does not decrease as fast as observed in Figure 3(a) for intermediate correlation lengths. Nor does Eq. (23) account for changes in θ. Although an analytic prediction for the mean strength of the constrained weakest path through a spatially random soil has not yet been determined, Eq. (23) can be improved by making the following empirical corrections for the worst case ($\theta \simeq B$),

$$\mu_{\ln M_c} \simeq 0.92 \ln N_c(\mu_\phi) - 0.7 \ln\left(1 + \frac{\sigma_c^2}{\mu_c^2}\right) \tag{30}$$

where the overall reduction with σ_c/μ_c is assumed to follow the same form as predicted in Eq. (23). Some portion of the above correction may be due to finite element model error (for example, the finite element model slightly underestimates the deterministic value of N_c, giving $N_c = 19.6$ instead of 20.7, a 2% relative error in $\ln N_c$), but most is attributed to the weakest path issue and model errors arising by relating a spatial geometric average to a failure which is actually taking place along a curve through the 2-D soil mass.

Figure 5 illustrates the agreement between the sample mean of $\ln M_c$ and that predicted by Eq. (30) and between the sample standard deviation of $\ln M_c$ and Eq. (28) for $\rho = 0$. The estimated mean is seen to be in quite good agreement with the sample mean for all θ when $\sigma_c/\mu_c < 2$, and with the worst case ($\theta = B$) for $\sigma_c/\mu_c > 2$.

The predicted standard deviation was obtained by assuming a geometric average over a region under the footing of depth equal to the mean wedge zone depth,

$$w \simeq \tfrac{1}{2} B \tan\left(\frac{\pi}{4} + \frac{\mu_\phi}{2}\right) \tag{31}$$

and width of about $5w$. This is a rough approximation to the area of the failure region within the mean log-spiral curve on either side of the footing. Thus, D used in the variance function of Eq. (28) is a region of size $5w \times w$.

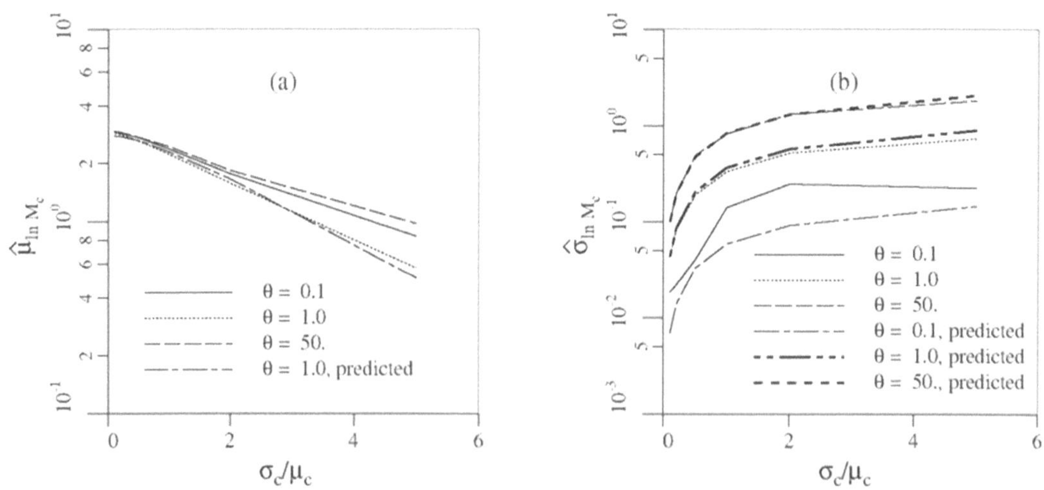

Figure 5. a) Sample and estimated mean (via Eq. 30) of $\ln M_c$, and b) its sample and estimated standard deviation (via Eq. 28).

Although Eq. (23) fails to reflect the effect of θ on the the reduction in the mean log-bearing capacity factor with increasing soil variability, the sample standard deviation is extremely well predicted by Eq. (28) – being only somewhat underpredicted for very small correlation lengths. To some extent the overall agreement in variance is as expected, since the variability along the weakest path will be similar to the variability along any nearby path through a statistically homogeneous medium.

The Monte Carlo simulation also allows the estimation of the probability density function of M_c. A Chi-Square goodness-of-fit test performed across all σ_c/μ_c, θ, and ρ parameter variations yields an average p-value of 33%. This is encouraging since large p-values indicate good agreement between the hypothesized distribution (lognormal) and the data. However, approximately 30% of the simulations had p-values less than 5%, indicating that a fair proportion of the runs had distributions that deviated from the lognormal to some extent. Some 10% of runs had p-values less than 0.01%. Figure 6(a) illustrates one of the better fits, with a p-value of 43% ($\sigma_c/\mu_c = 0.1$, $\theta = 4$, and $\rho = 0$), while Figure 6(b) illustrates one of the poorer fits, with a p-value of 0.01% ($\sigma_c/\mu_c = 5$, $\theta = 1$, and $\rho = 0$). It can be seen that even when the p-value is as low as 0.01%, the fit is still reasonable. There was no particular trend in degree of fit as far as the three parameters σ_c/μ_c, θ, and ρ was concerned. It appears, then, that M_c at least approximately follows a lognormal distribution. Note that if M_c does indeed arise from a geometric average of the underlying soil properties, c and N_c, then M_c will tend to a lognormal distribution by the Central Limit Theorem. It is also worth pointing out that this may be exactly why so many soil properties tend to follow a lognormal distribution.

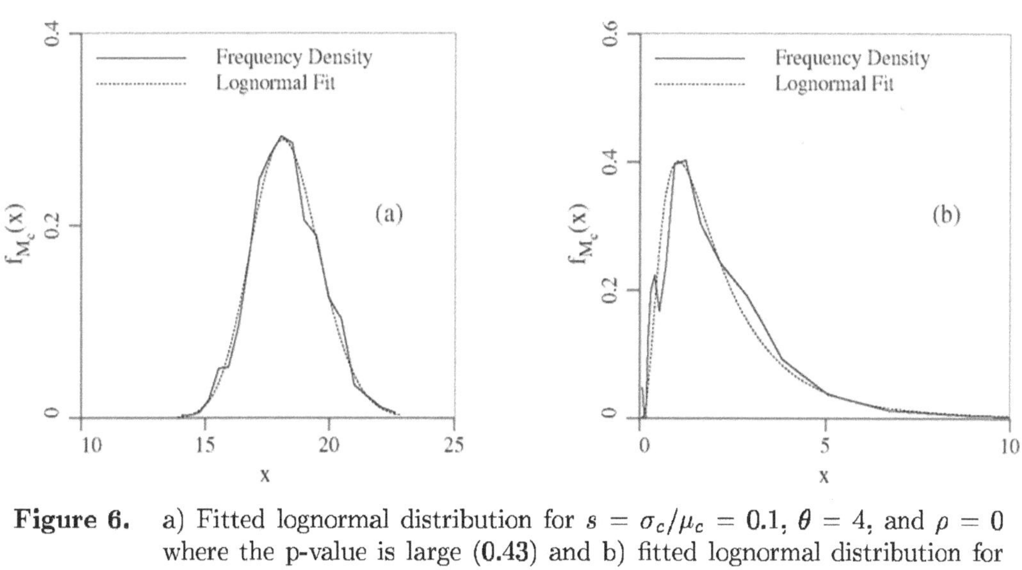

Figure 6. a) Fitted lognormal distribution for $s = \sigma_c/\mu_c = 0.1$, $\theta = 4$, and $\rho = 0$ where the p-value is large (0.43) and b) fitted lognormal distribution for $s = \sigma_c/\mu_c = 5$, $\theta = 1$, and $\rho = 0$ where the p-value is quite small (0.0001).

5 Probabilistic Interpretation

The results of the previous section indicated that Prandl's bearing capacity formula is still largely applicable in the case of spatially varying soil properties if geometrically averaged soil properties are used in the formula. The theoretical results presented above combined with the empirical correction to the mean proposed in the last section allows the approximate computation of probabilities associated with bearing capacity of a smooth strip footing. To illustrate this, consider an example strip footing of width $B = 2$ m founded on a weightless soil having $\mu_c = 75$ kPa, $\sigma_c = 50$ kPa, and $\theta = B = 2$ m (assuming the worst case correlation length). Assume also that the friction angle ϕ is independent of c (conservative assumption) and ranges from 5° to 35°, with mean 20° and $s = 1$. In this case, the deterministic value of N_c, based purely on μ_ϕ is

$$N_c(\mu_\phi) = \frac{e^{\pi \tan \mu_\phi} \tan^2\left(\frac{\pi}{4} + \frac{\mu_\phi}{2}\right) - 1}{\tan \mu_\phi} = 14.835 \tag{32}$$

so that, by Eq. (30),

$$\mu_{\ln \bar{M}_c} = 0.92 \ln(14.835) - 0.7 \ln\left(1 + \frac{50^2}{75^2}\right) = 2.2238 \tag{33}$$

For a footing width of $B = 2$, the wedge zone depth is

$$w = \tfrac{1}{2} B \tan\left(\frac{\pi}{4} + \frac{\mu_\phi}{2}\right) = \tan\left(\frac{\pi}{4} + \frac{20\pi}{360}\right) = 1.428 \tag{34}$$

Averaging over depth w by width $5w$ results in the variance reduction

$$\gamma(D) = \gamma(5w, w) = 0.1987$$

using the algorithm given in Appendix I for the Markov correlation function.

The slope of $\ln N_c$ at $\mu_\phi = 20^o$ is 3.62779 (rad^{-1}), using Eq. (27). These results applied to Eq. (28) give

$$\sigma^2_{\ln \bar{M}_c} = 0.1987 \left\{ \ln\left(1 + \frac{50^2}{75^2}\right) + \left[\frac{s}{4\pi}(\phi_{max} - \phi_{min})\beta(\mu_\phi\right]^2 \right\} = 0.07762 \tag{35}$$

so that $\sigma_{\ln \bar{M}_c} = 0.2778$.

The probability that M_c is less than half the deterministic value of N_c, based on μ_ϕ, is, then

$$\mathrm{P}\left[M_c \leq \frac{14.835}{2}\right] = \Phi\left(\frac{\ln(14.835/2) - \mu_{\ln M_c}}{\sigma_{\ln M_c}}\right) = \Phi(-0.79) = 0.215 \tag{36}$$

where Φ is the cumulative distribution function for the standard normal and where M_c is assumed lognormally distributed, as was found to be reasonable above. A simulation of the above problem yields $\mathrm{P}\left[M_c \leq \frac{14.835}{2}\right] = 0.2155$. Although this amazing agreement seems too good to be true, this is, in fact, the first example problem that the authors considered. The caveat, however, is that predictions derived from the results of a finite element program are being compared to the results of the same finite element program, albeit at different parameter values. Nevertheless, the fact that the agreement here is so good is encouraging since it indicates that the theoretical results given above may have some overall generality – namely that Prandtl's bearing capacity solution is applicable to spatially variable soils if the soil properties are taken from geometric averages, suitably modified to reflect weakest path issues. Inasmuch as the finite element method represents the actual soil behaviour, this observation seems reasonable.

6 Concluding Remarks

Most soil properties are local averages of some sort and are derived from measurements of properties over some finite volume. In the case of the shear resistance of a soil sample, tests involve determining the average shear resistance over some surface through the soil sample. Since this surface will tend to avoid the high strength areas in favour of low strength areas, the average will be less than a strictly arithmetic mean over a flat plane. Of the various common types of averages – arithmetic, geometric, and harmonic – the one that generally shows the best agreement with 'block' soil properties is the geometric average. The geometric average favours low strength areas, although not as drastically as does a harmonic average, lying between the arithmetic and harmonic averages.

The bearing capacity factor of Prandtl (1921) has been observed in practice to give reasonable agreement with test results, particularly under controlled conditions. When soil properties become spatially random, the failure surface migrates from the log-spiral surface to some nearby surface which is weaker. The results presented in this chapter indicate that the statistics of the resulting surface are well represented by geometrically averaging the soil properties over a domain of about the size of the plastically deformed bearing failure region (taken to be $5w \times w$ in this study). That is, that Prandtl's formula can be used to predict the statistics of bearing capacity if the soil properties used in the formula are based on geometric averages, with some empirical adjustment for the mean.

In this sense, the weakest path through the soil is what governs the stochastic bearing capacity behaviour. This means that the details of the distributions selected for c and ϕ are not particularly important, so long as they are physically reasonable, unimodal, and continuous. Although the lognormal distribution, for example, is mathematically convenient when dealing with geometric averages, very similar bearing capacity results are expected using other distributions, such as the normal distribution (suitably truncated to avoid negative strengths). The distribution selected for the friction angle basically resembles a truncated normal distribution over most values of s, but, for example, it is believed that a beta distribution could also have been used here without significantly affecting the results.

In the event that the soil is statistically anisotropic, that is that the correlation lengths differ in the vertical and horizontal directions, it is felt that the above results can still be used with some accuracy by using the algorithm of Appendex I with differing vertical and horizontal correlation lengths. However, some additional study is necessary to establish whether the mean bearing capacity in the anisotropic case is at least conservatively represented by Eq. (30).

Some limitations to this study are noted as follows;

1) The simulations were performed using a finite element analysis in which the values of the underlying normally distributed soil properties assigned to the elements are derived from arithmetic averages of the soil properties over each element domain. While this is believed to be a very realistic approach, intimately related to the

soil property measurement process, it is nevertheless an approach where geometric averaging is being performed at the element scale (at least for the cohesion – note that arithmetic averaging of a normally distributed field corresponds to geometric averaging of the associated lognormally distribution random field) in a method which is demonstrating that geometric averaging is applicable over the site scale. Although it is felt that the fine scale averaging assumptions should not significantly affect the large scale results through the finite element method, there is some possibility that there are effects that are not reflected in reality.

2) Model error has been entirely neglected in this analysis. That is, the ability of the finite element method to reflect the actual behaviour of an ideal soil, and the ability of Eq. (**3**) to do likewise have not been considered. It has been assumed that the finite element method and Eq. (**3**) are sufficiently reasonable approximations to the behaviour of soils to allow the investigation of the major features of stochastic soil behaviour under loading from a smooth strip footing. Note that the model error associated with traditional usage of Eq. (**3**) may be due in large part precisely to spatial variation of soil properties, so that this study may effectively be reducing, or at least quantifying, model error (although whether this is really true or not will have to wait until sufficient experimental evidence has been gathered).

The geometric averaging model has been shown to be a reasonable approach to estimating the statistics of bearing capacity. This is particularly true of the standard deviation. Some adjustment was required to the mean, since the geometric average was not able to completely account for the weakest path at intermediate correlation lengths. The proposed relationships for the mean and standard deviation, along with the simulation results indicating that the bearing capacity factor, M_c, is lognormally distributed, allow reasonably accurate calculations of probabilities associated with the bearing capacity. In the event that little is known about the cross-correlation of c and ϕ at a particular site, assuming that these properties are independent is deemed to be conservative (as long as the actual correlation is negative). In any case, the cross-correlation was not found to be a significant factor in the stochastic behaviour of bearing capacity.

Perhaps more importantly, since little is generally known about the correlation length at a site, the results of this study indicate that there exists a worst case correlation length of $\theta \simeq B$. Using this value, in the absence of improved information, allows conservative estimates of the probability of bearing failure. The estimate of the mean log-bearing capacity factor (Eq. **30**) is based on this conservative case.

Acknowledgements

The authors would like to thank the National Sciences and Engineering Research Council of Canada, under operating grant OPG0105445, and to the National Science Foundation of the United States of America, under grant CMS-9877189, for their essential support of this research. Any opinions, findings, conclusions or recommendations are those of the authors and do not necessarily reflect the views of the aforementioned organizations.

Notation

The following symbols are used in this paper:

$a = \tan\phi$ in Eq. (27)

$b=$ $e^{\pi \tan\phi}$ in Eq. (27)
$B=$ footing width
$c=$ cohesion
$\bar{c}=$ geometric average of cohesion field over domain D
$d=$ $\tan(\frac{\pi}{4}+\frac{\phi}{2})$ in Eq. (27)
$D=$ averaging domain ($5w \times w$)
$E=$ elastic modulus
$\mathrm{E}[\cdot]=$ expectation operator
$G_1(\underset{\sim}{x})=$ standard normal random field
$G_2(\underset{\sim}{x})=$ standard normal random field
$G_{\ln c}=$ standard normal random field (log-cohesion)
$G_\phi=$ standard normal random field (underlying friction angle)
$\bar{G}_{\ln c}=$ arithmetic average of $G_{\ln c}$ over domain D
$\bar{G}_\phi=$ arithmetic average of G_ϕ over domain D
$\underset{\approx}{L}=$ lower triangular matrix, square root of covariance matrix
$M_c=$ stochastic equivalent of the N_c factor
$M_{c_i}=$ ith realization of M_c
$N_c=$ N-factor associated with cohesion
$\bar{N}_c=$ cohesion N-factor based on a geometric average of cohesion
$N_q=$ N-factor associated with overburden
$N_\gamma=$ N-factor associated with the base width and unit weight
$q_f=$ ultimate bearing stress
$\bar{q}=$ overburden stress
$s=$ scale factor in distribution of ϕ
$\underset{\sim}{x}=$ spatial coordinate, (x_1, x_2) in 2-D
$\underset{\sim}{x}_i=$ spatial coordinate of the center of the ith element
$\beta(\phi)=$ derivated of N_c, with respect to ϕ, at ϕ
$\phi=$ friction angle (radians unless otherwise stated)
$\bar{\phi}=$ geometric average of ϕ over domain D
$\phi_{min}=$ minimum friction angle
$\phi_{max}=$ maximum friction angle
$\Phi=$ standard normal cumulative distribution function
$\gamma(D)=$ variance function giving variance reduction due to averaging over domain D
$\mu_c=$ cohesion mean
$\mu_{\ln c}=$ log-cohesion mean
$\mu_{\ln M_c}=$ mean of $\ln M_c$
$\hat{\mu}_{\ln M_c}=$ sample mean of $\ln M_c$ (from simulations)
$\mu_{\ln \bar{c}}=$ mean of the logarithm of $\bar{c}$
$\mu_{\ln \bar{N}_c}=$ mean of the logarithm of $\bar{N}_c$
$\mu_\phi=$ mean friction angle
$\mu_{\bar{\phi}}=$ mean of $\bar{\phi}$
$\nu=$ Poisson's ratio
$\theta=$ correlation length of the random fields
$\theta_{\ln c}=$ correlation length of the log-cohesion field
$\theta_\phi=$ correlation length of the G_ϕ field

ρ= correlation coefficient
$\rho_{\ln c}(\tau)$= correlation function giving correlation between two points in the log-cohesion field
$\underset{\approx}{\rho}$= correlation matrix
σ_c= cohesion standard deviation
$\sigma_{\ln c}$= log-cohesion standard deviation
$\sigma_{\ln \bar{c}}$= standard deviation of $\ln \bar{c}$
$\sigma_{\bar{\phi}}$= standard deviation of $\bar{\phi}$
σ_{G_ϕ}= standard deviation of G_ϕ (which is 1.0)
$\sigma_{\bar{G}_\phi}$= standard deviation of $\bar{G}_\phi$
$\sigma_{\ln M_c}$= standard deviation of $\ln M_c$
$\hat{\sigma}_{\ln M_c}$= sample standard deviation of $\ln M_c$ (from simulations)
τ= distance between two points in the soil domain
ψ= dilation angle

Bibliography

Bowles, J.E. . *Foundation Analysis and Design*, (5th Ed.), McGraw-Hill, New York, NY, 1996.

Cherubini, C. . Reliability evaluation of shallow foundation bearing capacity on c', ϕ' soils, *Canadian Geotechnical Journal*, **37**, 264–269, 2000.

Cherubini, C. . Data and considerations on the variability of geotechnical properties of soils., *Proceedings of the Int. Conf. on Safety and Reliability (ESREL) 97*, Vol. 2, Lisbon, 1583–1591, 1997.

Fenton, G.A. . Error evaluation of three random field generators, *ASCE Journal of Engineering Mechanics*, **120**(12), 2478–2497, 1994.

Fenton, G.A. and Vanmarcke, E.H. . Simulation of Random Fields via Local Average Subdivision, *ASCE Journal of Engineering Mechanics*, **116**(8), 1733–1749, 1990.

Fenton, G.A. . Estimation for stochastic soil models, *ASCE Journal of Geotechnical and Geoenvironmental Engineering*, **125**(6), 470–485, 1999.

Lumb, P. . Safety factors and the probability distribution of soil strength, *Canadian Geotechnical Journal*, **7**, 225–242, 1970.

Meyerhof, G. G. . Some recent research on the bearing capacity of foundations, *Canadian Geotechnical Journal*, 1(1), 16–26, 1963.

Meyerhof, G. G. . The ultimate bearing capacity of foundations, *Géotechnique*, **2**(4), 301–332, 1951.

Phoon, K-K. and Kulwawy, F.H. . Characterization of geotechnical variability, *Canadian Geotechnical Journal*, **36**, 612–624, 1999.

Prandtl, L. . Uber die Eindringungsfestigkeit (Harte) plastischer Baustoffe und die Festigkeit von Schneiden, *Zeitschrift fur angewandte Mathematik und Mechanik*, **1**(1), 15–20, 1921.

Press, W.H., Teukolsky, S.A., Vetterling, W.T. and Flannery, B.P. . *Numerical Recipes in C: The Art of Scientific Computing*, ((2nd Ed.)), Cambridge University Press, New York, 1997.

Smith, I.M. and Griffiths, D.V. . *Programming the Finite Element Method*, ((3rd Ed.)), John Wiley & Sons, New York, NY, 1998.

Sokolovski, V.V. . *Statics of Granular Media*, 270 pages, Pergamon Press, London, UK, 1965.
Terzaghi, K. . *Theoretical Soil Mechanics*, John Wiley & Sons, New York, NY, 1943.
Vanmarcke, E.H. . *Random Fields: Analysis and Synthesis*, The MIT Press, Cambridge, Massachusetts, 1984.
Wolff, T.H. . Analysis and design of embankment dam slopes: a probabilistic approach, Ph.D. Thesis, Purdue University, Lafayette, Indiana, 1985.
Yuceman, M.S., Tang, W.H. and Ang, A.H.S. . A probabilistic study of safety and design of earth slopes , Civil Engineering Studies, University of Illinois, Urbana, Structural Research Series 402, Urbana-Champagne, Illinois, 1973.

Appendix I

The variance reduction function $\gamma(D)$ gives the amount that the variance of a local average over the domain D is reduced from the point variance. If D is a rectangle of dimension $X \times Y$, then γ is actually a function of X and Y and is defined as

$$[A1] \qquad \gamma(X,Y) = \frac{1}{X^2Y^2} \int_0^X \int_0^X \int_0^Y \int_0^Y \rho(\xi_1 - \eta_1, \xi_2 - \eta_2)\, d\xi_1\, d\eta_1\, d\xi_1\, d\eta_2$$

where $\rho(\tau_1, \tau_2) = \rho_{\ln c}\left(\sqrt{\tau_1^2 + \tau_2^2}\right)$ (see Eq. 7). Since ρ is quandrant symmetric ($\rho(\tau_1,\tau_2) = \rho(-\tau_1,\tau_2) = \rho(\tau_1,-\tau_2) = \rho(-\tau_1,-\tau_2)$), the four-fold integration in Eq. (A1) can be reduced to a two-fold integration,

$$[A2] \qquad \gamma(X,Y) = \frac{4}{X^2Y^2} \int_0^X \int_0^Y (X - \tau_1)(Y - \tau_2)\, \rho(\tau_1, \tau_2)\, d\tau_1\, d\tau_2$$

which can be numerically calculated accurately and efficiently using a 5-point Gauss integration scheme as follows.

$$[A3] \qquad \gamma(X,Y) = \frac{1}{4} \sum_{i=1}^{5} w_i(1 - z_i) \sum_{j=1}^{5} w_j(1 - z_j)\rho(\xi_i, \eta_j)$$

where

$$\xi_i = \frac{X}{2}(1 + z_i)$$
$$\eta_j = \frac{Y}{2}(1 + z_j)$$

and the weights, w_i, and Gauss points, z_i, are as follows;

i	w_i	z_i
1	0.236926885056189	-0.906179845938664
2	0.478628670499366	-0.538469310105683
3	0.568888888888889	0.000000000000000
4	0.478628670499366	0.538469310105683
5	0.236926885056189	0.906179845938664

Appendix II

The cross-correlated random c and ϕ fields are obtained via Covariance Matrix Decomposition, as follows;

1) specify the cross-correlation coefficient, ρ $(-1 \leq \rho \leq 1)$, from statistical analyses. Three extreme cases are considered in this study: $\rho = -1$, 0 and 1, corresponding to completely negatively correlated, uncorrelated, and completely positively correlated, respectively.
2) form the correlation matrix between $G_{\ln c}(\underset{\sim}{x})$ and $G_{\phi}(\underset{\sim}{x})$, assumed to be stationary, ie. the same at all points $\underset{\sim}{x}$ in the field,

$$\underset{\approx}{\rho} = \begin{bmatrix} 1.0 & \rho \\ \rho & 1.0 \end{bmatrix}$$

3) compute the Cholesky decomposition of $\underset{\approx}{\rho}$. That is, find a lower triangular matrix $\underset{\approx}{L}$ such that $\underset{\approx}{L}\underset{\approx}{L}^T = \underset{\approx}{\rho}$. This is sometimes referred to as the square root of $\underset{\approx}{\rho}$. Note that when $\rho = \pm 1$, $\underset{\approx}{L}$ has the special form

$$\underset{\approx}{L} = \begin{bmatrix} 1.0 & 0.0 \\ \pm 1.0 & 0.0 \end{bmatrix}$$

4) generate two independent standard normally distributed random fields, $G_1(\underset{\sim}{x})$ and $G_2(\underset{\sim}{x})$, each having spatial correlation length θ (see Eq. 7).
5) at each spatial point, $\underset{\sim}{x}_i$, form the underlying point-wise correlated random fields

$$\begin{Bmatrix} G_{\ln c}(\underset{\sim}{x}_i) \\ G_{\phi}(\underset{\sim}{x}_i) \end{Bmatrix} = \begin{bmatrix} L_{11} & 0.0 \\ L_{21} & L_{22} \end{bmatrix} \begin{Bmatrix} G_1(\underset{\sim}{x}_i) \\ G_2(\underset{\sim}{x}_i) \end{Bmatrix}$$

6) use Eq's 5 and 8 to form the final c and ϕ random fields which are then mapped to the finite element mesh to specify the properties of each element.

The Random Finite Element Method (RFEM) in Slope Stability Analysis

D.V. Griffiths* and Gordon A. Fenton†

Division of Engineering, Colorado School of Mines, U.S.A.
Department of Engineering Mathematics, Dalhousie University, Canada

bstract The paper investigates the probability of failure of a cohesive slope using both simple and more advanced probabilistic analysis tools. The influence of local averaging on the probability of failure of a test problem is thoroughly investigated. In the simple approach, classical slope stability analysis techniques are used, and the shear strength is treated as a single random variable. The advanced method, called the random finite element method (RFEM), uses elasto-plasticity combined with random field theory. The RFEM method is shown to o er many advantages over traditional probabilistic slope stability techniques, because it enables slope failure to develop naturally by "seeking out" the most critical mechanism. Of particular importance in this work, is the conclusion that simplified probabilistic analysis, in which spatial variability is ignored by assuming perfect correlation, can lead to unconservative estimates of the probability of failure. This contradicts the findings of other investigators using classical slope stability analysis tools.

1 Introduction

Slope stability analysis is a branch of geotechnical engineering that is highly amenable to probabilistic treatment, and has received considerable attention in the literature. The earliest papers appeared in the 1970s (e.g. Matsuo and Kuroda 1974, Alonso 1976, Tang *et al.* 1976, Vanmarcke 1977) and have continued steadily (e.g. D'Andrea and Sangrey 1982, Li and Lumb 1987, Mostyn and Li 1993, Chowdhury and Tang 1987, Whitman 1984, Wolff 1996, Lacasse (1994), Christian *et al.* 1994, Christian 1996, Lacasse and Nadim (1996), Hassan and Wolff 1999, Duncan 2000). Most recently, El-Ramly *et al.* 2002 produced a useful review of the literature on this topic, and also noted that the geotechnical profession was slow to adopt probabilistic approaches to geotechnical design, especially in traditional problems such as slopes and foundations.

Two main observations can be made in relation to the existing body of work on this subject. First, the vast majority of probabilistic slope stability analyses, while using novel and sometimes quite sophisticated probabilistic methodologies, continue to use classical slope stability analysis techniques (e.g. Bishop 1955) that have changed little in decades, and were never intended for use with highly variable soil shear strength distributions. An obvious deficiency of the traditional slope stability approaches, is that the shape of the failure surface (e.g. circular) is often fixed by the method, thus the failure mechanism is not allowed to "seek out" the most critical path through the soil. Second, while the importance of spatial correlation (or auto-correlation) and local averaging of statistical geotechnical properties has long been recognized by some investigators (e.g. Mostyn and Soo 1990), it is still regularly omitted from many probabilistic slope stability analyses.

In recent years, the present authors have been pursuing a more rigorous method of probabilistic geotechnical analysis (e.g. Fenton and Griffiths 1993, Paice 1997, Griffiths and Fenton 2000), in which nonlinear finite element methods are combined with random field generation techniques. This method, called here the "Random Finite Element Method" (RFEM), fully accounts for spatial correlation and averaging, and is also a powerful slope stability analysis tool that does not require *a priori* assumptions relating to the shape or location of the failure mechanism.

In order to demonstrate the benefits of this method and to put it in context, this paper investigates the probabilistic stability characteristics of a cohesive slope using both the simple and more advanced methods. Initially, the slope is investigated using simple probabilistic concepts and classical slope stability techniques, followed by an investigation on the role of spatial correlation and local averaging. Finally, results are presented from a full-blown RFEM approach. Where possible throughout this paper, the Probability of Failure (p_f) is compared with the traditional Factor of Safety (FS) that would be obtained from charts or classical limit equilibrium methods.

The slope under consideration, known as the "Test Problem" is shown in Figure 1, and consists of undrained clay, with shear strength parameters $\phi_u = 0$ and c_u. In this study, the slope inclination and dimensions, given by β, H and D, and the saturated unit weight of the soil, γ_{sat} are held constant, while the undrained shear strength c_u is assumed to be a random variable. In the interests of generality, the undrained shear strength will be expressed in dimensionless form C, where $C = c_u/(\gamma_{sat}H)$.

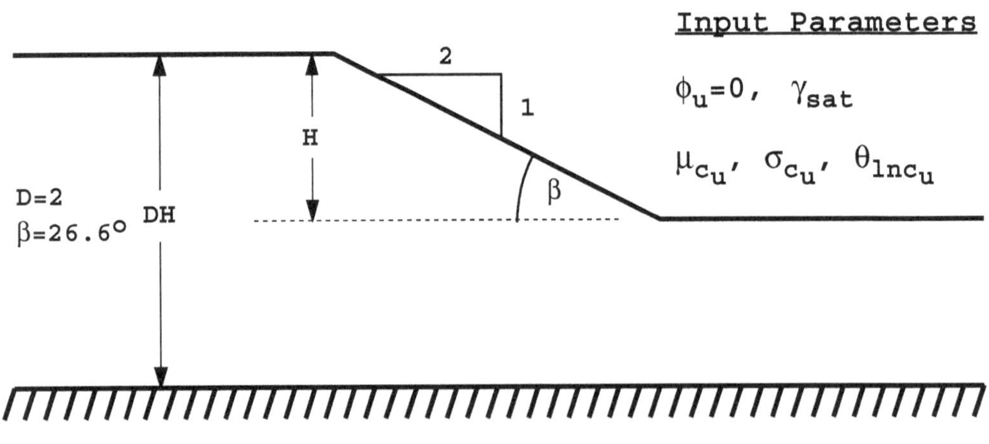

Figure 1 Cohesive slope test problem

2 Probabilistic description of shear strength

In this study, the shear strength C is assumed to be characterized statistically by a lognormal distribution defined by a mean, μ_C, and a standard deviation σ_C.

The probability density function of a lognormal distribution is given by,

$$f(C) = \frac{1}{C\,\sigma_{\ln C}\sqrt{2\pi}} \exp\left\{ \frac{1}{2} \left. \frac{\ln C \quad \mu_{\ln C}}{\sigma_{\ln C}} \right)^2 \right\} \tag{1}$$

shown in Figure 2 for a typical case with $\mu_C = 100$ kN/m^2 and $\sigma_C = 50$ kN/m^2. The function encloses an area of unity, thus the probability of the strength dropping below a given value is easily found from standard tables. The mean and standard deviation can conveniently be expressed in terms of the dimensionless coefficient of variation defined as

$$V_C = \frac{\sigma_C}{\mu_C} \tag{2}$$

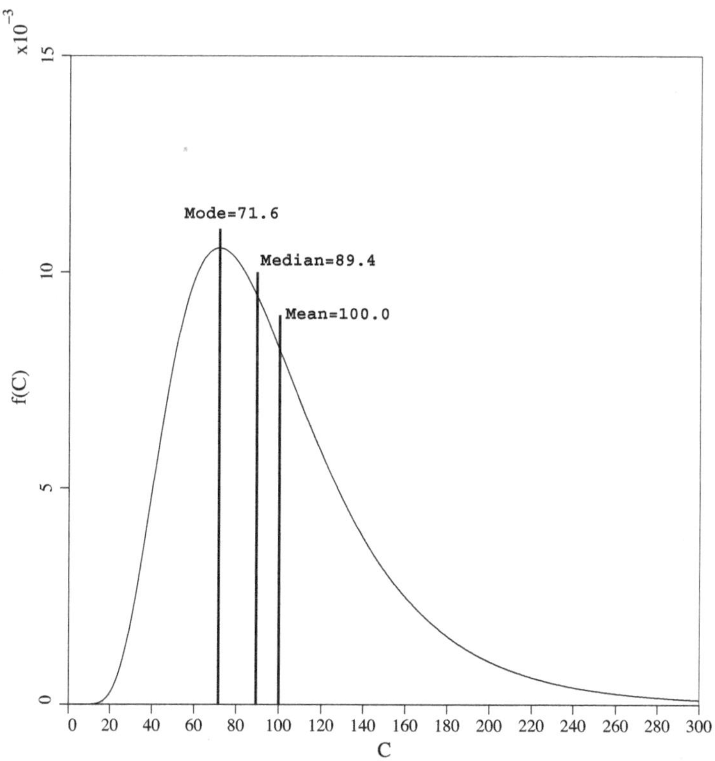

Figure 2 Typical log-normal distribution, with a mean of 100 and a standard deviation of 50 ($V_C = 0.5$)

Other useful relationships relating to the lognormal function include the standard deviation and mean of the underlying normal distribution as follows:

$$\sigma_{\ln C} = \sqrt{\ln\{1 + V_C^2\}} \tag{3}$$

$$\mu_{\ln C} = \ln \mu_C \quad \frac{1}{2}\sigma_{\ln C}^2 \tag{4}$$

Rearrangement of equations (3) and (4) gives the inverse relationships:

$$\mu_C = \exp \quad \mu_{\ln C} + \frac{1}{2}\sigma_{\ln C}^2\Big) \tag{5}$$

$$\sigma_C = \mu_C\sqrt{\exp(\sigma_{\ln C}^2) \quad 1} \tag{6}$$

Finally the median and mode of a lognormal distribution are given by:

$$\text{Median}_C = \exp(\mu_{\ln C}) \tag{7}$$

$$\text{Mode}_C = \exp(\mu_{\ln C} \quad \sigma_{\ln C}^2) \tag{8}$$

A third parameter, the spatial correlation length $\theta_{\ln C}$ will also be considered in this study. Since the actual undrained shear strength field is lognormally distributed, its logarithm yields an "underlying" normal distributed (or Gaussian) field. The spatial correlation length is measured with respect to this underlying field, that is, with respect to $\ln C$. In particular, the spatial correlation length ($\theta_{\ln C}$) describes the distance over which the spatially random values will tend to be significantly correlated in the underlying Gaussian field. Thus, a large value of $\theta_{\ln C}$ will imply a smoothly varying field, while a small value will imply a ragged field. The spatial correlation length can be estimated from a set of shear strength data taken over some spatial region simply by performing the statistical analyses on the log-data. In practice, however, $\theta_{\ln C}$ is not much different in magnitude from the correlation length in real space and, for most purposes, θ_C and $\theta_{\ln C}$ are interchangeable given their inherent uncertainty in the first place. In the current study, the spatial correlation length has been non-dimensionalized by dividing it by the height of the embankment H and will be expressed in the form,

$$_C = \theta_{\ln C}/H \tag{9}$$

It has been suggested (see e.g. Lee *et al* 1983, Kulhawy *et al* 1991) that typical V_C values for undrained shear strength lie in the range 0.1-0.5. The spatial correlation length however, is less well documented and may well exhibit anisotropy, especially in the horizontal direction. While the advanced analysis tools used later in this study have the capability of modeling an anisotropic spatial correlation field, the spatial correlation, when considered, will be assumed to be isotropic.

3 Preliminary Deterministic Study

To put the probabilistic analyses in context, an initial deterministic study has been performed assuming a homogeneous soil. For the simple slope shown in Figure 1, the

Factor of Safety can readily be obtained from Taylor's (1937) charts or simple limit equilibrium methods to give Table 1.

Table 1. Factors of Safety Assuming Homogeneous Soil

C	FS
0.15	0.88
0.17	1.00
0.20	1.18
0.25	1.47
0.30	1.77

These results, shown plotted in Figure 3, indicate the linear relationship between C and FS. The figure also shows that the test slope becomes unstable when the shear strength parameter falls below $C = 0.17$.

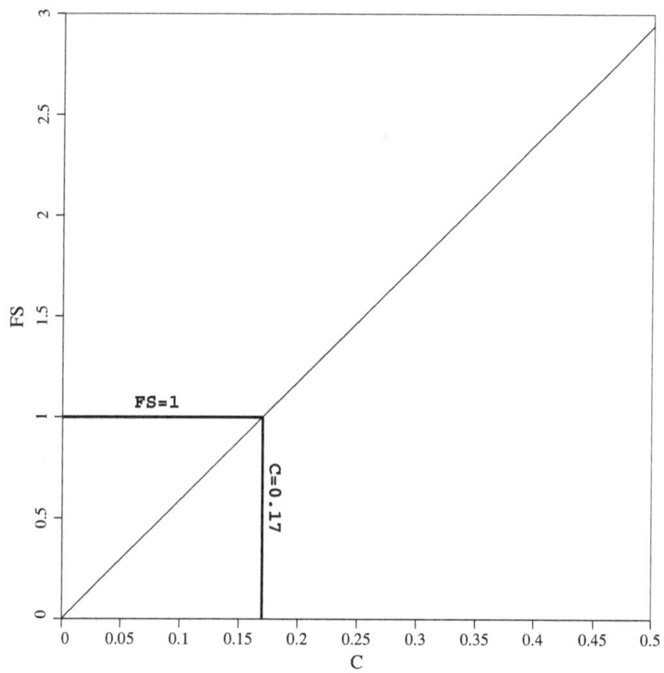

Figure 3 Linear relationship between FS and C for a cohesive slope with a slope angle of $\beta = 26.57^o$ and a depth ratio $D = 2$

4 Single random variable (SRV) approach

The first probabilistic analysis to be presented here investigates the influence of giving the shear strength C a lognormal probability density function similar to that shown in

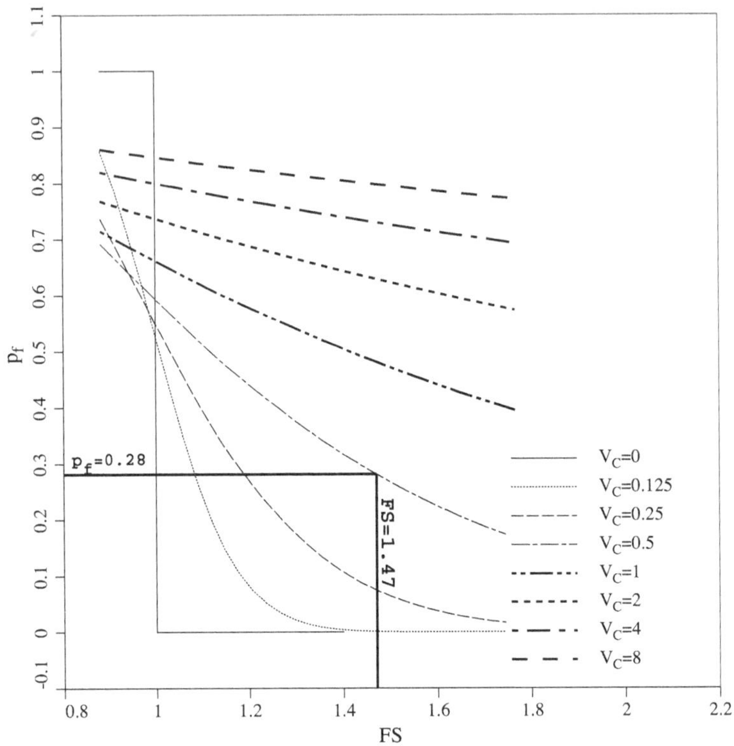

Figure 4 Probability of Failure vs. Factor of Safety (based on the mean) in a single random variable approach

Figure 2, based on a mean μ_C, and a standard deviation σ_C. The slope is assumed to have the same value of C everywhere, however the value of C is selected randomly from the lognormal distribution. Anticipating the random field analyses to be described later in this paper, this "single random variable approach" implies a spatial correlation length of $_C = \infty$, so no local averaging is applicable.

The Probability of Failure (p_f) in this case, is simply equal to the probability that the shear strength parameter C will be less than 0.17. Quantitatively, this equals the area of the probability density function corresponding to $C \leq 0.17$.

For example, if $\mu_C = 0.25$ and $\sigma_C = 0.125$ ($V_C = 0.5$), equations (3) and (4) give that the mean and standard deviation of the underlying normal distribution of the strength parameter are $\mu_{\ln C} =$ 1.498 and $\sigma_{\ln C} = 0.472$.

The Probability of Failure is therefore given by:

$$p_f = p[C < 0.17] = \Phi \quad \frac{\ln 0.17 \quad \mu_{\ln C}}{\sigma_{\ln C}} \Bigg) = 0.281 \tag{10}$$

where Φ is the cumulative standard normal distribution function.

This approach has been repeated for a range of μ_C and V_C values, for the slope under consideration, leading to Figure 4 which gives a direct relationship between the Factor of Safety and the Probability of Failure. It should be emphasized that the Factor of Safety in this plot is based on the value that would have been obtained if the slope had consisted of a homogeneous soil with a shear strength equal to the mean value μ_C from Figure 3.

From Figure 4, the Probability of Failure (p_f) clearly increases as the Factor of Safety decreases, however it is also shown that for $FS > 1$, the Probability of Failure increases as the V_C increases. The exception to this trend occurs when $FS < 1$. As shown in Figure 4, the Probability of Failure in such cases is understandably high, however the role of V_C is to have the opposite effect, with lower values of V_C tending to give the highest values of the probability of failure. This is explained by the "bunching up" of the shear strength distribution at low V_C rapidly excluding area to the right of the critical value of $C = 0.17$.

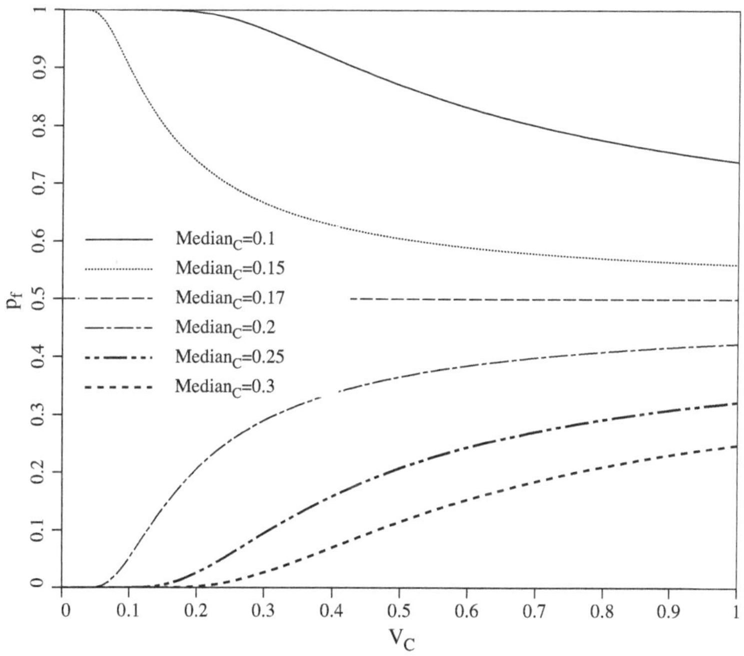

Figure 5 p_f vs. V_C for different Median$_C$ values

Figure 5 shows that the Median$_C$ is the key to understanding how the Probability of Failure changes in this analysis. When Median$_C < 0.17$, increasing V_C causes p_f to fall, whereas when Median$_C > 0.17$, increasing V_C causes p_f to rise.

While the single random variable approach described in this section leads to simple calculations, and useful qualitative comparisons between the Probability of Failure and the Factor of Safety, the quantitative value of the approach is more questionable. An important observation highlighted in Figure 4, is that a soil with a mean strength of $\mu_C = 0.25$ (implying $FS = 1.47$), would give a Probability of Failure as high as $p_f = 0.28$

for a soil with $V_C = 0.5$. Practical experience indicates that slopes with a Factor of Safety as high as $FS = 1.47$ rarely fail.

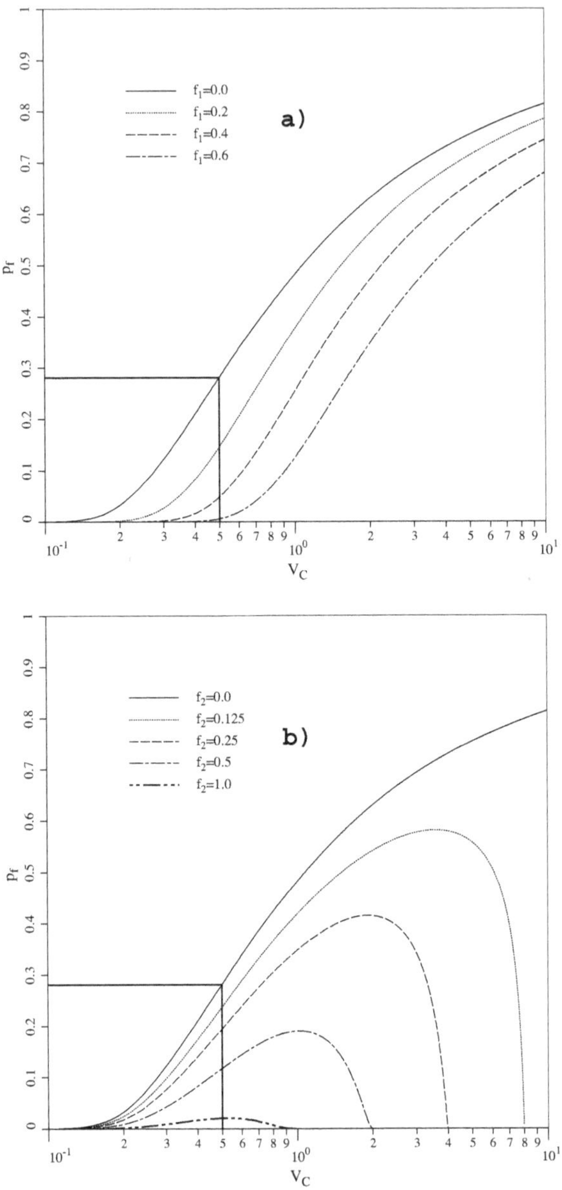

Figure 6 Influence of different mean strength factoring strategies on the Probability of Failure vs. Factor of Safety relationship. a) linear factoring b) standard deviation factoring. All curves assume FS=1.47 (based on $C_{des} = 0.25$)

An implication of this result is that either the perfectly correlated single random variable approach is entirely pessimistic in the prediction of the Probability of Failure, and/or it is unconservative to use the mean strength of a variable soil to estimate the Factor of Safety. Presented with a range of shear strengths at a given site, a geotechnical engineer would likely select a "pessimistic" or "lowest plausible" value for design, C_{des}, that would be lower than the mean. Assuming for the time being that the single random variable approach is reasonable, Figure 6 shows the influence on the Probability of Failure of two strategies for factoring the mean strength μ_C prior to calculating the Factor of Safety for the test problem.

In Figure 6a, a linear reduction in the mean strength has been proposed using a factor f_1, where:

$$C_{des} = \mu_C(1 \quad f_1) \tag{11}$$

and in Figure 6b, the mean strength has been reduced by a factor f_2 of the standard deviation, where:

$$C_{des} = \mu_C \quad f_2\sigma_C \tag{12}$$

All the results shown in Figure 6 assume that after factorization, $C_{des} = 0.25$, implying a Factor of Safety $FS = 1.47$. The Probability of Failure of $p_f = 0.28$ with no factorization $f_1 = f_2 = 0$, has also been highlighted for the case of $V_C = 0.5$. In both cases, an increase in the strength reduction factor reduces the Probability of Failure, which is to be expected, however the nature of the two sets of reduction curves is quite different, especially for higher values of V_C. From the linear mean strength reduction (equation 11), $f_1 = 0.6$ would result in a Probability of Failure of about 0.6%. By comparison, a mean strength reduction of one standard deviation given by $f_2 = 1$ (equation 12), would result in a Probability of Failure of about 2%. Figure 6a shows a gradual reduction of the Probability of Failure for all values of f_1 and V_C, however a quite different behavior is shown in Figure 6b, where standard deviation factoring results in a very rapid reduction in the Probability of Failure, especially for higher values of $V_C > 2$. This curious result is easily explained by the functional relationship between p_f and V_C, where the design strength can be written as:

$$C_{des} = 0.25 = \mu_C \quad f_2\sigma_C = \mu_C(1 \quad f_2V_C) \tag{13}$$

hence as $V_C \rightarrow 1/f_2$, $\mu_C \rightarrow \infty$. With the mean strength so much greater than the critical value of 0.17, the Probability of Failure falls very rapidly towards zero.

5 Spatial Correlation

Implicit in the single random variable approach described above, is that the spatial correlation length is infinite. In other words only homogeneous slopes are considered, in which the property assigned to the slope is taken at random from a lognormal distribution. A more realistic model would properly take account of smaller spatial correlation lengths in which the soil strength is allowed to vary spatially within the slope. The parameter that controls this is the spatial correlation length $\theta_{\ln C}$ as discussed previously. In this work, an exponentially decaying (Markovian) correlation function is used of the form:

$$\rho = e^{-\frac{\tau}{\ln C}} \tag{14}$$

where ρ is the familiar correlation coefficient, and τ is the absolute distance between two points in the random field. A plot of this function is given in Figure 7 and indicates, for example, that the strength at two points separated by $\theta_{\ln C}$ $(\tau/\theta_{\ln C} = 1)$ will have an expected correlation of $\rho = 0.135$. This correlation function is merely a way of representing the field observation that soil samples taken close together are more likely to have similar properties, than samples taken from far apart. There is also the issue of anisotropic spatial correlation, in that soil is likely to have longer spatial correlation lengths in the horizontal direction than in the vertical, due to the depositional history. While the tools described in this paper can take account of anisotropy, this refinement is left for future studies.

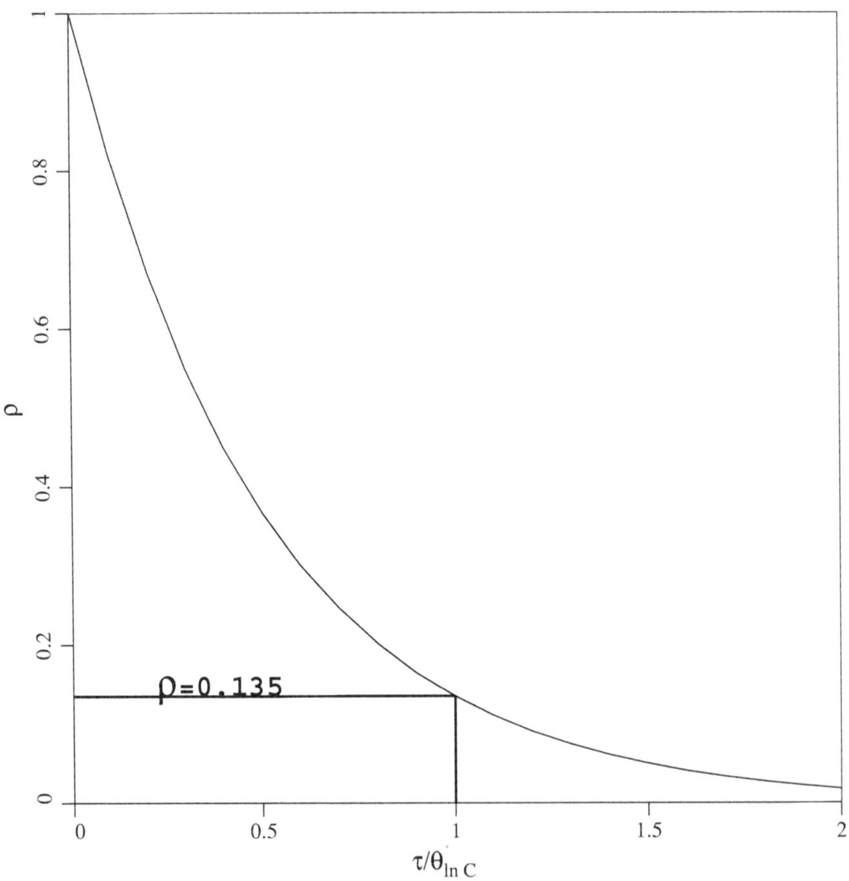

Figure 7 Markov correlation function

6 The Random Finite Element Method (RFEM)

A powerful and general method of accounting for spatially random shear strength parameters and spatial correlation, is the Random Finite Element Method (RFEM) which combines elasto-plastic finite element analysis with random field theory generated using the Local Average Subdivision Method (Fenton and Vanmarcke 1990). The methodology has been described in detail in other publications (e.g. Griffiths and Fenton 2001), so only a brief description will be repeated here.

A typical finite element mesh for the test problem considered in this paper is shown in Figure 8. The majority of the elements are square, however the elements adjacent to the slope are degenerated into triangles.

The code developed by the authors enables a random field of shear strength values to be generated and mapped onto the finite element mesh, taking full account of element size in the local averaging process. In a random field, the value assigned to each cell (or finite element in this case) is itself a random variable, thus the mesh of Figure 8, which has 910 finite elements, contains 910 random variables.

The random variables can be correlated to one another by controlling the spatial correlation length $\theta_{\ln C}$ as described previously, hence the single random variable approach discussed in the previous section where the spatial correlation length is implicitly set to infinity, can now be viewed as a special case of a much more powerful analytical tool. Figures 9a and b show typical meshes corresponding to different spatial correlation lengths. Figure 9a shows a relatively low spatial correlation length of $_C = 0.2$ and Figure 9b shows a relatively high spatial correlation length of $_C = 2$. Dark and light regions depict "weak" and "strong" soil respectively. It should be emphasized that both these shear strength distributions come from the same lognormal distribution, and it is only the spatial correlation length that is different.

In brief, the analyses involve the application of gravity loading, and the monitoring of stresses at all the Gauss points. The slope stability analyses use an elastic-perfectly plastic stress-strain law with a Tresca failure criterion which is appropriate for "undrained clays". If the Tresca criterion is violated, the program attempts to redistribute excess stresses to neighboring elements that still have reserves of strength. This is an iterative process which continues until the Tresca criterion and global equilibrium are satisfied at all points within the mesh under quite strict tolerances.

Plastic stress redistribution is accomplished using a viscoplastic algorithm with 8-node quadrilateral elements and reduced integration in both the stiffness and stress redistribution parts of the algorithm. The theoretical basis of the method is described more fully in Chapter 6 of the text by Smith and Griffiths (1998), and for a detailed discussion of the method applied to slope stability analysis, the reader is referred to Griffiths and Lane (1999).

For a given set of input shear strength parameters (mean, standard deviation and spatial correlation length), Monte-Carlo simulations are performed. This means that the slope stability analysis is repeated many times until the statistics of the output quantities of interest become stable. Each "realization" of the Monte-Carlo process differs in the locations at which the strong and weak zones are situated. For example, in one realization, weak soil may be situated in the locations where a critical failure

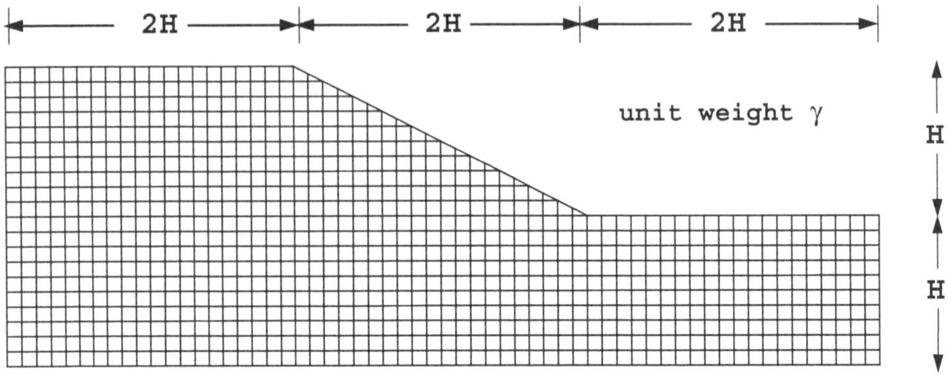

Figure 8 Mesh used for RFEM slope stability analyses

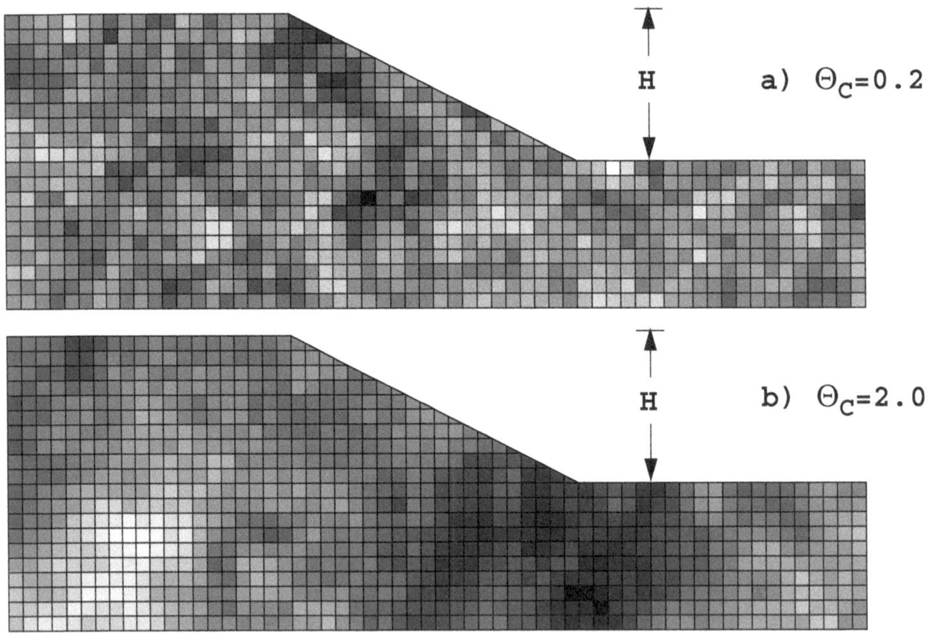

Figure 9 Influence of the scale of fluctuation in RFEM analysis

mechanism develops causing the slope to fail, whereas in another, strong soil in those locations means that the slope remains stable.

In this study, it was determined that 1000 realizations of the Monte-Carlo process for each parametric group, was sufficient to give reliable and reproducible estimates of the Probability of Failure, which was simply defined as the proportion of the 1000 Monte-Carlo slope stability analyses that failed.

In this study, "failure" was said to have occurred if, for any given realization, the algorithm was unable to converge within 500 iterations. While the choice of 500 as the iteration ceiling is subjective, Figure 10 confirms, for the case of $\mu_C = 0.25$ and $\ _C = 1$, that the Probability of Failure defined this way, is stable for iteration ceilings greater than about 200.

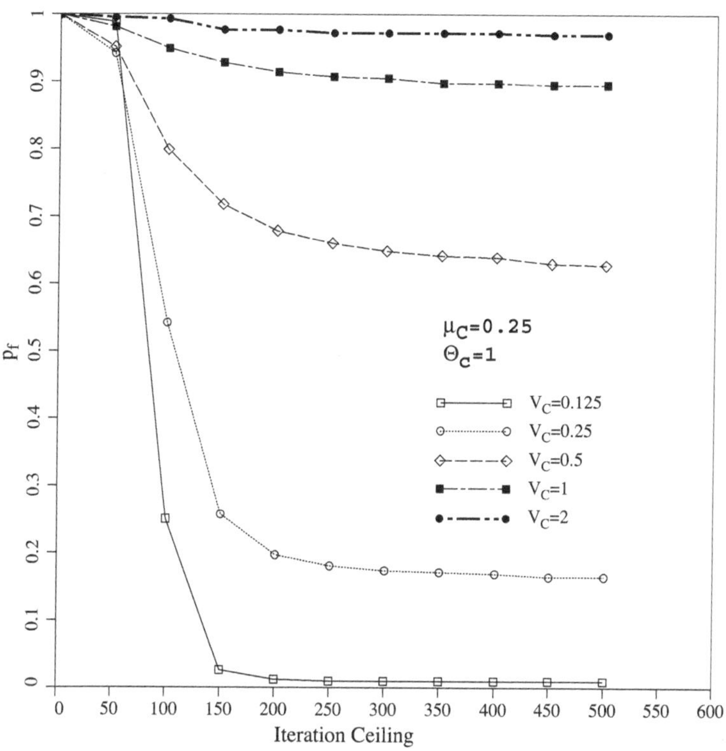

Figure 10 Influence the plastic iteration ceiling on the computed Probability of Failure

7 Local Averaging

The input parameters relating to the mean, standard deviation and spatial correlation length of the undrained strength, are assumed to be defined at the point level. While statistics at this resolution are obviously impossible to measure in practice, they represent a fundamental baseline of the inherent soil variability which can be corrected through local averaging to take account of the sample size.

In the context of the RFEM approach, each element is assigned a constant property at each realization of the Monte-Carlo process. The "sample" is represented by the size of each finite elements used to discretize the slope. If the point distribution is normal, local averaging results in a reduced variance but the mean is unaffected. In a lognormal distribution however, both the mean and the standard deviation are reduced by local averaging. This is because from equations (5) and (6), the mean of a lognormal relationship depends on both the mean *and* the variance of the underlying normal log-relationship. Thus the cruder the discretization of the slope stability problem and the larger the elements, the greater the influence of local averaging in the form of a reduced mean and standard deviation. These adjustments to the point statistics are fully accounted for in the RFEM, and are implemented before the elasto-plastic finite element slope stability analysis takes place.

8 Variance reduction over a square nite element

In this section, the algorithm used to compute the locally averaged statistics applied to the mesh is described.

A lognormal distribution of a random variable C, with point statistics given by a mean μ_C, a standard deviation σ_C and spatial correlation length $\theta_{\ln C}$, is to be mapped onto a mesh of square finite elements. Each element will be assigned a single value of the undrained strength parameter.

The locally averaged statistics over the elements will be referred to here as the "area" statistics with the subscript A. Thus, with reference to the underlying normal distribution of $\ln C$, the mean, which is unaffected by local averaging, is given by $\mu_{\ln C_A}$, and the standard deviation, which is affected by local averaging is given by $\sigma_{\ln C_A}$.

The variance reduction factor due to local averaging γ, is defined:

$$\gamma = \frac{\sigma_{\ln C_A}}{\sigma_{\ln C}}\Bigg)^2 \tag{15}$$

and is a function of the element size and the correlation function from equation (14), repeated here in the form,

$$\rho = \exp\left\{ \frac{2}{\theta_{\ln C}}\sqrt{\tau_x^2 + \tau_y^2}\right\} \tag{16}$$

where τ_x is the difference between the x coordinates of any two points in the random field, and τ_y is the difference between the y-coordinates.

For a square finite element of side length $\theta_{\ln C}$ as shown in Figure 11, it can be shown (Vanmarcke 1984) that for an isotropic spatial correlation field, the variance reduction factor is given by:

$$\gamma = \frac{4}{(\;\theta_{\ln C})^4}\int_0^{\theta_{\ln C}}\int_0^{\theta_{\ln C}} \exp\left\{ \frac{2}{\theta_{\ln C}}\sqrt{x^2+y^2}\right\}(\;\theta_{\ln C}\;\;x)(\;\theta_{\ln C}\;\;y)\,dx\,dy \tag{17}$$

Numerical integration of this function leads to the variance reduction values given in Table 2, and shown plotted in Figure 11.

Table 2. Variance reduction over a square element

	γ
0.01	0.9896
0.1	0.9021
1	0.3965
10	0.0138

The figure indicates that elements that are small relative to the correlation length ($\to 0$) lead to very little variance reduction ($\gamma \to 1$), whereas elements that are large relative to the correlation length can lead to very significant variance reduction ($\gamma \to 0$).

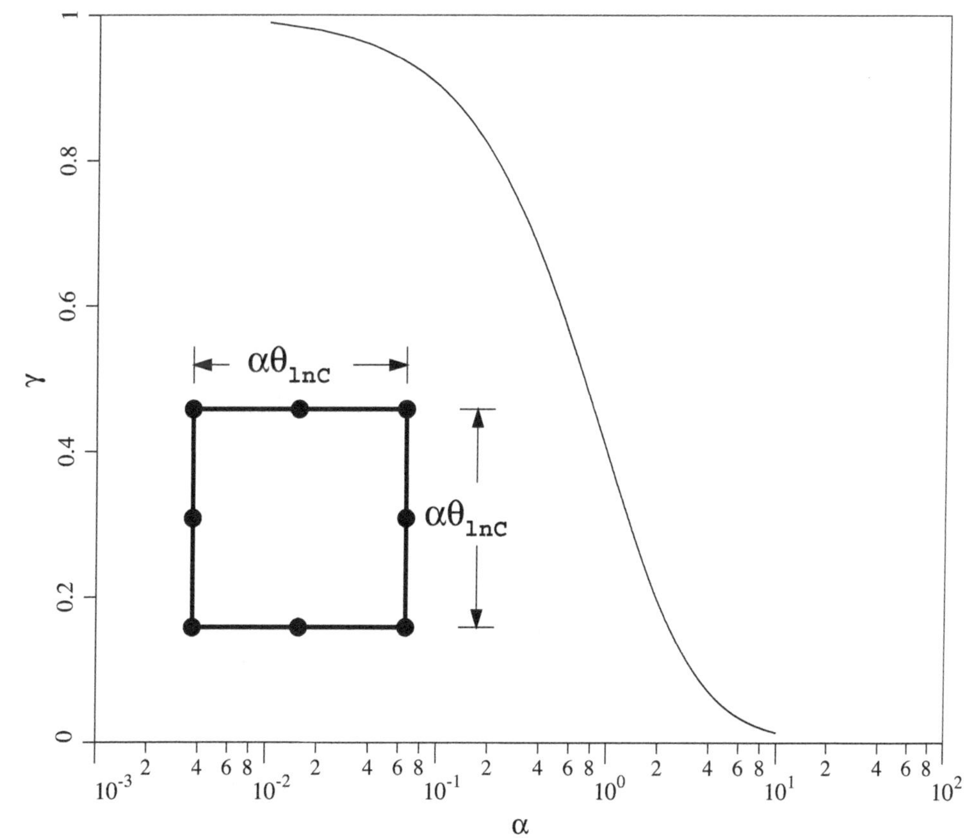

Figure 11 Variance reduction function over a square element of side length $\theta_{\ln C}$ with a Markov correlation function

The statistics of the underlying log-field, including local averaging, are therefore given by:

$$\sigma_{\ln C_A} = \sigma_{\ln C}\sqrt{\gamma} \tag{18}$$

and

$$\mu_{\ln C_A} = \mu_{\ln C} \tag{19}$$

which leads to the following statistics of the lognormal field, including local averaging, that is actually mapped onto the finite element mesh from equations (5) and (6), thus

$$\mu_{C_A} = \exp \quad \mu_{\ln C_A} + \frac{1}{2}\sigma^2_{\ln C_A}\Big) \tag{20}$$

$$\sigma_{C_A} = \mu_{C_A}\sqrt{\exp(\sigma^2_{\ln C_A}) \quad 1} \tag{21}$$

It is instructive to consider the range of locally averaged statistics, since this helps to explain the influence of the spatial correlation length $\quad_C(= \theta_{\ln C}/H)$ on the Probability of Failure in the RFEM slope analyses described in the next section.

Expressing the mean and the coefficient of variation of the locally averaged variable as a proportion of the point values of these quantities, leads to Figures 12a and 12b respectively. In both cases, there is virtually no reduction due to local averaging for elements that are small relative to the spatial correlation length ($\quad \rightarrow 0$). This is to be expected, since the elements are able to model the point field quite accurately. For larger elements relative to the spatial correlation length however, Figure 12a indicates that the average of the locally averaged field tends to a constant equal to the median, and Figure 12b indicates that the coefficient of variation of the locally averaged field tends to zero.

From equations (18) to (21), the expression plotted in Figure 12a for the mean can be written as,

$$\frac{\mu_{C_A}}{\mu_C} = \frac{1}{(1+V_C^2)^{\,1 \quad \gamma)/2}} \tag{22}$$

which gives that when $\gamma \rightarrow 0$, $\mu_{C_A}/\mu_C \rightarrow 1/(1+V_C^2)^{1/2}$, thus $\mu_{C_A} \rightarrow e^{\mu_{\ln C}} = \text{Median}_C$.

The expression plotted in Figure 12b for the coefficient of variation of the locally averaged variable can be written as,

$$\frac{V_{C_A}}{V_C} = \frac{\sqrt{(1+V_C^2)^{\gamma} \quad 1}}{V_C} \tag{23}$$

which gives that when $\gamma \rightarrow 0$, $V_{C_A}/V_C \rightarrow 0$, thus $V_{C_A} \rightarrow 0$.

Further examination of equations (22) and (23) shows that for all values of γ,

$$\text{Median}_{C_A} = \text{Median}_C \tag{24}$$

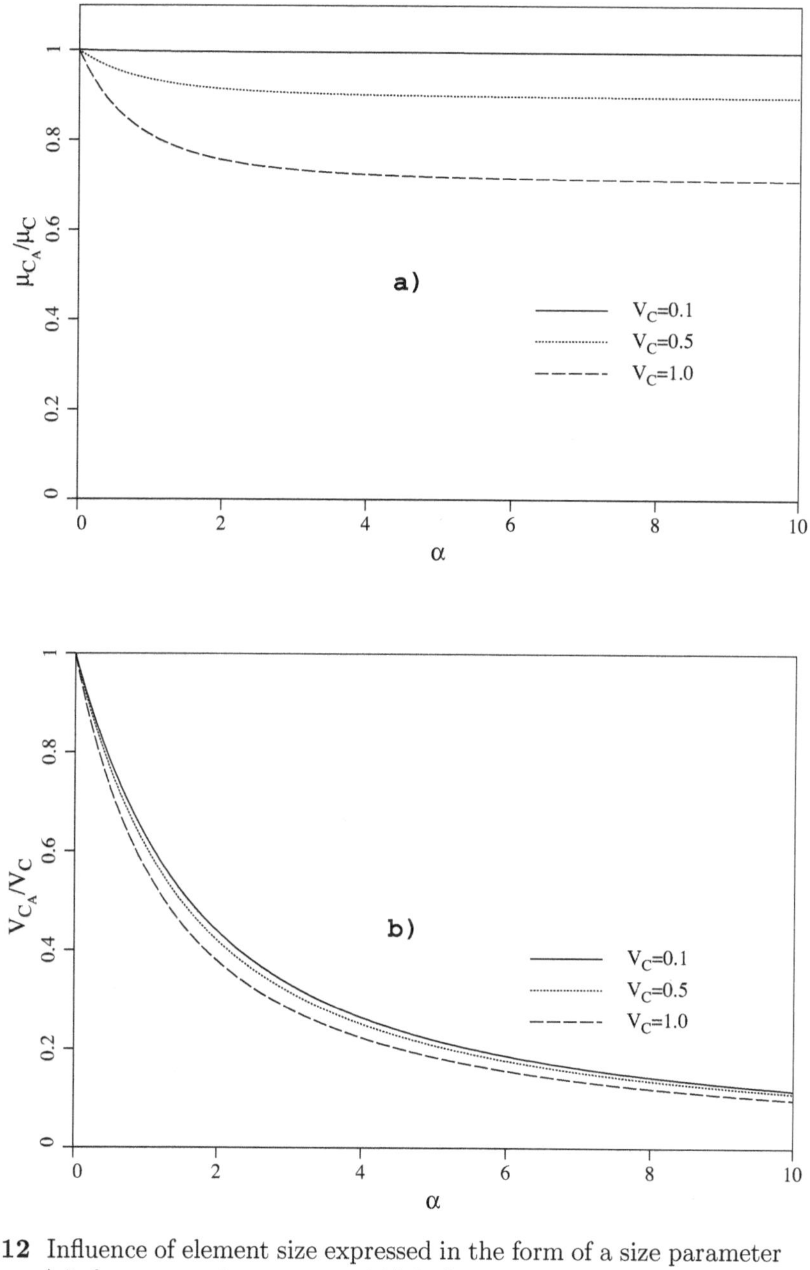

Figure 12 Influence of element size expressed in the form of a size parameter on local averaging. a) Influence on the mean, and b) Influence on the coefficient of variation

In Summary,

1. local averaging reduces both the mean and the variance of a lognormal point distribution.

2. local averaging preserves the Median of the point distribution, and

3. in the limit, local averaging removes all variance, and the mean tends to the Median.

9 Locally averaged single random variable approach

In this section the Probability of Failure is reworked with the single random variable approach using properties derived from local averaging over an individual finite element, termed "finite element locally averaged properties" throughout the rest of this paper. With reference to the mesh shown in Figure 8, the square elements have a side length of $0.1H$, thus $_C = 0.1/$. Figure 13 shows the probability of failure p_f as a function of $_C$ for a range of input point coefficients of variation, with the point mean fixed at $\mu_C = 0.25$. The Probability of Failure is defined, as before, by $p[C < 0.17]$, but this time the calculation is based on the finite element locally averaged properties, μ_{C_A} and σ_{C_A} from equations 20 and 21. The figure clearly shows two tails to the results, with $p_f \to 1$ as $_C \to 0$ for all $V_C > 1.0783$, and $p_f \to 0$ as $_C \to 0$ for all $V_C < 1.0783$. The horizontal line at $p_f = 0.5$ is given by $V_C = 1.0783$, which is the special value of the coefficient of variation that causes the $\text{Median}_C = 0.17$.

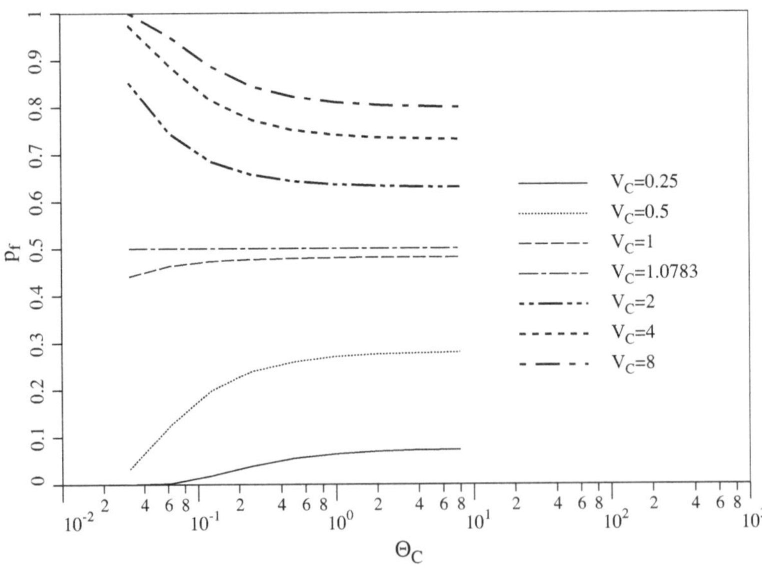

Figure 13 Probability of Failure vs. Spatial Correlation Length based on finite element locally averaged properties. The mean is fixed at $\mu_C = 0.25$

Recalling Table 1, this is the critical value of C that would give $FS = 1$ in the test slope. Higher values of V_C lead to Median$_C$ < 0.17 and a tendency for $p_f \rightarrow 1$ as $_C \rightarrow 0$. Conversely, lower values of V_C lead to Median$_C$ > 0.17 and a tendency for $p_f \rightarrow 0$. Figure 14 shows the same data plotted the other way round with V_C along the abscissa. This figure clearly shows the full influence of spatial correlation in the range $0 \leq \ _C < \infty$. All the curves cross over at the critical value of $V_C = 1.0783$, and it is of interest to note the step function corresponding to $\ _C = 0$ when p_f changes suddenly from zero to unity.

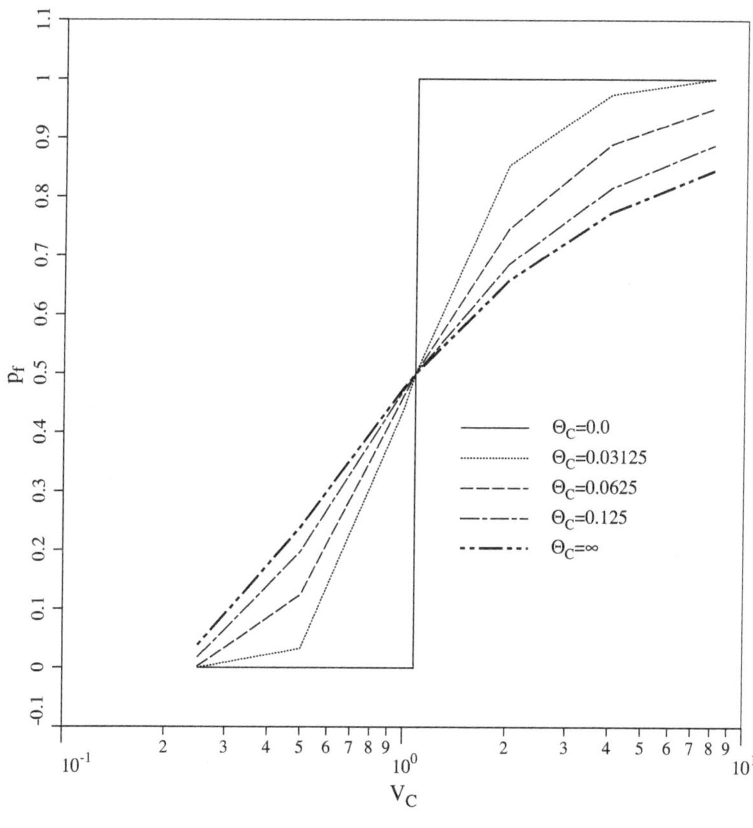

Figure 14 Probability of Failure vs. Coefficient of Variation based on finite element locally averaged properties. The mean is fixed at $\mu_C = 0.25$

It should be emphasized that the results presented in this section involved no finite element analysis, and were based solely on an SRV approach with statistical properties based on finite element locally averaged properties based on a typical finite element of the mesh in Figure 8.

10 Results of RFEM Analyses

In this section, the results of full nonlinear RFEM analyses with Monte-Carlo simulations are described, based on a range of parametric variations of μ_C, V_C and $_C$.

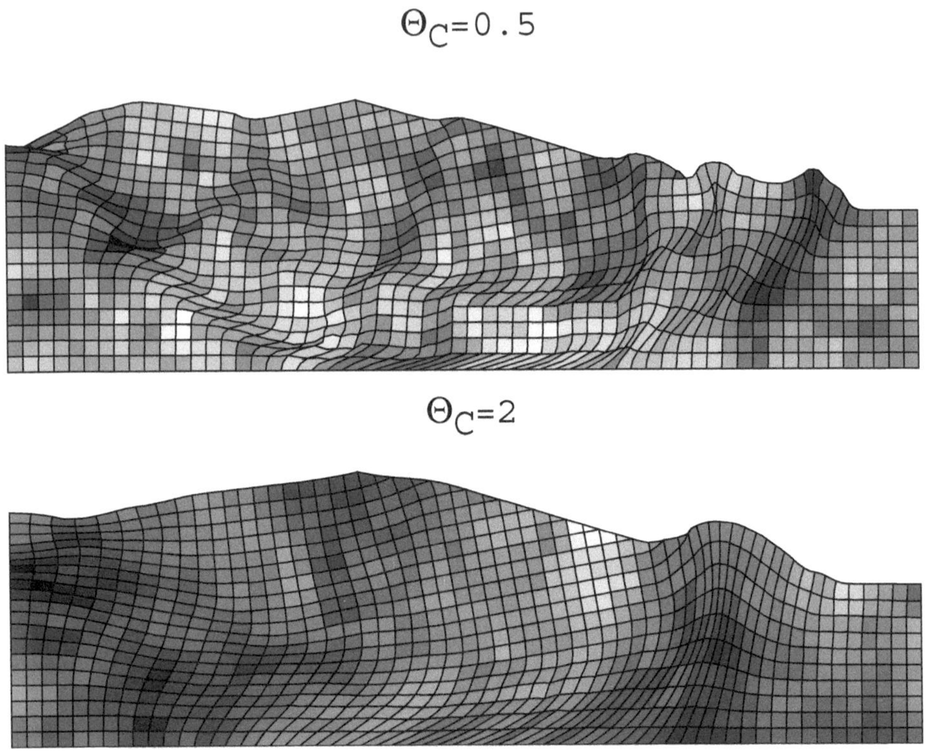

Figure 15 Typical random field realizations and deformed mesh at slope failure for two different spatial correlation lengths

In the elasto-plastic RFEM approach, the failure mechanism is free to "seek out" the weakest path through the soil. Figure 15 shows two typical random field realizations and the associated failure mechanisms for slopes with $_C = 0.5$ and $_C = 2$. The convoluted nature of the failure mechanisms, especially when $_C = 0.5$, would defy analysis by conventional slope stability analysis tools. While the mechanism is attracted to the weaker zones within the slope, it will inevitably pass through elements assigned many different strength values. This weakest path determination, and the strength averaging that goes with it, occurs quite naturally in the finite element slope stability method, and represents a very significant improvement over traditional limit equilibrium approaches to probabilistic slope stability, in which local averaging, if included at all, has to be computed over a failure mechanism that is pre-set the particular analysis method (e.g. a circular failure mechanism when using Bishop's Method).

Fixing the point mean strength at $\mu_C = 0.25$, Figures 16 and 17 show the effect of the spatial correlation length $_C$ and the coefficient of variation V_C on the probability of failure for the test problem. Figure 16 clearly indicates two branches, with the Probability of Failure tending to unity or zero for higher and lower values of V_C, respectively. This behavior is qualitatively similar to that observed in Figure 13, in which a single random variable approach was used to predict the Probability of Failure based solely on finite element locally averaged properties.

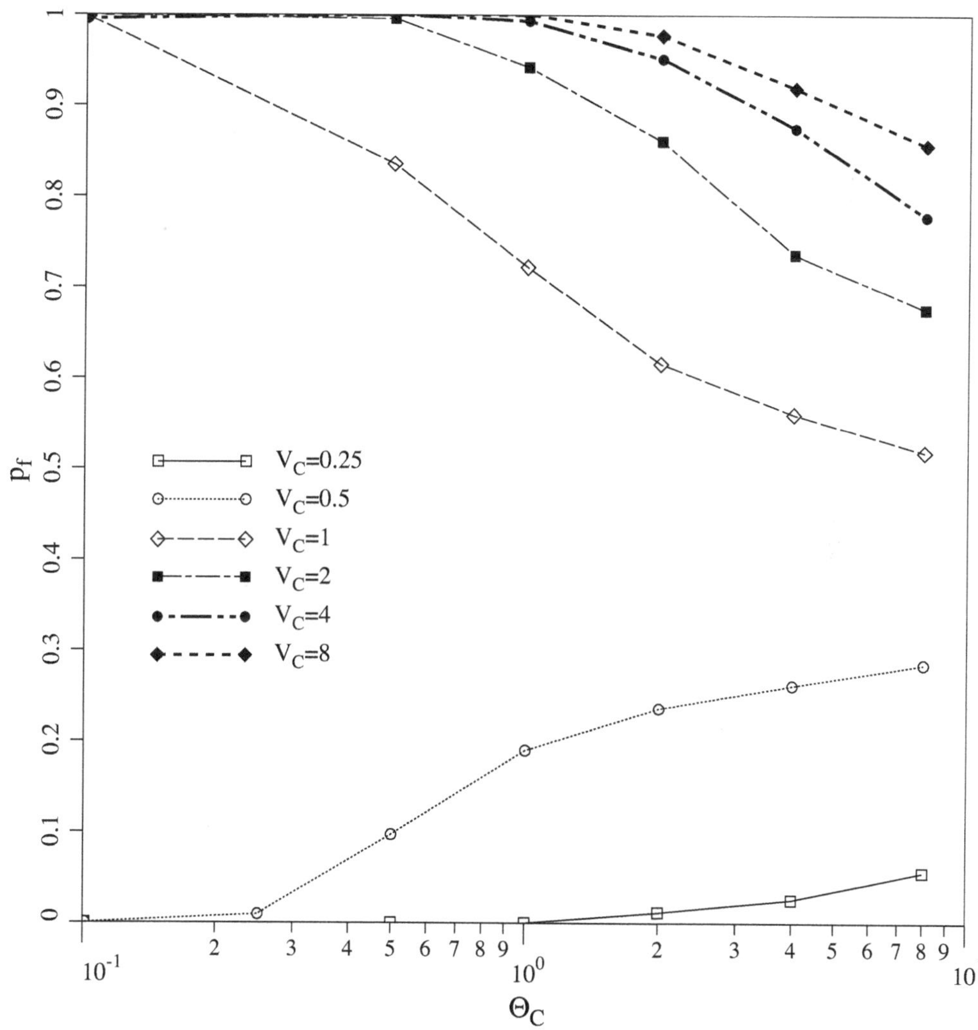

Figure 16 Probability of Failure vs. Spatial Correlation Length from RFEM. The mean is fixed at $\mu_C = 0.25$

Figure 17 shows the same results as Figure 16, but plotted the other way round with the coefficient of variation along the abscissa. Figure 17 also shows the theoretically obtained result corresponding to $\ _C = \infty$, indicating that a single random variable approach with no local averaging will overestimate the Probability of Failure (conservative) when the coefficient of variation is relatively small and underestimate the Probability of Failure (unconservative) when the coefficient of variation is relatively high. Figure 17 also confirms that the single random variable approach described earlier in the paper, which gave $p_f = 0.28$ corresponding to $\mu_C = 0.25$ and $V_C = 0.5$ with no local averaging, is indeed pessimistic. The RFEM results show that the inclusion of spatial correlation and local averaging in this case will always lead to a smaller Probability of Failure.

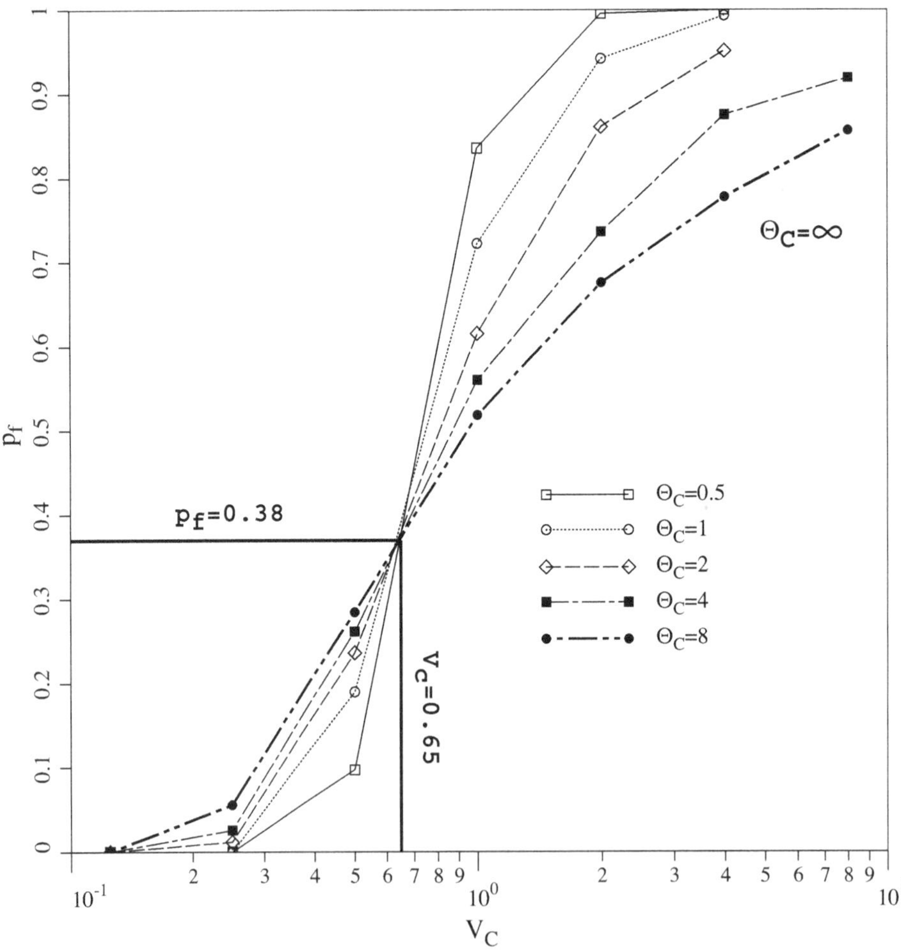

Figure 17 Probability of Failure vs. Coefficient of Variation from RFEM. The mean is fixed at $\mu_C = 0.25$

Comparison of Figures 13 and 14, with Figures 16 and 17, highlights the influence of the finite element approach to slope stability, where the failure mechanism is free to locate itself optimally within the mesh. From Figures 14 and 17, it is clear that the "weakest path" concept made possible by the RFEM approach has resulted in the crossover point falling to lower values of both V_C and p_f. With only finite element local averaging, the crossover occurred at $V_C = 1.0783$, whereas by the RFEM it occurred at $V_C \approx 0.65$. In terms of the Probability of Failure with only finite element local averaging, the crossover occurred at $p_f = 0.5$ whereas by the RFEM it occurred at $p_f \approx 0.38$. The RFEM solutions show that the single random variable approach becomes unconservative over a wider range of V_C values than would be indicated by finite element local averaging alone.

Figure 18 gives a direct comparison between Figures 13 and 16, indicating clearly that for higher values of V_C, RFEM always gives a higher probability of Failure than when using finite element local averaging alone. This is caused by the weaker elements in the distribution dominating the strength of the slope and the failure mechanism "seeking out" the weakest path through the soil.

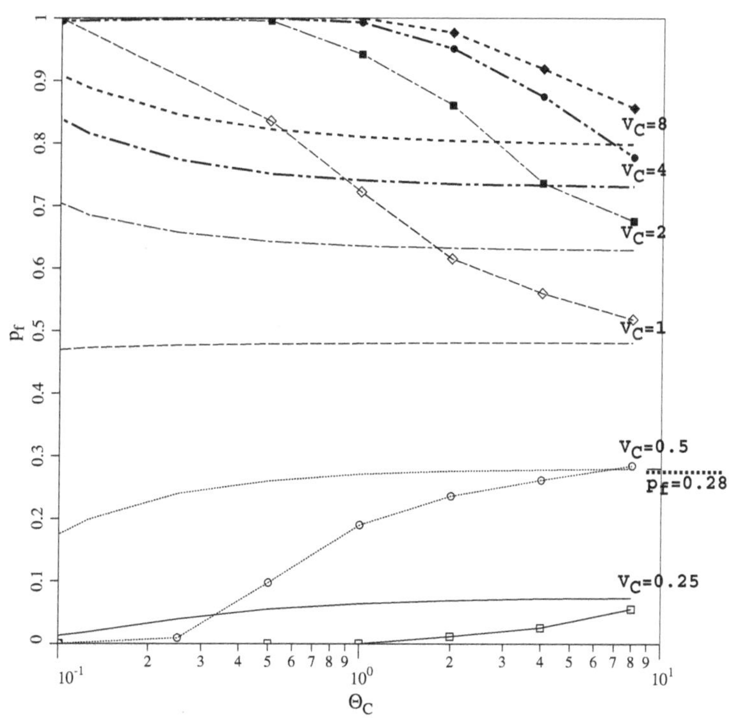

Figure 18 Comparison of the Probability of Failure predicted by RFEM and by finite element local averaging only. The curve with plotted points comes from the RFEM analyses.The mean is fixed at $\mu_C = 0.25$

At lower values of V_C, the locally averaged results tends to overestimate the Probability of Failure and give conservative results compared with RFEM. In this case the stronger elements of the slope are dominating the solution and the higher median combined with the "bunching up" of the locally averaged solution at low values of $_C$, means that potential failure mechanisms cannot readily find a weak path through the soil.

In all cases, as $_C$ increases, the RFEM and the locally averaged solutions converge on the single random variable solution corresponding to $_C = \infty$ with no local averaging. The $p_f = 0.28$ value, corresponding to $V_C = 0.5$, and discussed earlier in the paper is also indicated on Figure 18.

All of the above results and discussion in this section so far were applied to the test slope from Figure 1 with the mean strength fixed at $\mu_C = 0.25$ corresponding to a Factor of Safety (based on the mean) of 1.47. In the next set of results μ_C is varied while V_C is held constant at 0.5. Figure 19 shows the relationship between FS (based on the mean) and p_f assuming finite element local averaging only, and Figure 20 shows the same relationship as computed using RFEM.

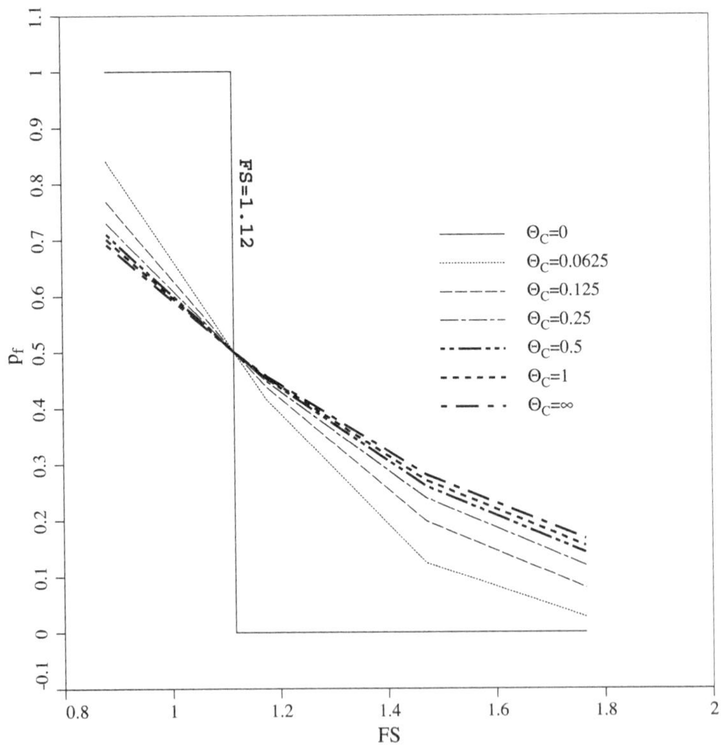

Figure 19 Probability of Failure vs. Factor of Safety (based on the mean) using finite element local averaging only for the test slope. The coefficient of variation is fixed at $V_C = 0.5$

Figure 19, based on finite element local averaging only, shows the full range of behavior for $0 \leq \;_C < \infty$. The figure shows that $_C$ only starts to have a significant influence on the FS vs. p_f relationship when the correlation length becomes significantly smaller than the slope height ($_C << 1$). The step function in which p_f jumps from zero to unity, occurs when $_C = 0$ and corresponds to a local average having zero variance. In this limiting case, the local average of the soil is deterministic, yielding a constant strength everywhere in the slope. With $V_C = 0.5$, the critical value of mean shear strength that would give $\mu_{C_A} = \text{Median}_C = 0.17$ is easily shown by equation 22 to be $\mu_C = 0.19$, which corresponds to a $FS = 1.12$. For higher values of $_C$, the relationship between FS and p_f is quite "bunched up", and generally insensitive to $_C$. For example, there is little difference between the curves corresponding to $_C = \infty$ and $_C = 0.5$. It should also be observed from Figure 19, that for $FS > 1.12$, failure to account for local averaging by assuming $_C = \infty$ is conservative, in that the predicted p_f is higher than it should be. When $FS < 1.12$ however, failure to account for local averaging is unconservative.

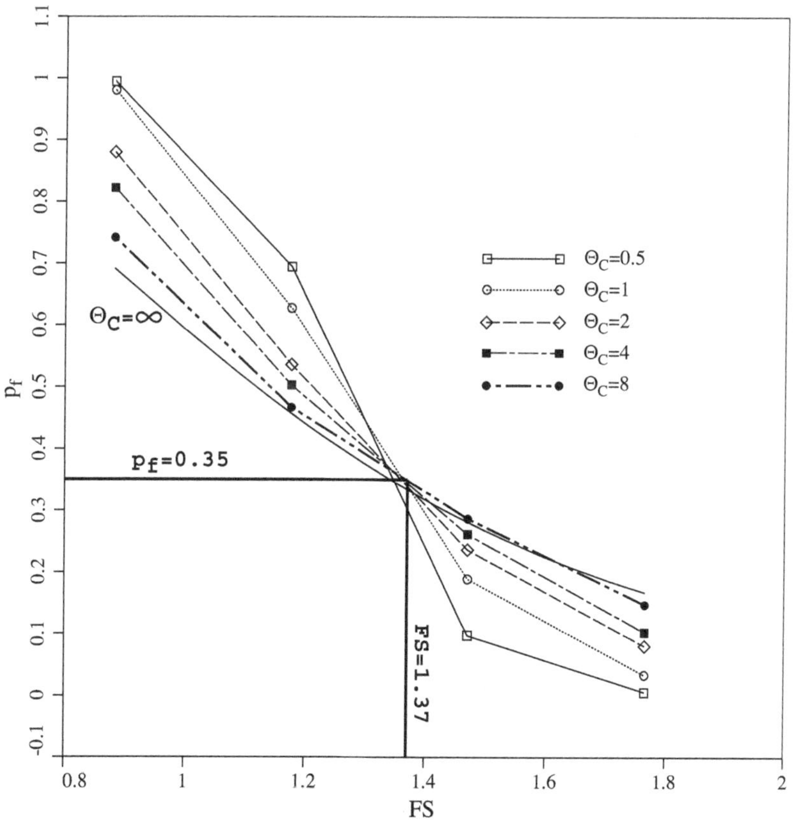

Figure 20 Probability of Failure vs. Factor of Safety (based on the mean) using RFEM for the test slope. The coefficient of variation is fixed at $V_C = 0.5$

Figure 20 gives the same relationships as computed using RFEM. By comparison with Figure 19, the RFEM results are more spread out, implying that the Probability of Failure is more sensitive to the spatial correlation length $_C$. Of greater significance, is that the crossover point has again shifted by RFEM as it seeks out the weakest path through the slope. In Figure 20, the crossover occurs at $FS \approx 1.37$ which is significantly higher and of greater practical significance than the crossover point of $FS \approx 1.12$ by finite element local averaging alone. The theoretical line corresponding to $_C = \infty$ is also shown in this plot. From a practical viewpoint, the RFEM analysis indicates that failure to properly account for local averaging is unconservative over a wider range of Factors of Safety than would be the case by finite element local averaging alone. To further highlight this difference, the particular results from Figures 19 and 20 corresponding to $_C = 0.5$ (spatial correlation length equal to half the embankment height) have been re plotted in Figure 21.

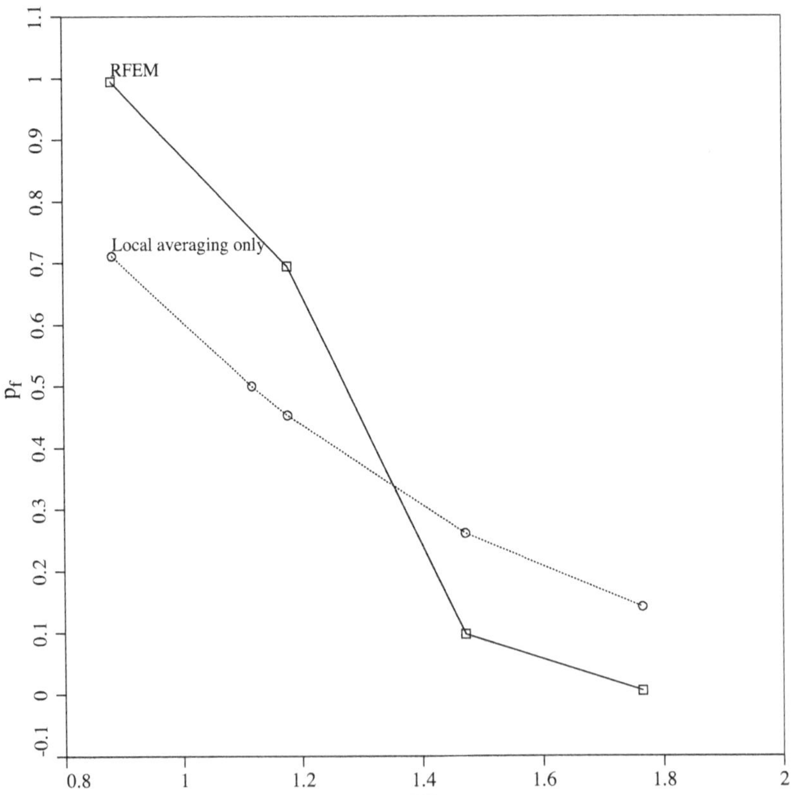

Figure 21 Probability of Failure vs. Factor of Safety (based on the mean) using finite element local averaging alone and RFEM for the test slope. $V_C = 0.5$ and $_C$=0.5

11 Concluding remarks

The paper has investigated the probability of failure of a cohesive slope using both simple and more advanced probabilistic analysis tools. The simple approach treated the strength of the entire slope as a single random variable, ignoring spatial correlation and local averaging. In the simple studies, the Probability of Failure was estimated as the probability that the shear strength would fall below a critical value based on a lognormal probability density function. These results led to a discussion on the appropriate choice of a design shear strength value suitable for deterministic analysis. Two factorization methods were proposed that were able to bring the Probability of Failure and the Factor of Safety more into line with practical experience.

The second half of the paper implemented the random finite element method (RFEM) on the same test problem. The non-linear elasto-plastic analyses with Monte-Carlo simulation were able to take full account of spatial correlation and local averaging, and observe their impact on the Probability of Failure using a parametric approach. The elasto-plastic finite element slope stability method makes no *a priori* assumptions about the shape or location of the critical failure mechanism, and therefore offers very significant benefits over traditional limit equilibrium methods in the analysis of highly variable soils. In the elasto-plastic RFEM, the failure mechanism is free to "seek out" the weakest path through the soil and it has been shown that this generality can lead to higher probabilities of failure than could be explained by finite element local averaging alone.

In summary, simplified probabilistic analysis, in which spatial variability is ignored by assuming perfect correlation, can lead to unconservative estimates of the probability of failure. This effect is most pronounced at relatively low factors of safety (Figure 20) or when the coefficient of variation of the soil strength is relatively high (Figure 18).

12 Acknowledgment

The writers wish to acknowledge the support of NSF Grant No. CMS-9877189. This is a modified version of a paper that first appeared as, D.V. Griffiths and Gordon A. Fenton (2004). "Probabilistic slope stability analysis by finite elements" *J. Geotech. Geoenv. Eng.* Vol. 130, No. 5, 507-518

13 Notation

c_u	Undrained shear strength
C	Dimensionless shear strength
C_{des}	Design value of C
D	Foundation depth ratio
f_1	Linear strength reduction factor
f_2	Strength reduction factor based on standard deviation
FS	Factor of Safety
H	Height of slope
p_f	Probability of Failure
V_C	Coefficient of variation of C
x	Cartesian x-coordinate
y	Cartesian y-coordinate
	Dimensionless element size parameter
β	Slope angle
γ	Variance reduction factor
γ_{sat}	Saturated unit weight
θ_C	Spatial correlation length of C
$\theta_{\ln C}$	Spatial correlation length of $\ln C$
$_C$	Dimensionless spatial correlation length of $\ln C$
μ_C	Mean of C
$\mu_{\ln C}$	Mean of $\ln C$
$\mu_{\ln C_A}$	Locally averaged mean of $\ln C$ over a square finite element
μ_{C_A}	Locally averaged mean of C over a square finite element
ρ	Correlation coefficient
σ_C	Standard deviation of C
σ_{C_A}	Locally averaged standard deviation of C over a square finite element
$\sigma_{\ln C}$	Standard deviation of $\ln C$
$\sigma_{\ln C_A}$	Locally averaged standard deviation of $\ln C$ over a square finite element
τ	Absolute distance between two points
τ_x	x-component of distance between two points
τ_y	y-component of distance between two points
ϕ_u	Undrained friction angle

Bibliography

E.E. Alonso. Risk analysis of slopes and its application to slopes in Canadian sensitive clays. *Géotechnique*, 26(453-472), 1976.

A.W. Bishop. The use of the slip circle in the stability analysis of slopes. *Géotechnique*, 5(1):7–17, 1955.

R.N. Chowdhury and W.H. Tang. Comparison of risk models for slopes. In *Proceedings of the Fifth International Conference on Applications of Statistics and Probability in Soil and Structural Engineering*, volume 2, pages 863–869. Vancouver, B.C., 1987.

J.T. Christian. Reliability methods for stability of existing slopes. In C.D. Shackelford *et al*, editor, *Uncertainty in the geologic environment: From theory to practice*, pages 409–419. GSP 58, ASCE, 1996. Wisconsin, August 1996.

J.T. Christian, C.C. Ladd, and G.B. Baecher. Reliability appled to slope stability analysis. *J Geotech Eng, ASCE*, 120(12):2180–2207, 1994.

R. A. D'Andrea and D. A. Sangrey. Safety factors for probabilistic slope design. *J Geotech Eng, ASCE*, 108(9):1108–1118, 1982.

J. M. Duncan. Factors of safety and reliability in geotechnical engineering. *J Geotech Geoenv Eng, ASCE*, 126(4):307–316, 2000.

H. El-Ramly, N.R. Morgenstern, and D.M. Cruden. Probabilistic slope stability analysis for practice. *Can Geotech J*, 39:665–683, 2002.

G. A. Fenton and D. V. Griffiths. Statistics of flow through a simple bounded stochastic medium. *Water Resour Res*, 29(6):1825–1830, 1993.

G.A. Fenton and E.H. Vanmarcke. Simulation of random fields via local average subdivision. *J Eng Mech, ASCE*, 116(8):1733–1749, 1990.

D. V. Griffiths and G. A. Fenton. Bearing capacity of spatially random soil: the undrained clay Prandtl problem revisited. *Géotechnique*, 51(4):351–359, 2001.

D. V. Griffiths and G. A. Fenton. Influence of soil strength spatial variability on the stability of an undrained clay slope by finite elements. In *Slope Stability 2000, Proceeding of GeoDenver 2000*, pages 184–193. ASCE, 2000.

D. V. Griffiths and P. A. Lane. Slope stability analysis by finite elements. *Géotechnique*, 49(3):387–403, 1999.

A.M. Hassan and T.F. Wolff. Effect of deterministic and probabilistic models on slope reliability index. In D. V. Griffiths *et al*, editor, *Slope Stability 2000*, pages 194–208. ASCE, 2000. GSP No. 101.

F.H. Kulhawy, M.J.S. Roth, and M.D. Grigoriu. Some statistical evaluations of geotechnical properties. In *Proc. ICASP6, 6th Int. Conf. Appl. Stats. Prob. Civ. Eng.* 1991.

S. Lacasse. Reliability and probabilistic methods. In *Proc 13th Int Conf Soil Mech Found Eng*, pages 225–227. 1994. New Delhi, India.

S. Lacasse and F. Nadim. Uncertainties in characterising soil properties. In C.D. Shackelford *et al*, editor, *Geotechnical Special Publication No 58, Proceedings of Uncertainity '96 held in Madison, Wisconson, July 31 - August 3, 1996*, pages 49–75. GSP 58, ASCE, 1996.

I. K. Lee, W. White, and O. G. Ingles. *Geotechnical Engineering.* Pitman, London, 1983.

K.S. Li and P. Lumb. Probabilistic design of slopes. *Can Geotech J*, 24:520–531, 1987.

M. Matsuo and K. Kuroda. Probabilistic approach to the design of embankments. *Soils Found*, 14(1):1–17, 1974.

G.R. Mostyn and K.S. Li. Probabilistic slope stability – State of play. In K.S. Li and S-C.R. Lo, editors, *Proc. Conf. Probabilistic Meths. Geotech. Eng.*, pages 89–110. A. A. Balkema, Rotterdam, 1993.

G.R. Mostyn and S. Soo. The effect of autocorrelation on the probability of failure of slopes. In *6th Australia, New Zealand Conference on Geomechanics: Geotechnical Risk*, pages 542–546. 1992.

G.M. Paice. *Finite element analysis of stochastic soils.* PhD thesis, University of Manchester, U.K., 1997.

I. M. Smith and D. V. Griffiths. *Programming the Finite Element Method.* John Wiley and Sons, Chichester, New York, 3rd edition, 1998.

W.H. Tang, M.S. Yucemen, and A.H.S. Ang. Probability based short-term design of slopes. *Can Geotech J*, 13:201–215, 1976.

D. W. Taylor. Stability of earth slopes. *J. Boston Soc. Civ. Eng.*, 24:197–246, 1937.

E.H. Vanmarcke. *Random elds: Analysis and synthesis.* The MIT Press, Cambridge, Mass., 1984.

E.H. Vanmarcke. Reliability of earth slopes. *J Geotech Eng, ASCE*, 103(11):1247–1265, 1977.

R.V. Whitman. Organizing and evaluating in geotechnical engineering. *J Geotech Geoenv Eng, ASCE*, 126(7):583–593, 2000.

T.F. Wolff. Probabilistic slope stability in theory and practice. In C.D. Shackelford *et al*, editor, *Uncertainty in the geologic environment: From theory to practice*, pages 419–433. GSP 58, ASCE, 1996. Wisconsin, August 1996.

Kolibri 360 GmbH
Ferdinand-Jühlke-Straße 7
99095 Erfurt, Deutschland
produktsicherheit@kolibri360.de